Introduction to Wavelets and Wavelet Transforms

A Primer

C. Sidney Burrus, Ramesh A. Gopinath, and Haitao Guo

with additional material and programs by

Jan E. Odegard and Ivan W. Selesnick

Electrical and Computer Engineering Department
and Computer and Information Technology Institute
Rice University
Houston, Texas
csb@rice.edu

Prentice Hall
Upper Saddle River, New Jersey 07458

Library of Congress Cataloging-in-Publication Data

BURRUS, C. S.
  Introduction to wavelets and wavelet transforms : a primer / C.
    Sidney Burrus, Ramesh A. Gopinath, and Haitao Guo ; with
  additional material and programs by Jan E. Odegard and Ivan W. Selesnick.
      p.   cm.
  Includes bibliographical references and index.
  ISBN   0-13-489600-9
  1. Wavelets (Mathematics)   2. Signal processing–Mathematics.
  I. Gopinath, Ramesh A.   II. Guo, Haitao.   III. Title.
  QA403.3.B87   1998   97-53263
  515'.2433–DC21   CIP

Acquisitions Editor: *Tom Robbins*
Production Editor: *Sharyn Vitrano*
Editor-in-Chief *Marcia Horton*
Managing Editor: *Bayani Mendoza DeLeon*
Copy Editor: *Irwin Zucker*
Cover Designer: *Bruce Kenselaar*
Director of Production and Manufacturing: *David W. Riccardi*
Manufacturing Buyer: *Donna Sullivan*
Editorial Assistant: *Nancy Garcia*
Composition: PRETEX, INC.

© 1998 by Prentice-Hall, Inc.
Upper Saddle River, New Jersey 07458

Printed in the United States of America

10 9 8 7 6 5

ISBN   0-13-489600-9

Prentice-Hall International (UK) Limited, *London*
Prentice-Hall of Australia Pty. Limited, *Sydney*
Prentice-Hall Canada Inc., *Toronto*
Prentice-Hall Hispanoamericana, S.A., *Mexico*
Prentice-Hall of India Private Limited, *New Delhi*
Prentice-Hall of Japan, Inc., *Tokyo*
Editora Prentice-Hall do Brasil, Ltda., *Rio de Janeiro*

To

Virginia Burrus and Charles Burrus,

Kalyani Narasimhan,

Yongtai Guo and Caijie Li

Contents

Preface

This book develops the ideas behind and properties of *wavelets* and shows how they can be used as analytical tools for signal processing, numerical analysis, and mathematical modeling. We try to present this in a way that is accessible to the engineer, scientist, and applied mathematician both as a theoretical approach and as a potentially practical method to solve problems. Although the roots of this subject go back some time, the modern interest and development have a history of only a few years.

The early work was in the 1980's by Morlet, Grossmann, Meyer, Mallat, and others, but it was the paper by Ingrid Daubechies [Dau88a] in 1988 that caught the attention of the larger applied mathematics communities in signal processing, statistics, and numerical analysis. Much of the early work took place in France [CGT89, Mey92a] and the USA [Dau88a, RBC*92, Dau92, RV91]. As in many new disciplines, the first work was closely tied to a particular application or traditional theoretical framework. Now we are seeing the theory abstracted from application and developed on its own and seeing it related to other parallel ideas. Our own background and interests in signal processing certainly influence the presentation of this book.

The goal of most modern wavelet research is to create a set of basis functions (or general expansion functions) and transforms that will give an informative, efficient, and useful description of a function or signal. If the signal is represented as a function of time, wavelets provide efficient localization in both time and frequency or scale. Another central idea is that of *multiresolution* where the decomposition of a signal is in terms of the resolution of detail.

For the Fourier series, sinusoids are chosen as basis functions, then the properties of the resulting expansion are examined. For wavelet analysis, one poses the desired properties and then derives the resulting basis functions. An important property of the wavelet basis is providing a multiresolution analysis. For several reasons, it is often desired to have the basis functions orthonormal. Given these goals, you will see aspects of correlation techniques, Fourier transforms, short-time Fourier transforms, discrete Fourier transforms, Wigner distributions, filter banks, subband coding, and other signal expansion and processing methods in the results.

Wavelet-based analysis is an exciting new problem-solving tool for the mathematician, scientist, and engineer. It fits naturally with the digital computer with its basis functions defined by summations not integrals or derivatives. Unlike most traditional expansion systems, the basis functions of the wavelet analysis are not solutions of differential equations. In some areas, it is the first truly new tool we have had in many years. Indeed, use of wavelets and wavelet transforms requires a new point of view and a new method of interpreting representations that we are still learning how to exploit.

More recently, work by Donoho, Johnstone, Coifman, and others have added theoretical reasons for why wavelet analysis is so versatile and powerful, and have given generalizations that are still being worked on. They have shown that wavelet systems have some inherent generic advantages and are near optimal for a wide class of problems [Don93b]. They also show that adaptive means can create special wavelet systems for particular signals and classes of signals.

The multiresolution decomposition seems to separate components of a signal in a way that is superior to most other methods for analysis, processing, or compression. Because of the ability of the discrete wavelet transform to decompose a signal at different independent scales and to do it in a very flexible way, Burke calls wavelets "The Mathematical Microscope" [Bur94, Hub96]. Because of this powerful and flexible decomposition, linear and nonlinear processing of signals in the wavelet transform domain offers new methods for signal detection, filtering, and compression [Don93b, Don95, Don93a, Sai94b, WTWB97, Guo97]. It also can be used as the basis for robust numerical algorithms.

You will also see an interesting connection and equivalence to filter bank theory from digital signal processing [Vai92, AH92]. Indeed, some of the results obtained with filter banks are the same as with discrete-time wavelets, and this has been developed in the signal processing community by Vetterli, Vaidyanathan, Smith and Barnwell, and others. Filter banks, as well as most algorithms for calculating wavelet transforms, are part of a still more general area of multirate and time-varying systems.

The presentation here will be as a tutorial or primer for people who know little or nothing about wavelets but do have a technical background. It assumes a knowledge of Fourier series and transforms and of linear algebra and matrix theory. It also assumes a background equivalent to a B.S. degree in engineering, science, or applied mathematics. Some knowledge of signal processing is helpful but not essential. We develop the ideas in terms of one-dimensional signals [RV91] modeled as real or perhaps complex functions of time, but the ideas and methods have also proven effective in image representation and processing [SA92, Mal89a] dealing with two, three, or even four dimensions. Vector spaces have proved to be a natural setting for developing both the theory and applications of wavelets. Some background in that area is helpful but can be picked up as needed. The study and understanding of wavelets is greatly assisted by using some sort of wavelet software system to work out examples and run experiments. MATLAB^{TM} programs are included at the end of this book and on our web site (noted at the end of the preface). Several other systems are mentioned in Chapter 10.

There are several different approaches that one could take in presenting wavelet theory. We have chosen to start with the representation of a signal or function of continuous time in a series expansion, much as a Fourier series is used in a Fourier analysis. From this series representation, we can move to the expansion of a function of a discrete variable (e.g., samples of a signal) and the theory of filter banks to efficiently calculate and interpret the expansion coefficients. This would be analogous to the discrete Fourier transform (DFT) and its efficient implementation, the fast Fourier transform (FFT). We can also go from the series expansion to an integral transform called the continuous wavelet transform, which is analogous to the Fourier transform or Fourier integral. We feel starting with the series expansion gives the greatest insight and provides ease in seeing both the similarities and differences with Fourier analysis.

This book is organized into sections and chapters, each somewhat self-contained. The earlier chapters give a fairly complete development of the discrete wavelet transform (DWT) as a series expansion of signals in terms of wavelets and scaling functions. The later chapters are short descriptions of generalizations of the DWT and of applications. They give references to other works, and serve as a sort of annotated bibliography. Because we intend this book as an introduction to wavelets which already have an extensive literature, we have included a rather long bibliography. However, it will soon be incomplete because of the large number of papers that are currently being published. Nevertheless, a guide to the other literature is essential to our goal of an introduction.

A good sketch of the philosophy of wavelet analysis and the history of its development can be found in a recent book published by the National Academy of Science in the chapter by Barbara Burke [Bur94]. She has written an excellent expanded version in [Hub96], which should be read by anyone interested in wavelets. Daubechies gives a brief history of the early research in [Dau96].

Many of the results and relationships presented in this book are in the form of theorems and proofs or derivations. A real effort has been made to ensure the correctness of the statements of theorems but the proofs are often only outlines of derivations intended to give insight into the result rather than to be a formal proof. Indeed, many of the derivations are put in the Appendix in order not to clutter the presentation. We hope this style will help the reader gain insight into this very interesting but sometimes obscure new mathematical signal processing tool.

We use a notation that is a mixture of that used in the signal processing literature and that in the mathematical literature. We hope this will make the ideas and results more accessible, but some uniformity and cleanness is lost.

The authors acknowledge AFOSR, ARPA, NSF, Nortel, Inc., Texas Instruments, Inc. and Aware, Inc. for their support of this work. We specifically thank H. L. Resnikoff, who first introduced us to wavelets and who proved remarkably accurate in predicting their power and success. We also thank W. M. Lawton, R. O. Wells, Jr., R. G. Baraniuk, J. E. Odegard, I. W. Selesnick, M. Lang, J. Tian, and members of the Rice Computational Mathematics Laboratory for many of the ideas and results presented in this book. The first named author would like to thank the Maxfield and Oshman families for their generous support. The students in EE-531 and EE-696 at Rice University provided valuable feedback as did Bruce Francis, Strela Vasily, Hans Schüssler, Peter Steffen, Gary Sitton, Jim Lewis, Yves Angel, Curt Michel, J. H. Husoy, Kjersti Engan, Ken Castleman, Jeff Trinkle, Katherine Jones, and other colleagues at Rice and elsewhere.

We also particularly want to thank Tom Robbins and his colleagues at Prentice Hall for their support and help. Their reviewers added significantly to the book.

We would appreciate learning of any errors or misleading statements that any readers discover. Indeed, any suggestions for improvement of the book would be most welcome. Send suggestions or comments via email to csb@rice.edu. Software, articles, errata for this book, and other information on the wavelet research at Rice can be found on the world-wide-web URL: http://www-dsp.rice.edu/ with links to other sites where wavelet research is being done.

C. Sidney Burrus, Ramesh A. Gopinath, and Haitao Guo
Houston, Texas, Yorktown Heights, New York, and Sunnyvale, California

Instructions to the Reader

Although this book in arranged in a somewhat progressive order, starting with basic ideas and definitions, moving to a rather complete discussion of the basic wavelet system, and then on to generalizations, one should skip around when reading or studying from it. Depending on the background of the reader, he or she should skim over most of the book first, then go back and study parts in detail. The Introduction at the beginning and the Summary at the end should be continually consulted to gain or keep a perspective; similarly for the Table of Contents and Index. The MATLAB programs in the Appendix or the Wavelet Toolbox from Mathworks or other wavelet software should be used for continual experimentation. The list of references should be used to find proofs or detail not included here or to pursue research topics or applications. The theory and application of wavelets are still developing and in a state of rapid growth. We hope this book will help open the door to this fascinating new subject.

Chapter 1

Introduction to Wavelets

This chapter will provide an overview of the topics to be developed in the book. Its purpose is to present the ideas, goals, and outline of properties for an understanding of and ability to use wavelets and wavelet transforms. The details and more careful definitions are given later in the book.

A *wave* is usually defined as an oscillating function of time or space, such as a sinusoid. Fourier analysis is wave analysis. It expands signals or functions in terms of sinusoids (or, equivalently, complex exponentials) which has proven to be extremely valuable in mathematics, science, and engineering, especially for periodic, time-invariant, or stationary phenomena. A *wavelet* is a "small wave", which has its energy concentrated in time to give a tool for the analysis of transient, nonstationary, or time-varying phenomena. It still has the oscillating wave-like characteristic but also has the ability to allow simultaneous time and frequency analysis with a flexible mathematical foundation. This is illustrated in Figure 1.1 with the wave (sinusoid) oscillating with equal amplitude over $-\infty \leq t \leq \infty$ and, therefore, having infinite energy and with the wavelet having its finite energy concentrated around a point.

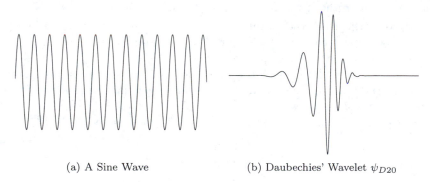

(a) A Sine Wave   (b) Daubechies' Wavelet ψ_{D20}

Figure 1.1. A Wave and a Wavelet

We will take wavelets and use them in a series expansion of signals or functions much the same way a Fourier series uses the wave or sinusoid to represent a signal or function. The signals are functions of a continuous variable, which often represents time or distance. From this series expansion, we will develop a discrete-time version similar to the discrete Fourier transform where the signal is represented by a string of numbers where the numbers may be samples of a signal, samples of another string of numbers, or inner products of a signal with some expansion set. Finally, we will briefly describe the continuous wavelet transform where both the signal and the transform are functions of continuous variables. This is analogous to the Fourier transform.

1.1   Wavelets and Wavelet Expansion Systems

Before delving into the details of wavelets and their properties, we need to get some idea of their general characteristics and what we are going to do with them [Swe96b].

What is a Wavelet Expansion or a Wavelet Transform?

A signal or function $f(t)$ can often be better analyzed, described, or processed if expressed as a linear decomposition by

$$f(t) = \sum_{\ell} a_\ell \, \psi_\ell(t) \tag{1.1}$$

where ℓ is an integer index for the finite or infinite sum, a_ℓ are the real-valued expansion coefficients, and $\psi_\ell(t)$ are a set of real-valued functions of t called the expansion set. If the expansion (1.1) is unique, the set is called a *basis* for the class of functions that can be so expressed. If the basis is orthogonal, meaning

$$\langle \psi_k(t), \psi_\ell(t) \rangle = \int \psi_k(t) \, \psi_\ell(t) \, dt = 0 \qquad k \neq \ell, \tag{1.2}$$

then the coefficients can be calculated by the *inner product*

$$a_k = \langle f(t), \psi_k(t) \rangle = \int f(t) \, \psi_k(t) \, dt. \tag{1.3}$$

One can see that substituting (1.1) into (1.3) and using (1.2) gives the single a_k coefficient. If the basis set is not orthogonal, then a dual basis set $\widetilde{\psi}_k(t)$ exists such that using (1.3) with the dual basis gives the desired coefficients. This will be developed in Chapter 2.

For a Fourier series, the orthogonal basis functions $\psi_k(t)$ are $\sin(k\omega_0 t)$ and $\cos(k\omega_0 t)$ with frequencies of $k\omega_0$. For a Taylor's series, the nonorthogonal basis functions are simple monomials t^k, and for many other expansions they are various polynomials. There are expansions that use splines and even fractals.

For the *wavelet expansion*, a two-parameter system is constructed such that (1.1) becomes

$$f(t) = \sum_k \sum_j a_{j,k} \, \psi_{j,k}(t) \tag{1.4}$$

where both j and k are integer indices and the $\psi_{j,k}(t)$ are the wavelet expansion functions that usually form an orthogonal basis.

The set of expansion coefficients $a_{j,k}$ are called the *discrete wavelet transform* (DWT) of $f(t)$ and (1.4) is the inverse transform.

What is a Wavelet System?

The wavelet expansion set is not unique. There are many different wavelets systems that can be used effectively, but all seem to have the following three general characteristics [Swe96b].

1. A wavelet system is a set of *building blocks* to construct or represent a signal or function. It is a two-dimensional expansion set (usually a basis) for some class of one- (or higher) dimensional signals. In other words, if the wavelet set is given by $\psi_{j,k}(t)$ for indices of $j, k = 1, 2, \cdots$, a linear expansion would be $f(t) = \sum_k \sum_j a_{j,k}\,\psi_{j,k}(t)$ for some set of coefficients $a_{j,k}$.

2. The wavelet expansion gives a time-frequency *localization* of the signal. This means most of the energy of the signal is well represented by a few expansion coefficients, $a_{j,k}$.

3. The calculation of the coefficients from the signal can be done *efficiently*. It turns out that many wavelet transforms (the set of expansion coefficients) can calculated with $O(N)$ operations. This means the number of floating-point multiplications and additions increase linearly with the length of the signal. More general wavelet transforms require $O(N \log(N))$ operations, the same as for the fast Fourier transform (FFT) [BP85].

Virtually all wavelet systems have these very general characteristics. Where the Fourier series maps a one-dimensional function of a continuous variable into a one-dimensional sequence of coefficients, the wavelet expansion maps it into a two-dimensional array of coefficients. We will see that it is this two-dimensional representation that allows localizing the signal in both time and frequency. A Fourier series expansion localizes in frequency in that if a Fourier series expansion of a signal has only one large coefficient, then the signal is essentially a single sinusoid at the frequency determined by the index of the coefficient. The simple time-domain representation of the signal itself gives the localization in time. If the signal is a simple pulse, the location of that pulse is the localization in time. A wavelet representation will give the location in both time and frequency simultaneously. Indeed, a wavelet representation is much like a musical score where the location of the notes tells when the tones occur and what their frequencies are.

More Specific Characteristics of Wavelet Systems

There are three additional characteristics [Swe96b, Dau92] that are more specific to wavelet expansions.

1. All so-called first-generation wavelet systems are generated from a single scaling function or wavelet by simple *scaling* and *translation*. The two-dimensional parameterization is achieved from the function (sometimes called the generating wavelet or mother wavelet) $\psi(t)$ by

$$\psi_{j,k}(t) = 2^{j/2}\,\psi(2^j t - k) \qquad\qquad j, k \in \mathbf{Z} \qquad\qquad (1.5)$$

where $\mathbf{Z}$ is the set of all integers and the factor $2^{j/2}$ maintains a constant norm independent of scale j. This parameterization of the time or space location by k and the frequency or scale (actually the logarithm of scale) by j turns out to be extraordinarily effective.

2. Almost all useful wavelet systems also satisfy the *multiresolution* conditions. This means that if a set of signals can be represented by a weighted sum of $\varphi(t - k)$, then a larger set (including the original) can be represented by a weighted sum of $\varphi(2t - k)$. In other words, if the basic expansion signals are made half as wide and translated in steps half as wide, they will represent a larger class of signals exactly or give a better approximation of any signal.

3. The lower resolution coefficients can be calculated from the higher resolution coefficients by a tree-structured algorithm called a *filter bank*. This allows a very efficient calculation of the expansion coefficients (also known as the discrete wavelet transform) and relates wavelet transforms to an older area in digital signal processing.

The operations of translation and scaling seem to be basic to many practical signals and signal-generating processes, and their use is one of the reasons that wavelets are efficient expansion functions. Figure 1.2 is a pictorial representation of the translation and scaling of a single mother wavelet described in (1.5). As the index k changes, the location of the wavelet moves along the horizontal axis. This allows the expansion to explicitly represent the location of events in time or space. As the index j changes, the shape of the wavelet changes in scale. This allows a representation of detail or resolution. Note that as the scale becomes finer (j larger), the steps in time become smaller. It is both the narrower wavelet and the smaller steps that allow representation of greater detail or higher resolution. For clarity, only every fourth term in the translation ($k = 1, 5, 9, 13, \cdots$) is shown, otherwise, the figure is a clutter. What is not illustrated here but is important is that the shape of the basic mother wavelet can also be changed. That is done during the design of the wavelet system and allows one set to well-represent a class of signals.

For the Fourier series and transform and for most signal expansion systems, the expansion functions (bases) are chosen, then the properties of the resulting transform are derived and

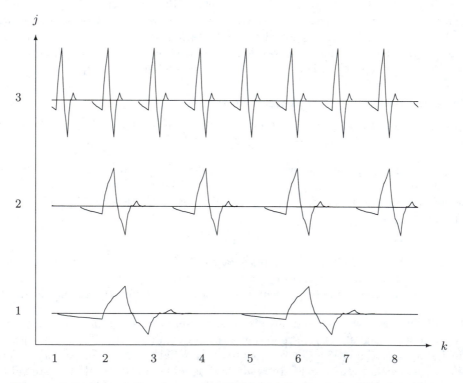

Figure 1.2. Translation (every fourth k) and Scaling of a Wavelet ψ_{D4}

analyzed. For the wavelet system, these properties or characteristics are mathematically required, then the resulting basis functions are derived. Because these constraints do not use all the degrees of freedom, other properties can be required to customize the wavelet system for a particular application. Once you decide on a Fourier series, the sinusoidal basis functions are completely set. That is not true for the wavelet. There are an infinity of very different wavelets that all satisfy the above properties. Indeed, the understanding and design of the wavelets is an important topic of this book.

Wavelet analysis is well-suited to transient signals. Fourier analysis is appropriate for periodic signals or for signals whose statistical characteristics do not change with time. It is the localizing property of wavelets that allow a wavelet expansion of a transient event to be modeled with a small number of coefficients. This turns out to be very useful in applications.

Haar Scaling Functions and Wavelets

The multiresolution formulation needs two closely related basic functions. In addition to the wavelet $\psi(t)$ that has been discussed (but not actually defined yet), we will need another basic function called the *scaling function* $\varphi(t)$. The reasons for needing this function and the details of the relations will be developed in the next chapter, but here we will simply use it in the wavelet expansion.

The simplest possible orthogonal wavelet system is generated from the Haar scaling function and wavelet. These are shown in Figure 1.3. Using a combination of these scaling functions and wavelets allows a large class of signals to be represented by

$$f(t) = \sum_{k=-\infty}^{\infty} c_k\, \varphi(t-k) + \sum_{k=-\infty}^{\infty} \sum_{j=0}^{\infty} d_{j,k}\, \psi(2^j t - k). \tag{1.6}$$

Haar [Haa10] showed this result in 1910, and we now know that wavelets are a generalization of his work. An example of a Haar system and expansion is given at the end of Chapter 2.

What do Wavelets Look Like?

All Fourier basis functions look alike. A high-frequency sine wave looks like a compressed low-frequency sine wave. A cosine wave is a sine wave translated by 90° or $\pi/2$ radians. It takes a

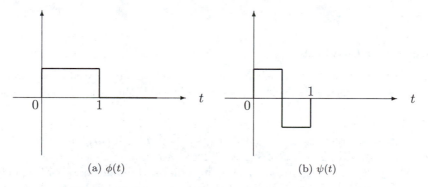

(a) $\phi(t)$ (b) $\psi(t)$

Figure 1.3. Haar Scaling Function and Wavelet

large number of Fourier components to represent a discontinuity or a sharp corner. In contrast, there are many different wavelets and some have sharp corners themselves.

To appreciate the special character of wavelets you should recognize that it was not until the late 1980's that some of the most useful basic wavelets were ever seen. Figure 1.4 illustrates four different scaling functions, each being zero outside of $0 < t < 6$ and each generating an orthogonal wavelet basis for all square integrable functions. This figure is also shown on the cover to this book.

Several more scaling functions and their associated wavelets are illustrated in later chapters, and the Haar wavelet is shown in Figure 1.3 and in detail at the end of Chapter 2.

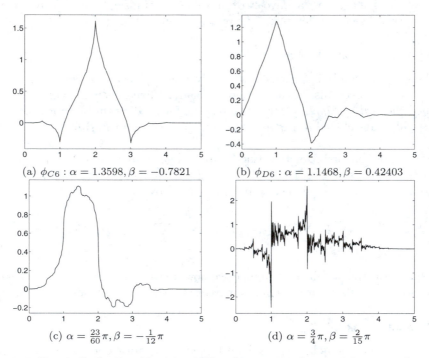

(a) $\phi_{C6} : \alpha = 1.3598, \beta = -0.7821$

(b) $\phi_{D6} : \alpha = 1.1468, \beta = 0.42403$

(c) $\alpha = \frac{23}{60}\pi, \beta = -\frac{1}{12}\pi$

(d) $\alpha = \frac{3}{4}\pi, \beta = \frac{2}{15}\pi$

Figure 1.4. Example Scaling Functions (See Section 5.8 for the Meaning of α and β)

Why is Wavelet Analysis Effective?

Wavelet expansions and wavelet transforms have proven to be very efficient and effective in analyzing a very wide class of signals and phenomena. Why is this true? What are the properties that give this effectiveness?

1. The size of the wavelet expansion coefficients $a_{j,k}$ in (1.4) or $d_{j,k}$ in (1.6) drop off rapidly with j and k for a large class of signals. This property is called being an *unconditional basis* and it is why wavelets are so effective in signal and image compression, denoising, and detection. Donoho [Don93b, DJKP95b] showed that wavelets are near optimal for a wide class of signals for compression, denoising, and detection.

2. The wavelet expansion allows a more accurate local description and separation of signal characteristics. A Fourier coefficient represents a component that lasts for all time and, therefore, temporary events must be described by a phase characteristic that allows cancellation or reinforcement over large time periods. A wavelet expansion coefficient represents a component that is itself local and is easier to interpret. The wavelet expansion may allow a separation of components of a signal that overlap in both time and frequency.

3. Wavelets are adjustable and adaptable. Because there is not just one wavelet, they can be designed to fit individual applications. They are ideal for adaptive systems that adjust themselves to suit the signal.

4. The generation of wavelets and the calculation of the discrete wavelet transform is well matched to the digital computer. We will later see that the defining equation for a wavelet uses no calculus. There are no derivatives or integrals, just multiplications and additions—operations that are basic to a digital computer.

While some of these details may not be clear at this point, they should point to the issues that are important to both theory and application and give reasons for the detailed development that follows in this and other books.

1.2   The Discrete Wavelet Transform

This two-variable set of basis functions is used in a way similar to the short-time Fourier transform, the Gabor transform, or the Wigner distribution for time-frequency analysis [Coh89, Coh95, HB92]. Our goal is to generate a set of expansion functions such that any signal in $L^2(\mathbf{R})$ (the space of square integrable functions) can be represented by the series

$$f(t) = \sum_{j,k} a_{j,k} \, 2^{j/2} \, \psi(2^j t - k) \tag{1.7}$$

or, using (1.5), as

$$f(t) = \sum_{j,k} a_{j,k} \, \psi_{j,k}(t) \tag{1.8}$$

where the two-dimensional set of coefficients $a_{j,k}$ is called the *discrete wavelet transform* (DWT) of $f(t)$. A more specific form indicating how the $a_{j,k}$'s are calculated can be written using inner products as

$$f(t) = \sum_{j,k} \langle \psi_{j,k}(t), f(t) \rangle \, \psi_{j,k}(t) \tag{1.9}$$

if the $\psi_{j,k}(t)$ form an orthonormal basis[1] for the space of signals of interest [Dau92]. The inner product is usually defined as

$$\langle x(t), y(t) \rangle = \int x^*(t) \, y(t) \, dt. \tag{1.10}$$

---

[1] Bases and tight frames are defined in Chapter 4.

The goal of most expansions of a function or signal is to have the coefficients of the expansion $a_{j,k}$ give more useful information about the signal than is directly obvious from the signal itself. A second goal is to have most of the coefficients be zero or very small. This is what is called a *sparse* representation and is extremely important in applications for statistical estimation and detection, data compression, nonlinear noise reduction, and fast algorithms.

Although this expansion is called the discrete wavelet transform (DWT), it probably should be called a wavelet series since it is a series expansion which maps a function of a continuous variable into a sequence of coefficients much the same way the Fourier series does. However, that is not the convention.

This wavelet series expansion is in terms of two indices, the time translation k and the scaling index j. For the Fourier series, there are only two possible values of k, zero and $\pi/2$, which give the sine terms and the cosine terms. The values j give the frequency harmonics. In other words, the Fourier series is also a two-dimensional expansion, but that is not seen in the exponential form and generally not noticed in the trigonometric form.

The DWT of a signal is somewhat difficult to illustrate because it is a function of two variables or indices, but we will show the DWT of a simple pulse in Figure 1.5 to illustrate the localization of the transform. Other displays will be developed in the next chapter.

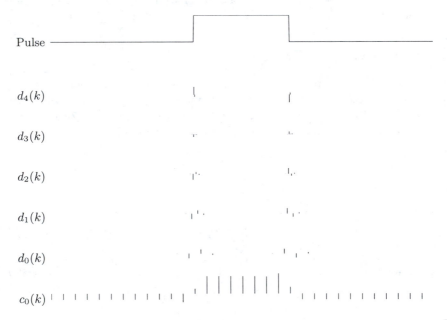

Figure 1.5. Discrete Wavelet Transform of a Pulse, using ψ_{D6} with a Gain of $\sqrt{2}$ for Each Higher Scale.

1.3   The Discrete-Time and Continuous Wavelet Transforms

If the signal is itself a sequence of numbers, perhaps samples of some function of a continuous variable or perhaps a set of inner products, the expansion of that signal is called a discrete-time wavelet transform (DTWT). It maps a sequence of numbers into a sequence of numbers much the same way the discrete Fourier transform (DFT) does. It does not, however, require the signal to be finite in duration or periodic as the DFT does. To be consistent with Fourier terminology, it probably should be called the discrete-time wavelet series, but this is not the convention. If the discrete-time signal is finite in length, the transform can be represented by a finite matrix. This formulation of a series expansion of a discrete-time signal is what filter bank methods accomplish [Vai92, VK95] and is developed in Chapter 8 of this book.

If the signal is a function of a continuous variable and a transform that is a function of two continuous variables is desired, the continuous wavelet transform (CWT) can be defined by

$$F(a,b) \; = \; \int f(t) \, w(\frac{t-a}{b}) \, dt \tag{1.11}$$

with an inverse transform of

$$f(t) \; = \; \iint F(a,b) \, w(\frac{t-a}{b}) \, da \, db \tag{1.12}$$

where $w(t)$ is the basic wavelet and $a, b \in \mathbf{R}$ are real continuous variables. Admissibility conditions for the wavelet $w(t)$ to support this invertible transform is discussed by Daubechies [Dau92], Heil and Walnut [HW89], and others and is briefly developed in Section 7.8 of this book. It is analogous to the Fourier transform or Fourier integral.

1.4 Exercises and Experiments

As the ideas about wavelets and wavelet transforms are developed in this book, it will be very helpful to experiment using the Matlab programs in the appendix of this book or in the MATLAB Toolbox [MMOP96]. An effort has been made to use the same notation in the programs in Appendix C as is used in the formulas in the book so that going over the programs can help in understanding the theory and vice versa.

1.5 This Chapter

This chapter has tried to set the stage for a careful introduction to both the theory and use of wavelets and wavelet transforms. We have presented the most basic characteristics of wavelets and tried to give a feeling of how and why they work in order to motivate and give direction and structure to the following material.

The next chapter will present the idea of multiresolution, out of which will develop the scaling function as well as the wavelet. This is followed by a discussion of how to calculate the wavelet expansion coefficients using filter banks from digital signal processing. Next, a more detailed development of the theory and properties of scaling functions, wavelets, and wavelet transforms is given followed by a chapter on the design of wavelet systems. Chapter 8 gives a detailed development of wavelet theory in terms of filter banks.

The earlier part of the book carefully develops the basic wavelet system and the later part develops several important generalizations, but in a less detailed form.

Chapter 2

A Multiresolution Formulation of Wavelet Systems

Both the mathematics and the practical interpretations of wavelets seem to be best served by using the concept of resolution [Mey93, Mal89b, Mal89c, Dau92] to define the effects of changing scale. To do this, we will start with a *scaling function* $\varphi(t)$ rather than directly with the wavelet $\psi(t)$. After the scaling function is defined from the concept of resolution, the wavelet functions will be derived from it. This chapter will give a rather intuitive development of these ideas, which will be followed by more rigorous arguments in Chapter 5.

This multiresolution formulation is obviously designed to represent signals where a single event is decomposed into finer and finer detail, but it turns out also to be valuable in representing signals where a time-frequency or time-scale description is desired even if no concept of resolution is needed. However, there are other cases where multiresolution is not appropriate, such as for the short-time Fourier transform or Gabor transform or for local sine or cosine bases or lapped orthogonal transforms, which are all discussed briefly later in this book.

2.1   Signal Spaces

In order to talk about the collection of functions or signals that can be represented by a sum of scaling functions and/or wavelets, we need some ideas and terminology from functional analysis. If these concepts are not familiar to you or the information in this section is not sufficient, you may want to skip ahead and read Chapter 5 or [VD95].

A *function space* is a linear vector space (finite or infinite dimensional) where the vectors are functions, the scalars are real numbers (sometime complex numbers), and scalar multiplication and vector addition are similar to that done in (1.1). The *inner product* is a scalar a obtained from two vectors, $f(t)$ and $g(t)$, by an integral. It is denoted

$$a = \langle f(t), g(t) \rangle = \int f^*(t)\, g(t)\, dt \qquad (2.1)$$

with the range of integration depending on the signal class being considered. This inner product defines a *norm* or "length" of a vector which is denoted and defined by

$$\|f\| = \sqrt{|\langle f, f \rangle|} \qquad (2.2)$$

which is a simple generalization of the geometric operations and definitions in three-dimensional Euclidean space. Two signals (vectors) with non-zero norms are called *orthogonal* if their inner product is zero. For example, with the Fourier series, we see that $\sin(t)$ is orthogonal to $\sin(2t)$.

A space that is particularly important in signal processing is call $L^2(\mathbf{R})$. This is the space of all functions $f(t)$ with a well defined integral of the square of the modulus of the function. The "L" signifies a Lebesque integral, the "2" denotes the integral of the square of the modulus of the function, and $\mathbf{R}$ states that the independent variable of integration t is a number over the whole real line. For a function $g(t)$ to be a member of that space is denoted: $g \in L^2(\mathbf{R})$ or simply $g \in L^2$.

Although most of the definitions and derivations are in terms of signals that are in L^2, many of the results hold for larger classes of signals. For example, polynomials are not in L^2 but can be expanded over any finite domain by most wavelet systems.

In order to develop the wavelet expansion described in (1.5), we will need the idea of an expansion set or a basis set. If we start with the vector space of signals, $\mathcal{S}$, then if any $f(t) \in \mathcal{S}$ can be expressed as $f(t) = \sum_k a_k \varphi_k(t)$, the set of functions $\varphi_k(t)$ are called an expansion set for the space $\mathcal{S}$. If the representation is unique, the set is a basis. Alternatively, one could start with the expansion set or basis set and define the space $\mathcal{S}$ as the set of all functions that can be expressed by $f(t) = \sum_k a_k \varphi_k(t)$. This is called the *span* of the basis set. In several cases, the signal spaces that we will need are actually the *closure* of the space spanned by the basis set. That means the space contains not only all signals that can be expressed by a linear combination of the basis functions, but also the signals which are the limit of these infinite expansions. The closure of a space is usually denoted by an over-line.

2.2   The Scaling Function

In order to use the idea of multiresolution, we will start by defining the scaling function and then define the wavelet in terms of it. As described for the wavelet in the previous chapter, we define a set of scaling functions in terms of integer translates of the basic scaling function by

$$\varphi_k(t) = \varphi(t - k) \qquad k \in \mathbf{Z} \qquad \varphi \in L^2. \tag{2.3}$$

The subspace of $L^2(\mathbf{R})$ spanned by these functions is defined as

$$\mathcal{V}_0 = \overline{\operatorname*{Span}_k \{\varphi_k(t)\}} \tag{2.4}$$

for all integers k from minus infinity to infinity. The over-bar denotes closure. This means that

$$f(t) = \sum_k a_k \varphi_k(t) \quad \text{for any} \quad f(t) \in \mathcal{V}_0. \tag{2.5}$$

One can generally increase the size of the subspace spanned by changing the time scale of the scaling functions. A two-dimensional family of functions is generated from the basic scaling function by scaling and translation by

$$\varphi_{j,k}(t) = 2^{j/2} \varphi(2^j t - k) \tag{2.6}$$

whose span over k is

$$V_j = \overline{\underset{k}{\text{Span}}\{\varphi_k(2^j t)\}} = \overline{\underset{k}{\text{Span}}\{\varphi_{j,k}(t)\}} \tag{2.7}$$

for all integers $k \in \mathbf{Z}$. This means that if $f(t) \in V_j$, then it can be expressed as

$$f(t) = \sum_k a_k \varphi(2^j t + k). \tag{2.8}$$

For $j > 0$, the span can be larger since $\varphi_{j,k}(t)$ is narrower and is translated in smaller steps. It, therefore, can represent finer detail. For $j < 0$, $\varphi_{j,k}(t)$ is wider and is translated in larger steps. So these wider scaling functions can represent only coarse information, and the space they span is smaller. Another way to think about the effects of a change of scale is in terms of resolution. If one talks about photographic or optical resolution, then this idea of scale is the same as resolving power.

Multiresolution Analysis

In order to agree with our intuitive ideas of scale or resolution, we formulate the basic requirement of multiresolution analysis (MRA) [Mal89c] by requiring a nesting of the spanned spaces as

$$\cdots \subset V_{-2} \subset V_{-1} \subset V_0 \subset V_1 \subset V_2 \subset \cdots \subset L^2 \tag{2.9}$$

or

$$V_j \subset V_{j+1} \quad \text{for all} \quad j \in \mathbf{Z} \tag{2.10}$$

with

$$V_{-\infty} = \{0\}, \qquad V_\infty = L^2. \tag{2.11}$$

The space that contains high resolution signals will contain those of lower resolution also. Because of the definition of V_j, the spaces have to satisfy a natural scaling condition

$$f(t) \in V_j \quad \Leftrightarrow \quad f(2t) \in V_{j+1} \tag{2.12}$$

which insures elements in a space are simply scaled versions of the elements in the next space. The relationship of the spanned spaces is illustrated in Figure 2.1.

The nesting of the spans of $\varphi(2^j t - k)$, denoted by V_j and shown in (2.9) and (2.12) and graphically illustrated in Figure 2.1, is achieved by requiring that $\varphi(t) \in V_1$, which means that if $\varphi(t)$ is in V_0, it is also in V_1, the space spanned by $\varphi(2t)$. This means $\varphi(t)$ can be expressed in terms of a weighted sum of shifted $\varphi(2t)$ as

$$\boxed{\varphi(t) = \sum_n h(n)\sqrt{2}\,\varphi(2t - n), \qquad n \in \mathbf{Z}} \tag{2.13}$$

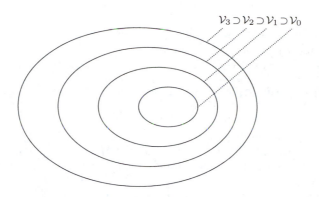

$$\mathcal{V}_3 \supset \mathcal{V}_2 \supset \mathcal{V}_1 \supset \mathcal{V}_0$$

Figure 2.1. Nested Vector Spaces Spanned by the Scaling Functions

where the coefficients $h(n)$ are a sequence of real or perhaps complex numbers called the scaling function coefficients (or the scaling filter or the scaling vector) and the $\sqrt{2}$ maintains the norm of the scaling function with the scale of two.

This recursive equation is fundamental to the theory of the scaling functions and is, in some ways, analogous to a differential equation with coefficients $h(n)$ and solution $\varphi(t)$ that may or may not exist or be unique. The equation is referred to by different names to describe different interpretations or points of view. It is called the refinement equation, the multiresolution analysis (MRA) equation, or the dilation equation.

The Haar scaling function is the simple unit-width, unit-height pulse function $\varphi(t)$ shown in Figure 2.2, and it is obvious that $\varphi(2t)$ can be used to construct $\varphi(t)$ by

$$\varphi(t) = \varphi(2t) + \varphi(2t - 1) \tag{2.14}$$

which means (2.13) is satisfied for coefficients $h(0) = 1/\sqrt{2}$, $h(1) = 1/\sqrt{2}$.

The triangle scaling function (also a first order spline) in Figure 2.2 satisfies (2.13) for $h(0) = \frac{1}{2\sqrt{2}}$, $h(1) = \frac{1}{\sqrt{2}}$, $h(2) = \frac{1}{\sqrt{2}}$, and the Daubechies scaling function shown in the first part of

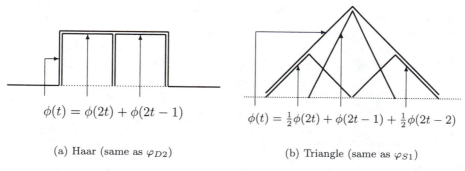

$$\phi(t) = \phi(2t) + \phi(2t - 1)$$

$$\phi(t) = \tfrac{1}{2}\phi(2t) + \phi(2t - 1) + \tfrac{1}{2}\phi(2t - 2)$$

(a) Haar (same as φ_{D2}) 　　　　　　　 (b) Triangle (same as φ_{S1})

Figure 2.2. Haar and Triangle Scaling Functions

Figure 6.1 satisfies (2.13) for $h = \{0.483, 0.8365, 0.2241, -0.1294\}$ as do all scaling functions for their corresponding scaling coefficients. Indeed, the design of wavelet systems is the choosing of the coefficients $h(n)$ and that is developed later.

2.3   The Wavelet Functions

The important features of a signal can better be described or parameterized, not by using $\varphi_{j,k}(t)$ and increasing j to increase the size of the subspace spanned by the scaling functions, but by defining a slightly different set of functions $\psi_{j,k}(t)$ that span the *differences* between the spaces spanned by the various scales of the scaling function. These functions are the wavelets discussed in the introduction of this book.

There are several advantages to requiring that the scaling functions and wavelets be orthogonal. Orthogonal basis functions allow simple calculation of expansion coefficients and have a Parseval's theorem that allows a partitioning of the signal energy in the wavelet transform domain. The orthogonal complement of $\mathcal{V}_j$ in $\mathcal{V}_{j+1}$ is defined as $\mathcal{W}_j$. This means that all members of $\mathcal{V}_j$ are orthogonal to all members of $\mathcal{W}_j$. We require

$$\langle \varphi_{j,k}(t), \psi_{j,\ell}(t) \rangle = \int \varphi_{j,k}(t) \, \psi_{j,\ell}(t) \, dt = 0 \tag{2.15}$$

for all appropriate $j, k, \ell \in \mathbf{Z}$.

The relationship of the various subspaces can be seen from the following expressions. From (2.9) we see that we may start at any $\mathcal{V}_j$, say at $j = 0$, and write

$$\mathcal{V}_0 \subset \mathcal{V}_1 \subset \mathcal{V}_2 \subset \cdots \subset L^2. \tag{2.16}$$

We now define the wavelet spanned subspace $\mathcal{W}_0$ such that

$$\mathcal{V}_1 = \mathcal{V}_0 \oplus \mathcal{W}_0 \tag{2.17}$$

which extends to

$$\mathcal{V}_2 = \mathcal{V}_0 \oplus \mathcal{W}_0 \oplus \mathcal{W}_1. \tag{2.18}$$

In general this gives

$$L^2 = \mathcal{V}_0 \oplus \mathcal{W}_0 \oplus \mathcal{W}_1 \oplus \cdots \tag{2.19}$$

when $\mathcal{V}_0$ is the initial space spanned by the scaling function $\varphi(t - k)$. Figure 2.3 pictorially shows the nesting of the scaling function spaces $\mathcal{V}_j$ for different scales j and how the wavelet spaces are the disjoint differences (except for the zero element) or, the orthogonal complements.

The scale of the initial space is arbitrary and could be chosen at a higher resolution of, say, $j = 10$ to give

$$L^2 = \mathcal{V}_{10} \oplus \mathcal{W}_{10} \oplus \mathcal{W}_{11} \oplus \cdots \tag{2.20}$$

or at a lower resolution such as $j = -5$ to give

$$L^2 = \mathcal{V}_{-5} \oplus \mathcal{W}_{-5} \oplus \mathcal{W}_{-4} \oplus \cdots \tag{2.21}$$

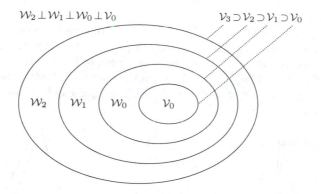

$$\mathcal{W}_2 \perp \mathcal{W}_1 \perp \mathcal{W}_0 \perp \mathcal{V}_0 \qquad\qquad \mathcal{V}_3 \supset \mathcal{V}_2 \supset \mathcal{V}_1 \supset \mathcal{V}_0$$

Figure 2.3. Scaling Function and Wavelet Vector Spaces

or at even $j = -\infty$ where (2.19) becomes

$$L^2 = \cdots \oplus \mathcal{W}_{-2} \oplus \mathcal{W}_{-1} \oplus \mathcal{W}_0 \oplus \mathcal{W}_1 \oplus \mathcal{W}_2 \oplus \cdots \tag{2.22}$$

eliminating the scaling space altogether and allowing an expansion of the form in (1.9).

Another way to describe the relation of $\mathcal{V}_0$ to the wavelet spaces is noting

$$\mathcal{W}_{-\infty} \oplus \cdots \oplus \mathcal{W}_{-1} = \mathcal{V}_0 \tag{2.23}$$

which again shows that the scale of the scaling space can be chosen arbitrarily. In practice, it is usually chosen to represent the coarsest detail of interest in a signal.

Since these wavelets reside in the space spanned by the next narrower scaling function, $\mathcal{W}_0 \subset \mathcal{V}_1$, they can be represented by a weighted sum of shifted scaling function $\varphi(2t)$ defined in (2.13) by

$$\psi(t) = \sum_n h_1(n)\,\sqrt{2}\,\varphi(2t - n), \qquad n \in \mathbf{Z} \tag{2.24}$$

for some set of coefficients $h_1(n)$. From the requirement that the wavelets span the "difference" or orthogonal complement spaces, and the orthogonality of integer translates of the wavelet (or scaling function), it is shown in the Appendix in (12.48) that the wavelet coefficients (modulo translations by integer multiples of two) are required by orthogonality to be related to the scaling function coefficients by

$$h_1(n) = (-1)^n h(1 - n). \tag{2.25}$$

One example for a finite even length-N $h(n)$ could be

$$h_1(n) = (-1)^n h(N - 1 - n). \tag{2.26}$$

The function generated by (2.24) gives the prototype or mother wavelet $\psi(t)$ for a class of expansion functions of the form

$$\psi_{j,k}(t) = 2^{j/2}\,\psi(2^j t - k) \tag{2.27}$$

where 2^j is the scaling of t (j is the $\log_2$ of the scale), $2^{-j}k$ is the translation in t, and $2^{j/2}$ maintains the (perhaps unity) L^2 norm of the wavelet at different scales.

The Haar and triangle wavelets that are associated with the scaling functions in Figure 2.2 are shown in Figure 2.4. For the Haar wavelet, the coefficients in (2.24) are $h_1(0) = 1/\sqrt{2}$, $h_1(1) = -1/\sqrt{2}$ which satisfy (2.25). The Daubechies wavelets associated with the scaling functions in Figure 6.1 are shown in Figure 6.2 with corresponding coefficients given later in the book in Tables 6.1 and 6.2.

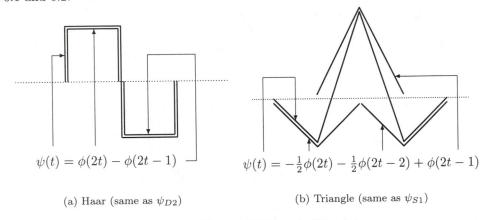

$$\psi(t) = \phi(2t) - \phi(2t - 1) \qquad\qquad \psi(t) = -\tfrac{1}{2}\phi(2t) - \tfrac{1}{2}\phi(2t - 2) + \phi(2t - 1)$$

(a) Haar (same as ψ_{D2})   (b) Triangle (same as ψ_{S1})

Figure 2.4. Haar and Triangle Wavelets

We have now constructed a set of functions $\varphi_k(t)$ and $\psi_{j,k}(t)$ that could span all of $L^2(\mathbf{R})$. According to (2.19), any function $g(t) \in L^2(\mathbf{R})$ could be written

$$g(t) = \sum_{k=-\infty}^{\infty} c(k)\,\varphi_k(t) + \sum_{j=0}^{\infty}\sum_{k=-\infty}^{\infty} d(j,k)\,\psi_{j,k}(t) \qquad (2.28)$$

as a series expansion in terms of the scaling function and wavelets.

In this expansion, the first summation in (2.28) gives a function that is a low resolution or coarse approximation of $g(t)$. For each increasing index j in the second summation, a higher or finer resolution function is added, which adds increasing detail. This is somewhat analogous to a Fourier series where the higher frequency terms contain the detail of the signal.

Later in this book, we will develop the property of having these expansion functions form an orthonormal basis or a tight frame, which allows the coefficients to be calculated by inner products as

$$c(k) = c_0(k) = \langle g(t), \varphi_k(t) \rangle = \int g(t)\,\varphi_k(t)\,dt \qquad (2.29)$$

and

$$d_j(k) = d(j,k) = \langle g(t), \psi_{j,k}(t) \rangle = \int g(t)\,\psi_{j,k}(t)\,dt. \qquad (2.30)$$

The coefficient $d(j,k)$ is sometimes written as $d_j(k)$ to emphasize the difference between the time translation index k and the scale parameter j. The coefficient $c(k)$ is also sometimes written as $c_j(k)$ or $c(j,k)$ if a more general "starting scale" other than $j = 0$ for the lower limit on the sum in (2.28) is used.

It is important at this point to recognize the relationship of the scaling function part of the expansion (2.28) to the wavelet part of the expansion. From the representation of the nested spaces in (2.19) we see that the scaling function can be defined at any scale j. Equation (2.28) uses $j = 0$ to denote the family of scaling functions.

You may want to examine the Haar system example at the end of this chapter just now to see these features illustrated.

2.4   The Discrete Wavelet Transform

Since

$$L^2 = \mathcal{V}_{j_0} \oplus \mathcal{W}_{j_0} \oplus \mathcal{W}_{j_0+1} \oplus \cdots \tag{2.31}$$

using (2.6) and (2.27), a more general statement of the expansion (2.28) can be given by

$$g(t) = \sum_k c_{j_0}(k) \, 2^{j_0/2} \, \varphi(2^{j_0}t - k) + \sum_k \sum_{j=j_0}^{\infty} d_j(k) \, 2^{j/2} \, \psi(2^j t - k) \tag{2.32}$$

or

$$g(t) = \sum_k c_{j_0}(k) \, \varphi_{j_0,k}(t) + \sum_k \sum_{j=j_0}^{\infty} d_j(k) \, \psi_{j,k}(t) \tag{2.33}$$

where j_0 could be zero as in (2.19) and (2.28), it could be ten as in (2.20), or it could be negative infinity as in (1.8) and (2.22) where no scaling functions are used. The choice of j_0 sets the coarsest scale whose space is spanned by $\varphi_{j_0,k}(t)$. The rest of $L^2(\mathbf{R})$ is spanned by the wavelets which provide the high resolution details of the signal. In practice where one is given only the samples of a signal, not the signal itself, there is a highest resolution when the finest scale is the sample level.

The coefficients in this wavelet expansion are called the *discrete wavelet transform* (DWT) of the signal $g(t)$. If certain conditions described later are satisfied, these wavelet coefficients completely describe the original signal and can be used in a way similar to Fourier series coefficients for analysis, description, approximation, and filtering. If the wavelet system is orthogonal, these coefficients can be calculated by inner products

$$c_j(k) = \langle g(t), \varphi_{j,k}(t) \rangle = \int g(t) \, \varphi_{j,k}(t) \, dt \tag{2.34}$$

and

$$d_j(k) = \langle g(t), \psi_{j,k}(t) \rangle = \int g(t) \, \psi_{j,k}(t) \, dt. \tag{2.35}$$

If the scaling function is well-behaved, then at a high scale, the scaling is similar to a Dirac delta function and the inner product simply samples the function. In other words, at high enough resolution, samples of the signal are very close to the scaling coefficients. More is said about this later. It has been shown [Don93b] that wavelet systems form an unconditional basis for a large class of signals. That is discussed in Chapter 5 but means that even for the worst case signal in the class, the wavelet expansion coefficients drop off rapidly as j and k increase. This is why the DWT is efficient for signal and image compression.

The DWT is similar to a Fourier series but, in many ways, is much more flexible and informative. It can be made periodic like a Fourier series to represent periodic signals efficiently.

However, unlike a Fourier series, it can be used directly on non-periodic transient signals with excellent results. An example of the DWT of a pulse was illustrated in Figure 3.3. Other examples are illustrated just after the next section.

2.5   A Parseval's Theorem

If the scaling functions and wavelets form an orthonormal basis[1], there is a Parseval's theorem that relates the energy of the signal $g(t)$ to the energy in each of the components and their wavelet coefficients. That is one reason why orthonormality is important.

For the general wavelet expansion of (2.28) or (2.33), Parseval's theorem is

$$
\int |g(t)|^2 \, dt = \sum_{l=-\infty}^{\infty} |c(l)|^2 + \sum_{j=0}^{\infty} \sum_{k=-\infty}^{\infty} |d_j(k)|^2
\tag{2.36}
$$

with the energy in the expansion domain partitioned in time by k and in scale by j. Indeed, it is this partitioning of the time-scale parameter plane that describes the DWT. If the expansion system is a tight frame, there is a constant multiplier in (2.36) caused by the redundancy.

Daubechies [Dau88a, Dau92] showed that it is possible for the scaling function and the wavelets to have compact support (i.e., be nonzero only over a finite region) and to be orthonormal. This makes possible the time localization that we desire. We now have a framework for describing signals that has features of short-time Fourier analysis and of Gabor-based analysis but using a new variable, scale. For the short-time Fourier transform, orthogonality and good time-frequency resolution are incompatible according to the Balian-Low-Coifman-Semmes theorem [Dau90, Sie86]. More precisely, if the short-time Fourier transform is orthogonal, either the time or the frequency resolution is poor and the trade-off is inflexible. This is not the case for the wavelet transform. Also, note that there is a variety of scaling functions and wavelets that can be obtained by choosing different coefficients $h(n)$ in (2.13).

Donoho [Don93b] has noted that wavelets are an unconditional basis for a very wide class of signals. This means wavelet expansions of signals have coefficients that drop off rapidly and therefore the signal can be efficiently represented by a small number of them.

We have first developed the basic ideas of the discrete wavelet system using a scaling multiplier of 2 in the defining equation (2.13). This is called a *two-band wavelet system* because of the two channels or bands in the related filter banks discussed in Chapters 3 and 8. It is also possible to define a more general discrete wavelet system using $\varphi(t) = \sum_n h(n) \sqrt{M} \, \varphi(Mt - n)$ where M is an integer [SHGB93]. This is discussed in Section 7.2. The details of numerically calculating the DWT are discussed in Chapter 9 where special forms for periodic signals are used.

2.6   Display of the Discrete Wavelet Transform and the Wavelet Expansion

It is important to have an informative way of displaying or visualizing the wavelet expansion and transform. This is complicated in that the DWT is a real-valued function of two integer indices and, therefore, needs a two-dimensional display or plot. This problem is somewhat analogous to plotting the Fourier transform, which is a complex-valued function.

There seem to be **five** displays that show the various characteristics of the DWT well:

---

[1]or a tight frame defined in Chapter 4

1. The most basic time-domain description of a signal is the signal itself (or, for most cases, samples of the signal) but it gives no frequency or scale information. A very interesting property of the DWT (and one different from the Fourier series) is for a high starting scale j_0 in (2.33), samples of the signal are the DWT at that scale. This is an extreme case, but it shows the flexibility of the DWT and will be explained later.

2. The most basic wavelet-domain description is a three-dimensional plot of the expansion coefficients or DWT values $c(k)$ and $d_j(k)$ over the j, k plane. This is difficult to do on a two-dimensional page or display screen, but we show a form of that in Figures 2.5 and 2.8.

3. A very informative picture of the effects of scale can be shown by generating time functions $f_j(t)$ at each scale by summing (2.28) over k so that

$$f(t) = f_{j_0} + \sum_j f_j(t) \tag{2.37}$$

where

$$f_{j_0} = \sum_k c(k)\,\varphi(t-k) \tag{2.38}$$

and

$$f_j(t) = \sum_k d_j(k)\,2^{j/2}\,\psi(2^j t - k). \tag{2.39}$$

This illustrates the components of the signal at each scale and is shown in Figures 2.7 and 2.10.

4. Another illustration that shows the time localization of the wavelet expansion is obtained by generating time functions $f_k(t)$ at each translation by summing (2.28) over k so that

$$f(t) = \sum_k f_k(t) \tag{2.40}$$

where

$$f_k(t) = c(k)\,\varphi(t-k) + \sum_j d_j(k)\,2^{j/2}\,\psi(2^j t - k). \tag{2.41}$$

This illustrates the components of the signal at each integer translation.

5. There is another rather different display based on a partitioning of the time-scale plane as if the time translation index and scale index were continuous variables. This display is called "tiling the time-frequency plane." Because it is a different type of display and is developed and illustrated in Chapter 9, it will not be illustrated here.

Experimentation with these displays can be very informative in terms of the properties and capabilities of the wavelet transform, the effects of particular wavelet systems, and the way a wavelet expansion displays the various attributes or characteristics of a signal.

2.7 Examples of Wavelet Expansions

In this section, we will try to show the way a wavelet expansion decomposes a signal and what the components look like at different scales. These expansions use what is called a length-8 Daubechies basic wavelet (developed in Chapter 6), but that is not the main point here. The local nature of the wavelet decomposition is the topic of this section.

These examples are rather standard ones, some taken from David Donoho's papers and web page. The first is a decomposition of a piecewise linear function to show how edges and constants are handled. A characteristic of Daubechies systems is that low order polynomials are completely contained in the scaling function spaces V_j and need no wavelets. This means that when a section of a signal is a section of a polynomial (such as a straight line), there are no wavelet expansion coefficients $d_j(k)$, but when the calculation of the expansion coefficients overlaps an edge, there is a wavelet component. This is illustrated well in Figure 2.6 where the high resolution scales gives a very accurate location of the edges and this spreads out over k at the lower scales. This gives a hint of how the DWT could be used for edge detection and how the large number of small or zero expansion coefficients could be used for compression.

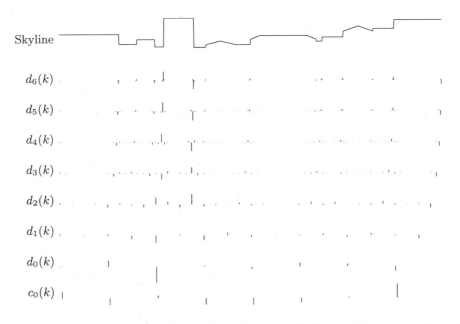

Figure 2.5. Discrete Wavelet Transform of the Houston Skyline, using $\psi_{D8'}$ with a Gain of $\sqrt{2}$ for Each Higher Scale

Figure 2.6 shows the approximations of the skyline signal in the various scaling function spaces V_j. This illustrates just how the approximations progress, giving more and more resolution at higher scales. The fact that the higher scales give more detail is similar to Fourier methods, but the localization is new. Figure 2.7 illustrates the individual wavelet decomposition by showing the components of the signal that exist in the wavelet spaces $\mathcal{W}_j$ at different scales j. This shows the same expansion as Figure 2.6, but with the wavelet components given separately rather than being cumulatively added to the scaling function. Notice how the large objects show up at the lower resolution. Groups of buildings and individual buildings are resolved according to their width. The edges, however, are located at the higher resolutions and are located very accurately.

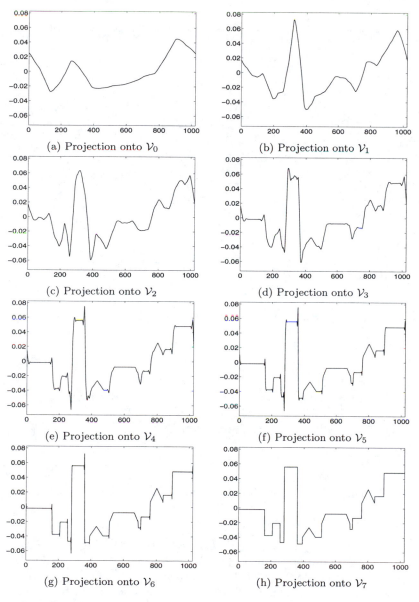

(a) Projection onto $\mathcal{V}_0$

(b) Projection onto $\mathcal{V}_1$

(c) Projection onto $\mathcal{V}_2$

(d) Projection onto $\mathcal{V}_3$

(e) Projection onto $\mathcal{V}_4$

(f) Projection onto $\mathcal{V}_5$

(g) Projection onto $\mathcal{V}_6$

(h) Projection onto $\mathcal{V}_7$

Figure 2.6. Projection of the Houston Skyline Signal onto $\mathcal{V}$ Spaces using $\phi_{D8'}$

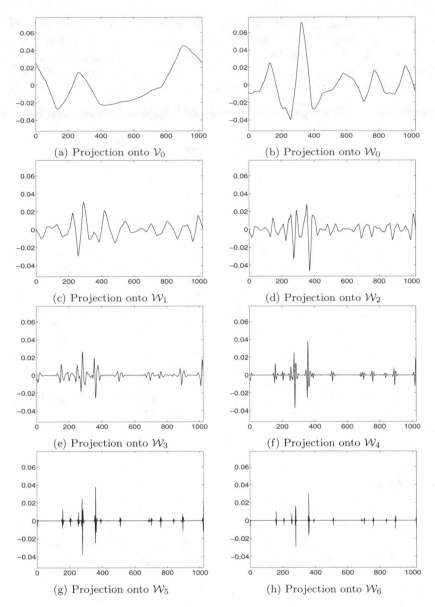

Figure 2.7. Projection of the Houston Skyline Signal onto $\mathcal{W}$ Spaces using $\psi_{D8'}$

The second example uses a chirp or doppler signal to illustrate how a time-varying frequency is described by the scale decomposition. Figure 2.8 gives the coefficients of the DWT directly as a function of j and k. Notice how the location in k tracks the frequencies in the signal in a way the Fourier transform cannot. Figures 2.9 and 2.10 show the scaling function approximations and the wavelet decomposition of this chirp signal. Again, notice in this type of display how the "location" of the frequencies are shown.

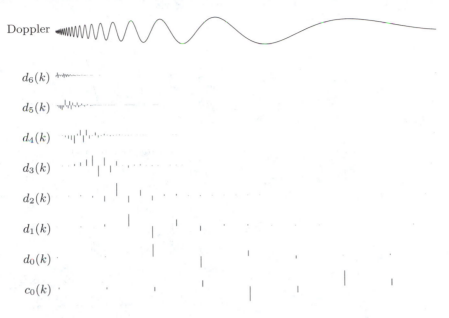

Figure 2.8. Discrete Wavelet Transform of a Doppler, using $\psi_{D8'}$ with a gain of $\sqrt{2}$ for each higher scale.

2.8   An Example of the Haar Wavelet System

In this section, we can illustrate our mathematical discussion with a more complete example. In 1910, Haar [Haa10] showed that certain square wave functions could be translated and scaled to create a basis set that spans L^2. This is illustrated in Figure 2.11. Years later, it was seen that Haar's system is a particular wavelet system.

If we choose our scaling function to have compact support over $0 \leq t \leq 1$, then a solution to (2.13) is a scaling function that is a simple rectangle function

$$\varphi(t) = \begin{cases} 1 & \text{if } 0 < t < 1 \\ 0 & \text{otherwise} \end{cases} \tag{2.42}$$

with only two nonzero coefficients $h(0) = h(1) = 1/\sqrt{2}$ and (2.24) and (2.25) require the wavelet to be

$$\psi(t) = \begin{cases} 1 & \text{for } 0 < t < 0.5 \\ -1 & \text{for } 0.5 < t < 1 \\ 0 & \text{otherwise} \end{cases} \tag{2.43}$$

with only two nonzero coefficients $h_1(0) = 1/\sqrt{2}$ and $h_1(1) = -1/\sqrt{2}$.

$\mathcal{V}_0$ is the space spanned by $\varphi(t - k)$ which is the space of piecewise constant functions over integers, a rather limited space, but nontrivial. The next higher resolution space $\mathcal{V}_1$ is spanned by $\varphi(2t - k)$ which allows a somewhat more interesting class of signals which does include $\mathcal{V}_0$. As we consider higher values of scale j, the space $\mathcal{V}_j$ spanned by $\varphi(2^j t - k)$ becomes better able to approximate arbitrary functions or signals by finer and finer piecewise constant functions.

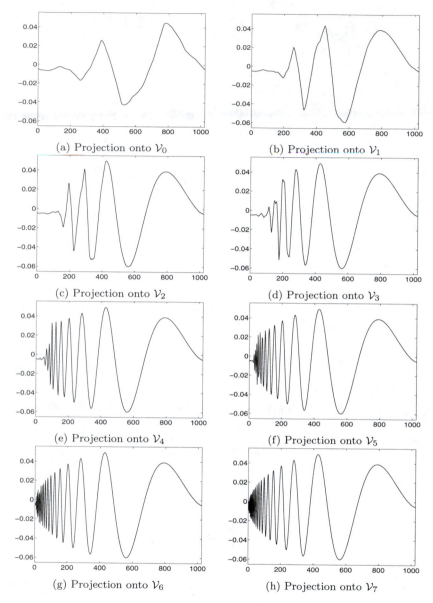

Figure 2.9. Projection of the Doppler Signal onto $\mathcal{V}$ Spaces using $\phi_{D8'}$

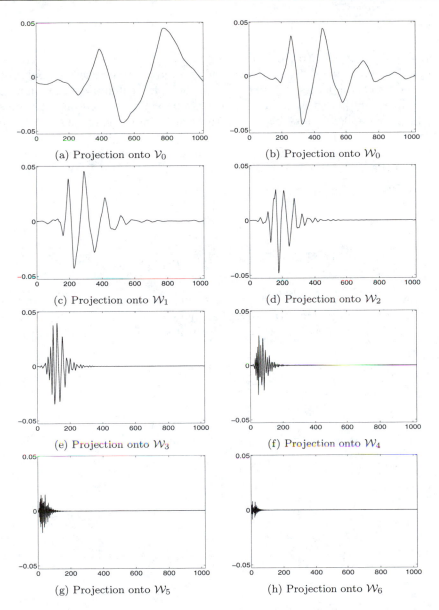

Figure 2.10. Projection of the Doppler Signal onto $\mathcal{W}$ Spaces using $\psi_{D8'}$

Haar showed that as $j \to \infty$, $\mathcal{V}_j \to L^2$. We have an approximation made up of step functions approaching any square integrable function.

The Haar functions are illustrated in Figure 2.11 where the first column contains the simple constant basis function that spans $\mathcal{V}_0$, the second column contains the unit pulse of width one half and the one translate necessary to span $\mathcal{V}_1$. The third column contains four translations of a pulse of width one fourth and the fourth contains eight translations of a pulse of width one eighth. This shows clearly how increasing the scale allows greater and greater detail to be realized. However, using only the scaling function does not allow the decomposition described in the introduction. For that we need the wavelet. Rather than use the scaling functions $\varphi(8t - k)$ in $\mathcal{V}_3$, we will use the orthogonal decomposition

$$\mathcal{V}_3 \; = \; \mathcal{V}_2 \oplus \mathcal{W}_2 \tag{2.44}$$

which is the same as

$$\overline{\underset{k}{\text{Span}}\{\varphi(8t - k)\}} \; = \; \overline{\underset{k}{\text{Span}}\{\varphi(4t - k)\}} \oplus \overline{\underset{k}{\text{Span}}\{\psi(4t - k)\}} \tag{2.45}$$

which means there are two sets of orthogonal basis functions that span $\mathcal{V}_3$, one in terms of $j = 3$ scaling functions, and the other in terms of half as many coarser $j = 2$ scaling functions plus the details contained in the $j = 2$ wavelets. This is illustrated in Figure 2.12.

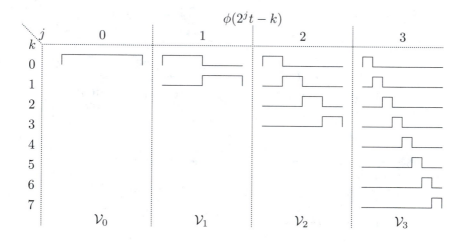

Figure 2.11. Haar Scaling Functions that Span $\mathcal{V}_j$

The $\mathcal{V}_2$ can be further decomposed into

$$\mathcal{V}_2 \; = \; \mathcal{V}_1 \oplus \mathcal{W}_1 \tag{2.46}$$

which is the same as

$$\overline{\underset{k}{\text{Span}}\{\varphi(4t - k)\}} \; = \; \overline{\underset{k}{\text{Span}}\{\varphi(2t - k)\}} \oplus \overline{\underset{k}{\text{Span}}\{\psi(2t - k)\}} \tag{2.47}$$

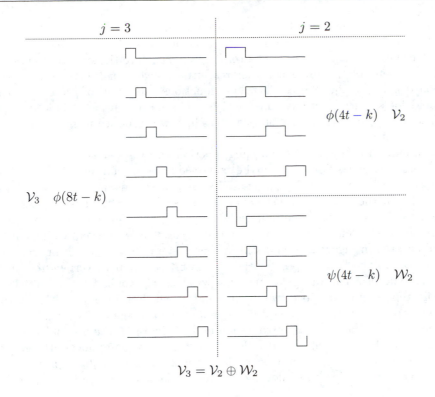

Figure 2.12. Haar Scaling Functions and Wavelets Decomposition of $\mathcal{V}_3$

and this is illustrated in Figure 2.14. This gives $\mathcal{V}_1$ also to be decomposed as

$$\mathcal{V}_1 \quad = \quad \mathcal{V}_0 \oplus \mathcal{W}_0 \tag{2.48}$$

which is shown in Figure 2.13. By continuing to decompose the space spanned by the scaling function until the space is one constant, the complete decomposition of $\mathcal{V}_3$ is obtained. This is symbolically shown in Figure 2.16.

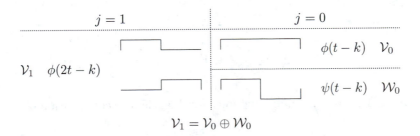

Figure 2.13. Haar Scaling Functions and Wavelets Decomposition of $\mathcal{V}_1$

Finally we look at an approximation to a smooth function constructed from the basis elements in $V_3 = V_0 \oplus W_0 \oplus W_1 \oplus W_2$. Because the Haar functions form an orthogonal basis in each subspace, they can produce an optimal least squared error approximation to the smooth function. One can easily imagine the effects of adding a higher resolution "layer" of functions to W_3 giving an approximation residing in V_4. Notice that these functions satisfy all of the conditions that we have considered for scaling functions and wavelets. The basic wavelet is indeed an oscillating function which, in fact, has an average of zero and which will produce finer and finer detail as it is scaled and translated.

The multiresolution character of the scaling function and wavelet system is easily seen from Figure 2.12 where a signal residing in V_3 can be expressed in terms of a sum of eight shifted scaling functions at scale $j = 3$ or a sum of four shifted scaling functions and four shifted wavelets at a scale of $j = 2$. In the second case, the sum of four scaling functions gives a low resolution approximation to the signal with the four wavelets giving the higher resolution "detail". The four shifted scaling functions could be further decomposed into coarser scaling functions and wavelets as illustrated in Figure 2.14 and still further decomposed as shown in Figure 2.13.

Figure 2.15 shows the Haar approximations of a test function in various resolutions. The signal is an example of a mixture of a pure sine wave which would have a perfectly localized Fourier domain representation and a two discontinuities which are completely localized in time domain. The component at the coarsest scale is simply the average of the signal. As we include more and more wavelet scales, the approximation becomes close to the original signal.

This chapter has skipped over some details in an attempt to communicate the general idea of the method. The conditions that can or must be satisfied and the resulting properties, together with examples, are discussed in the following chapters and/or in the references.

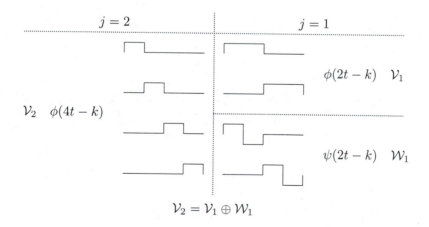

Figure 2.14. Haar Scaling Functions and Wavelets Decomposition of V_2

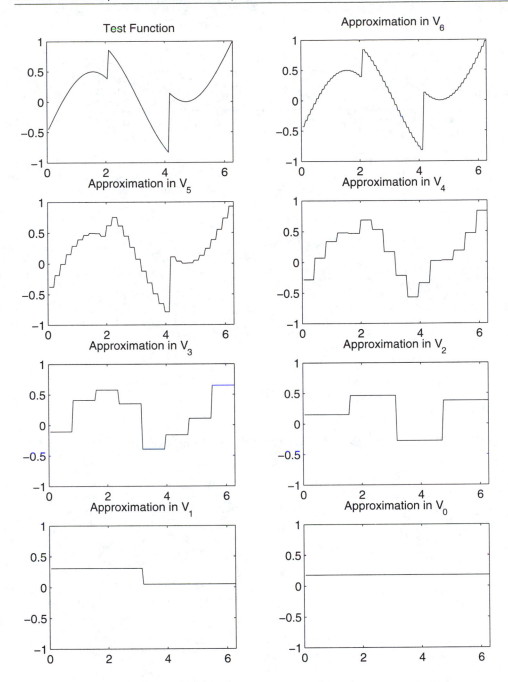

Figure 2.15. Haar Function Approximation in $\mathcal{V}_j$

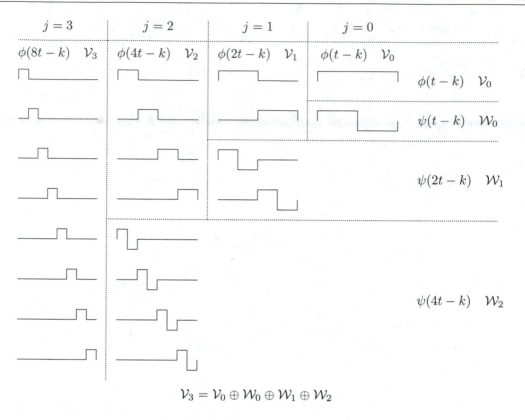

$$\mathcal{V}_3 = \mathcal{V}_0 \oplus \mathcal{W}_0 \oplus \mathcal{W}_1 \oplus \mathcal{W}_2$$

Figure 2.16. Haar Scaling Functions and Wavelets Decomposition of $\mathcal{V}_3$

Chapter 3

Filter Banks and the Discrete Wavelet Transform

In many applications, one never has to deal directly with the scaling functions or wavelets. Only the coefficients $h(n)$ and $h_1(n)$ in the defining equations (2.13) and (2.24) and $c(k)$ and $d_j(k)$ in the expansions (2.28), (2.29), and (2.30) need be considered, and they can be viewed as digital filters and digital signals respectively [GB92c, Vai92]. While it is possible to develop most of the results of wavelet theory using only filter banks, we feel that both the signal expansion point of view and the filter bank point of view are necessary for a real understanding of this new tool.

3.1   Analysis – From Fine Scale to Coarse Scale

In order to work directly with the wavelet transform coefficients, we will derive the relationship between the expansion coefficients at a lower scale level in terms of those at a higher scale. Starting with the basic recursion equation from (2.13)

$$\varphi(t) = \sum_n h(n) \sqrt{2}\, \varphi(2t - n) \tag{3.1}$$

and assuming a unique solution exists, we scale and translate the time variable to give

$$\varphi(2^j t - k) = \sum_n h(n) \sqrt{2}\, \varphi(2(2^j t - k) - n) = \sum_n h(n) \sqrt{2}\, \varphi(2^{j+1}t - 2k - n) \tag{3.2}$$

which, after changing variables $m = 2k + n$, becomes

$$\varphi(2^j t - k) = \sum_m h(m - 2k) \sqrt{2}\, \varphi(2^{j+1}t - m). \tag{3.3}$$

If we denote $\mathcal{V}_j$ as

$$\mathcal{V}_j = \underset{k}{\mathrm{Span}}\{2^{j/2}\, \varphi(2^j t - k)\} \tag{3.4}$$

then

$$f(t) \in \mathcal{V}_{j+1} \quad \Rightarrow \quad f(t) = \sum_k c_{j+1}(k)\, 2^{(j+1)/2}\, \varphi(2^{j+1}t - k) \tag{3.5}$$

is expressible at a scale of $j + 1$ with scaling functions only and no wavelets. At one scale lower resolution, wavelets are necessary for the "detail" not available at a scale of j. We have

$$f(t) = \sum_k c_j(k)\, 2^{j/2}\, \varphi(2^j t - k) + \sum_k d_j(k)\, 2^{j/2}\, \psi(2^j t - k) \tag{3.6}$$

where the $2^{j/2}$ terms maintain the unity norm of the basis functions at various scales. If $\varphi_{j,k}(t)$ and $\psi_{j,k}(t)$ are orthonormal or a tight frame, the j level scaling coefficients are found by taking the inner product

$$c_j(k) = \langle f(t), \varphi_{j,k}(t) \rangle = \int f(t)\, 2^{j/2}\, \varphi(2^j t - k)\, dt \tag{3.7}$$

which, by using (3.3) and interchanging the sum and integral, can be written as

$$c_j(k) = \sum_m h(m - 2k) \int f(t)\, 2^{(j+1)/2}\, \varphi(2^{j+1} t - m)\, dt \tag{3.8}$$

but the integral is the inner product with the scaling function at a scale of $j + 1$ giving

$$\boxed{c_j(k) = \sum_m h(m - 2k)\, c_{j+1}(m).} \tag{3.9}$$

The corresponding relationship for the wavelet coefficients is

$$\boxed{d_j(k) = \sum_m h_1(m - 2k)\, c_{j+1}(m).} \tag{3.10}$$

Filtering and Down-Sampling or Decimating

In the discipline of digital signal processing, the "filtering" of a sequence of numbers (the input signal) is achieved by convolving the sequence with another set of numbers called the filter coefficients, taps, weights, or impulse response. This makes intuitive sense if you think of a moving average with the coefficients being the weights. For an input sequence $x(n)$ and filter coefficients $h(n)$, the output sequence $y(n)$ is given by

$$y(n) = \sum_{k=0}^{N-1} h(k)\, x(n - k) \tag{3.11}$$

There is a large literature on digital filters and how to design them [PB87, OS89]. If the number of filter coefficients N is finite, the filter is called a Finite Impulse Response (FIR) filter. If the number is infinite, it is called an Infinite Impulse (IIR) filter. The design problem is the choice of the $h(n)$ to obtain some desired effect, often to remove noise or separate signals [OS89, PB87].

     In multirate digital filters, there is an assumed relation between the integer index n in the signal $x(n)$ and time. Often the sequence of numbers are simply evenly spaced samples of a function of time. Two basic operations in multirate filters are the down-sampler and the up-sampler. The down-sampler (sometimes simply called a sampler or a decimator) takes a signal

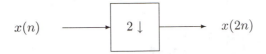

Figure 3.1. The Down Sampler or Decimator

$x(n)$ as an input and produces an output of $y(n) = x(2n)$. This is symbolically shown in Figure 3.1. In some cases, the down-sampling is by a factor other than two and in some cases, the output is the odd index terms $y(n) = x(2n+1)$, but this will be explicitly stated if it is important.

In down-sampling, there is clearly the possibility of losing information since half of the data is discarded. The effect in the frequency domain (Fourier transform) is called aliasing which states that the result of this loss of information is a mixing up of frequency components [PB87, OS89]. Only if the original signal is band-limited (half of the Fourier coefficients are zero) is there no loss of information caused by down-sampling.

We talk about digital filtering and down-sampling because that is exactly what (3.9) and (3.10) do. These equations show that the scaling and wavelet coefficients at different levels of scale can be obtained by convolving the expansion coefficients at scale j by the time-reversed recursion coefficients $h(-n)$ and $h_1(-n)$ then down-sampling or decimating (taking every other term, the even terms) to give the expansion coefficients at the next level of $j-1$. In other words, the scale-j coefficients are "filtered" by two FIR digital filters with coefficients $h(-n)$ and $h_1(-n)$ after which down-sampling gives the next coarser scaling and wavelet coefficients. These structures implement Mallat's algorithm [Mal89b, Mal89c] and have been developed in the engineering literature on filter banks, quadrature mirror filters (QMF), conjugate filters, and perfect reconstruction filter banks [SB86a, SB87, Vet86, VG89, Vet87, Vai87a, Vai92] and are expanded somewhat in Chapter 8 of this book. Mallat, Daubechies, and others showed the relation of wavelet coefficient calculation and filter banks. The implementation of equations (3.9) and (3.10) is illustrated in Figure 3.2 where the down-pointing arrows denote a decimation or down-sampling by two and the other boxes denote FIR filtering or a convolution by $h(-n)$ or $h_1(-n)$. To ease notation, we use both $h(n)$ and $h_0(n)$ to denote the scaling function coefficients for the dilation equation (2.13).

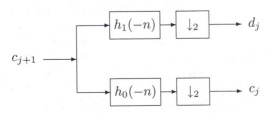

Figure 3.2. Two-Band Analysis Bank

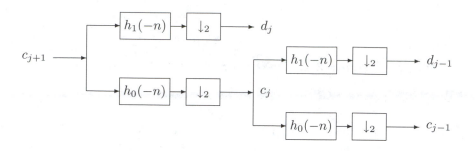

Figure 3.3. Two-Stage Two-Band Analysis Tree

As we will see in Chapter 5, the FIR filter implemented by $h(-n)$ is a lowpass filter, and the one implemented by $h_1(-n)$ is a highpass filter. Note the average number of data points out of this system is the same as the number in. The number is doubled by having two filters; then it is halved by the decimation back to the original number. This means there is the possibility that no information has been lost and it will be possible to completely recover the original signal. As we shall see, that is indeed the case. The aliasing occurring in the upper bank can be "undone" or cancelled by using the signal from the lower bank. This is the idea behind perfect reconstruction in filter bank theory [Vai92, Fli94].

This splitting, filtering, and decimation can be repeated on the scaling coefficients to give the two-scale structure in Figure 3.3. Repeating this on the scaling coefficients is called *iterating the filter bank*. Iterating the filter bank again gives us the three-scale structure in Figure 3.4.

The frequency response of a digital filter is the discrete-time Fourier transform of its impulse response (coefficients) $h(n)$. That is given by

$$H(\omega) = \sum_{n=-\infty}^{\infty} h(n)\, e^{i\omega n}. \tag{3.12}$$

The magnitude of this complex-valued function gives the ratio of the output to the input of the filter for a sampled sinusoid at a frequency of ω in radians per seconds. The angle of $H(\omega)$ is the phase shift between the output and input.

The first stage of two banks divides the spectrum of $c_{j+1}(k)$ into a lowpass and highpass band, resulting in the scaling coefficients and wavelet coefficients at lower scale $c_j(k)$ and $d_j(k)$. The second stage then divides that lowpass band into another lower lowpass band and a bandpass band. The first stage divides the spectrum into two equal parts. The second stage divides the lower half into quarters and so on. This results in a logarithmic set of bandwidths as illustrated in Figure 3.5. These are called "constant-Q" filters in filter bank language because the ratio of the band width to the center frequency of the band is constant. It is also interesting to note that a musical scale defines octaves in a similar way and that the ear responds to frequencies in a similar logarithmic fashion.

For any practical signal that is bandlimited, there will be an upper scale $j = J$, above which the wavelet coefficients, $d_j(k)$, are negligibly small [GOB94]. By starting with a high resolution description of a signal in terms of the scaling coefficients c_J, the analysis tree calculates the DWT

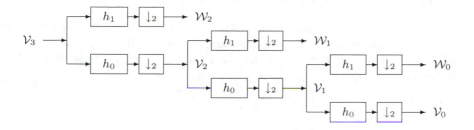

Figure 3.4. Three-Stage Two-Band Analysis Tree

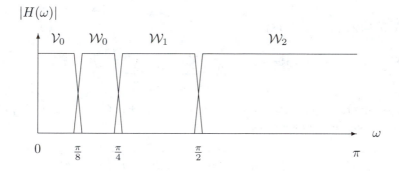

Figure 3.5. Frequency Bands for the Analysis Tree

down to as low a resolution, $j = j_0$, as desired by having $J - j_0$ stages. So, for $f(t) \in \mathcal{V}_J$, using (2.8) we have

$$f(t) = \sum_k c_J(k)\, \varphi_{J,k}(t) \tag{3.13}$$

$$= \sum_k c_{J-1}(k)\, \varphi_{J-1,k}(t) + \sum_k d_{J-1}(k)\, \psi_{J-1,k}(t) \tag{3.14}$$

$$f(t) = \sum_k c_{J-2}(k)\, \varphi_{J-2,k}(t) + \sum_k \sum_{j=J-2}^{J-1} d_j(k)\, \psi_{j,k}(t) \tag{3.15}$$

$$f(t) = \sum_k c_{j_0}(k)\, \varphi_{j_0,k}(t) + \sum_k \sum_{j=j_0}^{J-1} d_j(k)\, \psi_{j,k}(t) \tag{3.16}$$

which is a finite scale version of (2.33). We will discuss the choice of j_0 and J further in Chapter 9.

3.2   Synthesis – From Coarse Scale to Fine Scale

As one would expect, a reconstruction of the original fine scale coefficients of the signal can be made from a combination of the scaling function and wavelet coefficients at a coarse resolution. This is derived by considering a signal in the $j + 1$ scaling function space $f(t) \in V_{j+1}$. This function can be written in terms of the scaling function as

$$f(t) = \sum_k c_{j+1}(k) \, 2^{(j+1)/2} \, \varphi(2^{j+1}t - k) \tag{3.17}$$

or in terms of the next scale (which also requires wavelets) as

$$f(t) = \sum_k c_j(k) \, 2^{j/2} \, \varphi(2^j t - k) + \sum_k d_j(k) \, 2^{j/2} \, \psi(2^j t - k). \tag{3.18}$$

Substituting (3.1) and (2.24) into (3.18) gives

$$f(t) = \sum_k c_j(k) \sum_n h(n) \, 2^{(j+1)/2} \, \varphi(2^{j+1}t - 2k - n) + \sum_k d_j(k) \sum_n h_1(n) \, 2^{(j+1)/2} \, \varphi(2^{j+1}t - 2k - n).$$
$$\tag{3.19}$$

Because all of these functions are orthonormal, multiplying (3.17) and (3.19) by $\varphi(2^{j+1}t - k')$ and integrating evaluates the coefficient as

$$\boxed{c_{j+1}(k) = \sum_m c_j(m) \, h(k - 2m) + \sum_m d_j(m) \, h_1(k - 2m).} \tag{3.20}$$

Filtering and Up-Sampling or Stretching

For synthesis in the filter bank we have a sequence of first up-sampling or stretching, then filtering. This means that the input to the filter has zeros inserted between each of the original terms. In other words,

$$y(2n) = x(n) \quad \text{and} \quad y(2n + 1) = 0 \tag{3.21}$$

where the input signal is stretched to twice its original length and zeros are inserted. Clearly this up-sampling or stretching could be done with factors other than two, and the two equation above could have the $x(n)$ and 0 reversed. It is also clear that up-sampling does not lose any information. If you first up-sample then down-sample, you are back where you started. However, if you first down-sample then up-sample, you are not generally back where you started.

Our reason for discussing filtering and up-sampling here is that is exactly what the synthesis operation (3.20) does. This equation is evaluated by up-sampling the j scale coefficient sequence $c_j(k)$, which means double its length by inserting zeros between each term, then convolving it with the scaling coefficients $h(n)$. The same is done to the j level wavelet coefficient sequence and the results are added to give the $j + 1$ level scaling function coefficients. This structure is illustrated in Figure 3.6 where $g_0(n) = h(n)$ and $g_1(n) = h_1(n)$. This combining process can be continued to any level by combining the appropriate scale wavelet coefficients. The resulting two-scale tree is shown in Figure 3.7.

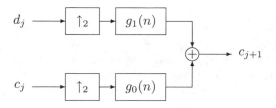

Figure 3.6. Two-Band Synthesis Bank

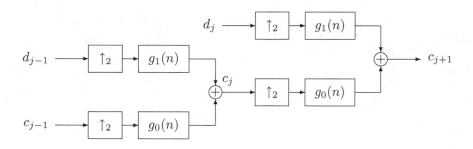

Figure 3.7. Two-Stage Two-Band Synthesis Tree

3.3 Input Coefficients

One might wonder how the input set of scaling coefficients c_{j+1} are obtained from the signal to use in the systems of Figures 3.2 and 3.3. For high enough scale, the scaling functions act as "delta functions" with the inner product to calculate the high scale coefficients as simply a sampling of $f(t)$ [GOB92, OGB92]. If the samples of $f(t)$ are above the Nyquist rate, they are good approximations to the scaling coefficients at that scale, meaning no wavelet coefficients are necessary at that scale. This approximation is particularly good if moments of the scaling function are zero or small. These ideas are further explained in Section 6.8 and Chapter 9.

An alternative approach is to "prefilter" the signal samples to make them a better approximation to the expansion coefficients. This is discussed in [Str86].

This set of analysis and synthesis operations is known as Mallat's algorithm [Mal89b, Mal89c]. The analysis filter bank efficiently calculates the DWT using banks of digital filters and downsamplers, and the synthesis filter bank calculates the inverse DWT to reconstruct the signal from the transform. Although presented here as a method of calculating the DWT, the filter bank description also gives insight into the transform itself and suggests modifications and generalizations that would be difficult to see directly from the wavelet expansion point of view. Filter banks will be used more extensively in the remainder of this book. A more general development of filter banks is presented in Section 7.2.

Although a pure wavelet expansion is possible as indicated in (1.7) and (2.22), properties of the wavelet are best developed and understood through the scaling function. This is certainly true if the scaling function has compact support because then the wavelet is composed of a finite sum of scaling functions given in (2.24).

In a practical situation where the wavelet expansion or transform is being used as a computational tool in signal processing or numerical analysis, the expansion can be made finite. If the basis functions have finite support, only a finite number of additions over k are necessary. If the scaling function is included as indicated in (2.28) or (3.6), the lower limit on the summation over j is finite. If the signal is essentially bandlimited, there is a scale above which there is little or no energy and the upper limit can be made finite. That is described in Chapter 9.

3.4   Lattices and Lifting

An alternative to using the basic two-band tree-structured filter bank is a lattice-structured filter bank. Because of the relationship between the scaling filter $h(n)$ and the wavelet filter $h_1(h)$ given in (2.25), some of the calculation can be done together with a significant savings in arithmetic. This is developed in Chapter 9 [Vai92].

Still another approach to the calculation of discrete wavelet transforms and to the calculations of the scaling functions and wavelets themselves is called "lifting." Although it is related to several other schemes [Mar92, Mar93, DM93, KS92], this idea was first explained by Wim Sweldens as a time-domain construction based on interpolation [Swe95]. Lifting does not use Fourier methods and can be applied to more general problems (e.g., nonuniform sampling) than the approach in this chapter. It was first applied to the biorthogonal system [Swe96a] and then extended to orthogonal systems [DS96a]. The application of lifting to biorthogonal is introduced in Section 7.4 later in this book. Implementations based on lifting also achieve the same improvement in arithmetic efficiency as the lattice structure do.

3.5   Different Points of View

Multiresolution versus Time-Frequency Analysis

The development of wavelet decomposition and the DWT has thus far been in terms of multiresolution where the higher scale wavelet components are considered the "detail" on a lower scale signal or image. This is indeed a powerful point of view and an accurate model for many signals and images, but there are other cases where the components of a composite signal at different scales and/or time are independent or, at least, not details of each other. If you think of a musical score as a wavelet decomposition, the higher frequency notes are not details on a lower frequency note; they are independent notes. This second point of view is more one of the time-frequency or time-scale analysis methods [Coh89, Coh95, HB92, Boa92, LP89] and may be better developed with wavelet packets (see Section 7.3), M-band wavelets (see Section 7.2), or a redundant representation (see Section 7.6), but would still be implemented by some sort of filter bank.

Periodic versus Nonperiodic Discrete Wavelet Transforms

Unlike the Fourier series, the DWT can be formulated as a periodic or a nonperiodic transform. Up until now, we have considered a nonperiodic series expansion (2.33) over $-\infty < t < \infty$ with the calculations made by the filter banks being an on-going string of coefficients at each of the scales. If the input to the filter bank has a certain rate, the output at the next lower scale will be two sequences, one of scaling function coefficients $c_{j-1,k-1}$ and one of wavelet coefficients $d_{j-1,k-1}$, each, after down-sampling, being at half the rate of the input. At the next lower scale, the same process is done on the scaling coefficients to give a total output of three strings, one at half rate and two at quarter rate. In other words, the calculation of the wavelet transform coefficients is a multirate filter bank producing sequences of coefficients at different rates but with the average number at any stage being the same. This approach can be applied to any signal, finite or infinite in length, periodic or nonperiodic. Note that while the average output rate is the same as the average input rate, the number of output coefficients is greater than the number of input coefficients because the length of the output of convolution is greater than the length of the input.

An alternative formulation that can be applied to finite duration signals or periodic signals (much as the Fourier series) is to make all of the filter bank filters cyclic or periodic convolution which is defined by

$$y(n) = \sum_{\ell=0}^{N-1} h(\ell)\, x(n - \ell),\tag{3.22}$$

for $n, \ell = 0, 1, \cdots, N - 1$ and all indices and arguments are evaluated modulo N. For a length N input at scale $j = J$, we have after one stage two length $N/2$ sequences, after two stages, one length $N/2$ and two length $N/4$ sequences, and so on. If $N = 2^J$, this can be repeated J times with the last stage being length one; one scaling function coefficient and one wavelet coefficient. An example of how the periodic DWT of a length 8 can be seen Figure 3.8.

$c_j(k)$	$d_j(k)$	$d_{j+1}(k)$	$d_{j+1}(k+1)$	$d_{j+2}(k)$	$d_{j+2}(k+1)$	$d_{j+2}(k+2)$	$d_{j+2}(k+3)$

Figure 3.8. The length-8 DWT vector

The details of this periodic approach are developed in Chapter 9 showing the aliasing that takes place in this system because of the cyclic convolution (3.22). This formulation is particularly clean because there are the same number of terms in the transform as in the signal. It can be represented by a square matrix with a simple inverse that has interesting structure. It can be efficiently calculated by an FFT although that is not needed for most applications.

For most of the theoretical developments or for conceptual purposes, there is little difference in these two formulations. However, for actual calculations and in applications, you should make sure you know which one you want or which one your software package calculates. As for the Fourier case, you can use the periodic form to calculate the nonperiodic transform by padding the signal with zeros but that wastes some of the efficiency that the periodic formulation was set up to provide.

The Discrete Wavelet Transform versus the Discrete-Time Wavelet Transform

Two more points of view concern looking at the signal processing methods in this book as based on an expansion of a signal or on multirate digital filtering. One can look at Mallat's algorithm either as a way of calculating expansion coefficients at various scales or as a filter bank for processing discrete-time signals. The first is analogous to use of the Fourier series (FS) where a continuous function is transformed into a discrete sequence of coefficients. The second is analogous to the discrete Fourier transform (DFT) where a discrete function is transformed into a discrete function. Indeed, the DFT (through the FFT) is often used to calculate the Fourier series coefficients, but care must be taken to avoid or minimize aliasing. The difference in these views comes partly from the background of the various researchers (i.e., whether they are "wavelet people" or "filter bank people"). However, there are subtle differences between using the series expansion of the signal (using the discrete wavelet transform (DWT)) and using a multirate digital filter bank on samples of the signal (using the discrete-time wavelet transform (DTWT)). Generally, using both views gives more insight into a problem than either achieves alone. The series expansion is the main approach of this book but filter banks and the DTWT are also developed in Chapters 7.8 and 8.

Numerical Complexity of the Discrete Wavelet Transform

Analysis of the number of mathematical operations (floating-point multiplications and additions) shows that calculating the DTWT of a length-N sequence of numbers using Mallat's algorithm with filter banks requires $O(N)$ operations. In other words, the number of operations is linear with the length of the signal. What is more, the constant of linearity is relatively small. This is in contrast to the FFT algorithm for calculating the DFT where the complexity is $O(N \log(N))$ or calculating a DFT directly requires $O(N^2)$ operations. It is often said that the FFT algorithm is based on a "divide and conquer" scheme, but that is misleading. The process is better described as a "organize and share" scheme. The efficiency (in fact, optimal efficiency) is based on organizing the calculations so that redundant operations can be shared. The cascaded filtering (convolution) and down-sampling of Mallat's algorithm do the same thing.

One should not make too much of this difference between the complexity of the FFT and DTWT. It comes from the DTWT having a logarithmic division of frequency bands and the FFT having a uniform division. This logarithmic scale is appropriate for many signals but if a uniform division is used for the wavelet system such as is done for wavelet packets (see Section 7.3) or the redundant DWT (see Chapter 7.6), the complexity of the wavelet system becomes $O(N \log(N))$.

If you are interested in more details of the discrete wavelet transform and the discrete-time wavelet transform, relations between them, methods of calculating them, further properties of them, or examples, see Section 7.8 and Chapter 9.

Chapter 4

Bases, Orthogonal Bases, Biorthogonal Bases, Frames, Tight Frames, and Unconditional Bases

Most people with technical backgrounds are familiar with the ideas of expansion vectors or basis vectors and of orthogonality; however, the related concepts of biorthogonality or of frames and tight frames are less familiar but also important. In the study of wavelet systems, we find that frames and tight frames are needed and should be understood, at least at a superficial level. One can find details in [You80, Dau92, Dau90, HW89]. Another perhaps unfamiliar concept is that of an unconditional basis used by Donoho, Daubechies, and others [Don93b, Mey90, Dau92] to explain why wavelets are good for signal compression, detection, and denoising [GOL*94b, GOL*94a]. In this chapter, we will very briefly define and discuss these ideas. At this point, you may want to skip these sections and perhaps refer to them later when they are specifically needed.

4.1 Bases, Orthogonal Bases, and Biorthogonal Bases

A set of vectors or functions $f_k(t)$ *spans* a vector space $\mathcal{F}$ (or $\mathcal{F}$ is the *Span* of the set) if any element of that space can be expressed as a linear combination of members of that set, meaning: Given the finite or infinite set of functions $f_k(t)$, we define $\text{Span}_k\{f_k\} = \mathcal{F}$ as the vector space with all elements of the space of the form

$$g(t) = \sum_k a_k \, f_k(t) \tag{4.1}$$

with $k \in \mathbf{Z}$ and $t, a \in \mathbf{R}$. An *inner product* is usually defined for this space and is denoted $\langle f(t), g(t) \rangle$. A *norm* is defined and is denoted by $\|f\| = \sqrt{\langle f, f \rangle}$.

We say that the set $f_k(t)$ is a *basis set* or a *basis* for a given space $\mathcal{F}$ if the set of $\{a_k\}$ in (4.1) are unique for any particular $g(t) \in \mathcal{F}$. The set is called an *orthogonal basis* if $\langle f_k(t), f_\ell(t) \rangle = 0$ for all $k \neq \ell$. If we are in three dimensional Euclidean space, orthogonal basis vectors are coordinate vectors that are at right (90°) angles to each other. We say the set is an *orthonormal basis* if $\langle f_k(t), f_\ell(t) \rangle = \delta(k - \ell)$ i.e. if, in addition to being orthogonal, the basis vectors are normalized to unity norm: $\|f_k(t)\| = 1$ for all k.

From these definitions it is clear that if we have an orthonormal basis, we can express any element in the vector space, $g(t) \in \mathcal{F}$, written as (4.1) by

$$g(t) = \sum_k \langle g(t),\, f_k(t) \rangle\, f_k(t) \tag{4.2}$$

since by taking the inner product of $f_k(t)$ with both sides of (4.1), we get

$$a_k = \langle g(t),\, f_k(t) \rangle \tag{4.3}$$

where this inner product of the signal $g(t)$ with the basis vector $f_k(t)$ "picks out" the corresponding coefficient a_k. This expansion formulation or representation is extremely valuable. It expresses (4.2) as an identity operator in the sense that the inner product operates on $g(t)$ to produce a set of coefficients that, when used to linearly combine the basis vectors, gives back the original signal $g(t)$. It is the foundation of Parseval's theorem which says the norm or energy can be partitioned in terms of the expansion coefficients a_k. It is why the interpretation, storage, transmission, approximation, compression, and manipulation of the coefficients can be very useful. Indeed, (4.2) is the form of all Fourier type methods.

Although the advantages of an orthonormal basis are clear, there are cases where the basis system dictated by the problem is not and cannot (or should not) be made orthogonal. For these cases, one can still have the expression of (4.1) and one similar to (4.2) by using a *dual basis set* $\tilde{f}_k(t)$ whose elements are not orthogonal to each other, but to the corresponding element of the expansion set

$$\langle f_\ell(t),\, \tilde{f}_k(t) \rangle \;=\; \delta(\ell - k) \tag{4.4}$$

Because this type of "orthogonality" requires two sets of vectors, the expansion set and the dual set, the system is called *biorthogonal*. Using (4.4) with the expansion in (4.1) gives

$$g(t) = \sum_k \langle g(t),\, \tilde{f}_k(t) \rangle\, f_k(t) \tag{4.5}$$

Although a biorthogonal system is more complicated in that it requires, not only the original expansion set, but the finding, calculating, and storage of a dual set of vectors, it is very general and allows a larger class of expansions. There may, however, be greater numerical problems with a biorthogonal system if some of the basis vectors are strongly correlated.

The calculation of the expansion coefficients using an inner product in (4.3) is called the *analysis* part of the complete process, and the calculation of the signal from the coefficients and expansion vectors in (4.1) is called the *synthesis* part.

In finite dimensions, analysis and synthesis operations are simply matrix–vector multiplications. If the expansion vectors in (4.1) are a basis, the synthesis matrix has these basis vectors as columns and the matrix is square and non singular. If the matrix is orthogonal, its rows and columns are orthogonal, its inverse is its transpose, and the identity operator is simply the matrix multiplied by its transpose. If it is not orthogonal, then the identity is the matrix multiplied by its inverse and the dual basis consists of the rows of the inverse. If the matrix is singular, then its columns are not independent and, therefore, do not form a basis.

Matrix Examples

Using a four dimensional space with matrices to illustrate the ideas of this chapter, the synthesis formula $g(t) = \sum_k a_k\, f_k(t)$ becomes

$$
\begin{bmatrix} g(0) \\ g(1) \\ g(2) \\ g(3) \end{bmatrix} = a_0 \begin{bmatrix} f_0(0) \\ f_0(1) \\ f_0(2) \\ f_0(3) \end{bmatrix} + a_1 \begin{bmatrix} f_1(0) \\ f_1(1) \\ f_1(2) \\ f_1(3) \end{bmatrix} + a_2 \begin{bmatrix} f_2(0) \\ f_2(1) \\ f_2(2) \\ f_2(3) \end{bmatrix} + a_3 \begin{bmatrix} f_3(0) \\ f_3(1) \\ f_3(2) \\ f_3(3) \end{bmatrix}
\tag{4.6}
$$

which can be compactly written in matrix form as

$$
\begin{bmatrix} g(0) \\ g(1) \\ g(2) \\ g(3) \end{bmatrix} = \begin{bmatrix} f_0(0) & f_1(0) & f_2(0) & f_3(0) \\ f_0(1) & f_1(1) & f_2(1) & f_3(1) \\ f_0(2) & f_1(2) & f_2(2) & f_3(2) \\ f_0(3) & f_1(3) & f_2(3) & f_3(3) \end{bmatrix} \begin{bmatrix} a_0 \\ a_1 \\ a_2 \\ a_3 \end{bmatrix}
\tag{4.7}
$$

The synthesis or expansion equation (4.1) or (4.7) becomes

$$
\mathbf{g} = \mathbf{F}\,\mathbf{a},
\tag{4.8}
$$

with the left-hand column vector $\mathbf{g}$ being the signal vector, the matrix $\mathbf{F}$ formed with the basis vectors $\mathbf{f_k}$ as columns, and the right-hand vector $\mathbf{a}$ containing the four expansion coefficients, a_k.

The equation for calculating the kth expansion coefficient in (4.6) is

$$
a_k = \langle g(t),\ \tilde{f}_k(t) \rangle = \tilde{\mathbf{f}}_k^T \mathbf{g}
\tag{4.9}
$$

which can be written in vector form as

$$
\begin{bmatrix} a_0 \\ a_1 \\ a_2 \\ a_3 \end{bmatrix} = \begin{bmatrix} \tilde{f}_0(0) & \tilde{f}_0(1) & \tilde{f}_0(2) & \tilde{f}_0(3) \\ \tilde{f}_1(0) & \tilde{f}_1(1) & \tilde{f}_1(2) & \tilde{f}_1(3) \\ \tilde{f}_2(0) & \tilde{f}_2(1) & \tilde{f}_2(2) & \tilde{f}_2(3) \\ \tilde{f}_3(0) & \tilde{f}_3(1) & \tilde{f}_3(2) & \tilde{f}_3(3) \end{bmatrix} \begin{bmatrix} g(0) \\ g(1) \\ g(2) \\ g(3) \end{bmatrix}
\tag{4.10}
$$

where each a_k is an inner product of the kth row of $\tilde{\mathbf{F}}^{\mathbf{T}}$ with $\mathbf{g}$ and analysis or coefficient equation (4.3) or (4.10) becomes

$$
\mathbf{a} = \tilde{\mathbf{F}}^{\mathbf{T}}\, \mathbf{g}
\tag{4.11}
$$

which together are (4.2) or

$$
\mathbf{g} = \mathbf{F}\, \tilde{\mathbf{F}}^{\mathbf{T}}\, \mathbf{g}.
\tag{4.12}
$$

Therefore,

$$
\tilde{\mathbf{F}}^{\mathbf{T}} = \mathbf{F}^{-1}
\tag{4.13}
$$

is how the dual basis in (4.4) is found.

If the columns of $\mathbf{F}$ are orthogonal and normalized, then

$$
\mathbf{F}\,\mathbf{F}^{\mathbf{T}} = \mathbf{I}.
\tag{4.14}
$$

This means the basis and dual basis are the same, and (4.12) and (4.13) become

$$\mathbf{g} = \mathbf{F}\,\mathbf{F^T}\,\mathbf{g} \tag{4.15}$$

and

$$\mathbf{\tilde{F}^T} = \mathbf{F^T} \tag{4.16}$$

which are both simpler and more numerically stable than (4.13).

The discrete Fourier transform (DFT) is an interesting example of a finite dimensional Fourier transform with orthogonal basis vectors where matrix and vector techniques can be informative as to the DFT's characteristics and properties. That can be found developed in several signal processing books.

Fourier Series Example

The Fourier Series is an excellent example of an infinite dimensional composition (synthesis) and decomposition (analysis). The expansion formula for an even function $g(t)$ over $0 < x < 2\pi$ is

$$g(t) = \sum_k a_k \, \cos(kt) \tag{4.17}$$

where the basis vectors (functions) are

$$f_k(t) = \cos(kt) \tag{4.18}$$

and the expansion coefficients are obtained as

$$a_k = \langle g(t),\, f_k(t) \rangle = \frac{2}{\pi} \int_0^\pi g(t) \, \cos(kt)\, dx. \tag{4.19}$$

The basis vector set is easily seen to be orthonormal by verifying

$$\langle f_\ell(t),\, f_k(t) \rangle = \delta(k - \ell). \tag{4.20}$$

These basis functions span an infinite dimensional vector space and the convergence of (4.17) must be examined. Indeed, it is the robustness of that convergence that is discussed in this section under the topic of unconditional bases.

Sinc Expansion Example

Another example of an infinite dimensional orthogonal basis is Shannon's sampling expansion [Mar91]. If $f(t)$ is band limited, then

$$f(t) = \sum_k f(Tk) \frac{\sin(\frac{\pi}{T}t - \pi k)}{\frac{\pi}{T}t - \pi k} \tag{4.21}$$

for a sampling interval $T < \frac{\pi}{W}$ if the spectrum of $f(t)$ is zero for $|\omega| > W$. In this case the basis functions are the sinc functions with coefficients which are simply samples of the original function. This means the inner product of a sinc basis function with a bandlimited function will give a sample of that function. It is easy to see that the sinc basis functions are orthogonal by taking the inner product of two sinc functions which will sample one of them at the points of value one or zero.

4.2 Frames and Tight Frames

While the conditions for a set of functions being an orthonormal basis are sufficient for the representation in (4.2) and the requirement of the set being a basis is sufficient for (4.5), they are not necessary. To be a basis requires uniqueness of the coefficients. In other words it requires that the set be *independent*, meaning no element can be written as a linear combination of the others.

If the set of functions or vectors is dependent and yet does allow the expansion described in (4.5), then the set is called a *frame*. Thus, a frame is a *spanning set*. The term frame comes from a definition that requires finite limits on an inequality bound [Dau92, You80] of inner products.

If we want the coefficients in an expansion of a signal to represent the signal well, these coefficients should have certain properties. They are stated best in terms of energy and energy bounds. For an orthogonal basis, this takes the form of Parseval's theorem. To be a frame in a signal space, an expansion set $\varphi_k(t)$ must satisfy

$$A\|g\|^2 \leq \sum_k |\langle \varphi_k, g \rangle|^2 \leq B\|g\|^2 \tag{4.22}$$

for some $0 < A$ and $B < \infty$ and for all signals $g(t)$ in the space. Dividing (4.22) by $\|g\|^2$ shows that A and B are bounds on the normalized energy of the inner products. They "frame" the normalized coefficient energy. If

$$A \ = \ B \tag{4.23}$$

then the expansion set is called a *tight frame*. This case gives

$$A\|g\|^2 \ = \ \sum_k |\langle \varphi_k, g \rangle|^2 \tag{4.24}$$

which is a generalized Parseval's theorem for tight frames. If $A = B = 1$, the tight frame becomes an orthogonal basis. From this, it can be shown that for a tight frame [Dau92]

$$g(t) = A^{-1} \sum_k \langle \varphi_k(t), g(t) \rangle \ \varphi_k(t) \tag{4.25}$$

which is the same as the expansion using an orthonormal basis except for the A^{-1} term which is a measure of the redundancy in the expansion set.

If an expansion set is a non tight frame, there is no strict Parseval's theorem and the energy in the transform domain cannot be exactly partitioned. However, the closer A and B are, the better an approximate partitioning can be done. If $A = B$, we have a tight frame and the partitioning can be done exactly with (4.24). Daubechies [Dau92] shows that the tighter the frame bounds in (4.22) are, the better the analysis and synthesis system is conditioned. In other words, if A is near or zero and/or B is very large compared to A, there will be numerical problems in the analysis–synthesis calculations.

Frames are an over-complete version of a basis set, and tight frames are an over-complete version of an orthogonal basis set. If one is using a frame that is neither a basis nor a tight frame, a dual frame set can be specified so that analysis and synthesis can be done as for a non-orthogonal basis. If a tight frame is being used, the mathematics is very similar to using an orthogonal basis. The Fourier type system in (4.25) is essentially the same as (4.2), and (4.24) is essentially a Parseval's theorem.

The use of frames and tight frames rather than bases and orthogonal bases means a certain amount of redundancy exists. In some cases, redundancy is desirable in giving a robustness to the representation so that errors or faults are less destructive. In other cases, redundancy is an inefficiency and, therefore, undesirable. The concept of a frame originates with Duffin and Schaeffer [DS52] and is discussed in [You80, Dau90, Dau92]. In finite dimensions, vectors can always be removed from a frame to get a basis, but in infinite dimensions, that is not always possible.

An example of a frame in finite dimensions is a matrix with more columns than rows but with independent rows. An example of a tight frame is a similar matrix with orthogonal rows. An example of a tight frame in infinite dimensions would be an over-sampled Shannon expansion. It is informative to examine this example.

Matrix Examples

An example of a frame of four expansion vectors f_k in a three-dimensional space would be

$$
\begin{bmatrix} g(0) \\ g(1) \\ g(2) \end{bmatrix} = \begin{bmatrix} f_0(0) & f_1(0) & f_2(0) & f_3(0) \\ f_0(1) & f_1(1) & f_2(1) & f_3(1) \\ f_0(2) & f_1(2) & f_2(2) & f_3(2) \end{bmatrix} \begin{bmatrix} a_0 \\ a_1 \\ a_2 \\ a_3 \end{bmatrix} \tag{4.26}
$$

which corresponds to the basis shown in the square matrix in (4.7). The corresponding analysis equation is

$$
\begin{bmatrix} a_0 \\ a_1 \\ a_2 \\ a_3 \end{bmatrix} = \begin{bmatrix} \tilde{f}_0(0) & \tilde{f}_0(1) & \tilde{f}_0(2) \\ \tilde{f}_1(0) & \tilde{f}_1(1) & \tilde{f}_1(2) \\ \tilde{f}_2(0) & \tilde{f}_2(1) & \tilde{f}_2(2) \\ \tilde{f}_3(0) & \tilde{f}_3(1) & \tilde{f}_3(2) \end{bmatrix} \begin{bmatrix} g(0) \\ g(1) \\ g(2) \end{bmatrix}. \tag{4.27}
$$

which corresponds to (4.10). One can calculate a set of dual frame vectors by temporarily appending an arbitrary independent row to (4.26), making the matrix square, then using the first three columns of the inverse as the dual frame vectors. This clearly illustrates the dual frame is not unique. Daubechies [Dau92] shows how to calculate an "economical" unique dual frame.

The tight frame system occurs in wavelet infinite expansions as well as other finite and infinite dimensional systems. A numerical example of a frame which is a normalized tight frame with four vectors in three dimensions is

$$
\begin{bmatrix} g(0) \\ g(1) \\ g(2) \end{bmatrix} = \frac{1}{A}\frac{1}{\sqrt{3}} \begin{bmatrix} 1 & 1 & -1 & -1 \\ 1 & -1 & 1 & -1 \\ 1 & 1 & 1 & 1 \end{bmatrix} \begin{bmatrix} a_0 \\ a_1 \\ a_2 \\ a_3 \end{bmatrix} \tag{4.28}
$$

which includes the redundancy factor from (4.25). Note the rows are orthogonal and the columns are normalized, which gives

$$
\mathbf{F}\,\mathbf{F}^{\mathbf{T}} = \frac{1}{\sqrt{3}} \begin{bmatrix} 1 & 1 & -1 & -1 \\ 1 & -1 & 1 & -1 \\ 1 & 1 & 1 & 1 \end{bmatrix} \frac{1}{\sqrt{3}} \begin{bmatrix} 1 & 1 & 1 \\ 1 & -1 & 1 \\ -1 & 1 & 1 \\ -1 & -1 & 1 \end{bmatrix} = \frac{4}{3} \begin{bmatrix} 1 & 0 & 0 \\ 0 & 1 & 0 \\ 0 & 0 & 1 \end{bmatrix} = \frac{4}{3}\mathbf{I} \tag{4.29}
$$

or

$$\mathbf{g} = \frac{1}{A} \mathbf{F} \mathbf{F}^{\mathbf{T}} \mathbf{g} \tag{4.30}$$

which is the matrix form of (4.25). The factor of $A = 4/3$ is the measure of redundancy in this tight frame using four expansion vectors in a three-dimensional space.

The identity for the expansion coefficients is

$$\mathbf{a} = \frac{1}{A} \mathbf{F}^{\mathbf{T}} \mathbf{F} \, \mathbf{a} \tag{4.31}$$

which for the numerical example gives

$$\mathbf{F^T\,F} = \frac{1}{\sqrt{3}} \begin{bmatrix} 1 & 1 & 1 \\ 1 & -1 & 1 \\ -1 & 1 & 1 \\ -1 & -1 & 1 \end{bmatrix} \frac{1}{\sqrt{3}} \begin{bmatrix} 1 & 1 & -1 & -1 \\ 1 & -1 & 1 & -1 \\ 1 & 1 & 1 & 1 \end{bmatrix} = \begin{bmatrix} 1 & 1/3 & 1/3 & -1/3 \\ 1/3 & 1 & -1/3 & 1/3 \\ 1/3 & -1/3 & 1 & 1/3 \\ -1/3 & 1/3 & 1/3 & 1 \end{bmatrix}.$$
$$\tag{4.32}$$

Although this is not a general identity operator, it is an identity operator over the three-dimensional subspace that $\mathbf{a}$ is in and it illustrates the unity norm of the rows of $\mathbf{F^T}$ and columns of $\mathbf{F}$.

If the redundancy measure A in (4.25) and (4.29) is one, the matrices must be square and the system has an orthonormal basis.

Frames are over-complete versions of non-orthogonal bases and tight frames are over-complete versions of orthonormal bases. Tight frames are important in wavelet analysis because the restrictions on the scaling function coefficients discussed in Chapter 5 guarantee not that the wavelets will be a basis, but a tight frame. In practice, however, they are usually a basis.

Sinc Expansion as a Tight Frame Example

An example of an infinite-dimensional tight frame is the generalized Shannon's sampling expansion for the over-sampled case [Mar91]. If a function is over-sampled but the sinc functions remains consistent with the upper spectral limit W, the sampling theorem becomes

$$g(t) = \frac{TW}{\pi} \sum_n g(Tn) \frac{\sin((t - Tn)W)}{(t - Tn)W} \tag{4.33}$$

or using R as the amount of over-sampling

$$RW = \frac{\pi}{T}, \quad \text{for } R \geq 1 \tag{4.34}$$

we have

$$g(t) = \frac{1}{R} \sum_n g(Tn) \frac{\sin(\frac{\pi}{RT}(t - Tn))}{\frac{\pi}{RT}(t - Tn)} \tag{4.35}$$

where the sinc functions are no longer orthogonal now. In fact, they are no longer a basis as they are not independent. They are, however, a tight frame and, therefore, act as though they were an orthogonal basis but now there is a "redundancy" factor R as a multiplier in the formula.

Notice that as R is increased from unity, (4.35) starts as (4.21) where each sample occurs where the sinc function is one or zero but becomes an expansion with the shifts still being $t = Tn$, however, the sinc functions become wider so that the samples are no longer at the zeros. If the signal is over-sampled, either the expression (4.21) or (4.35) could be used. They both are over-sampled but (4.21) allows the spectrum of the signal to increase up to the limit without distortion while (4.35) does not. The generalized sampling theorem (4.35) has a built-in filtering action which may be an advantage or it may not.

The application of frames and tight frames to what is called a redundant discrete wavelet transform (RDWT) is discussed later in Section 7.6 and their use in Section 10.3. They are also needed for certain adaptive descriptions discussed at the end of Section 7.6 where an independent subset of the expansion vectors in the frame are chosen according to some criterion to give an optimal basis.

4.3   Conditional and Unconditional Bases

A powerful point of view used by Donoho [Don93b] gives an explanation of which basis systems are best for a particular class of signals and why the wavelet system is good for a wide variety of signal classes.

Donoho defines an unconditional basis as follows. If we have a function class $\mathcal{F}$ with a norm defined and denoted $|| \cdot ||_{\mathcal{F}}$ and a basis set f_k such that every function $g \in \mathcal{F}$ has a unique representation $g = \sum_k a_k f_k$ with equality defined as a limit using the norm, we consider the infinite expansion

$$g(t) = \sum_k m_k \, a_k \, f_k(t). \tag{4.36}$$

If for all $g \in \mathcal{F}$, the infinite sum converges for all $|m_k| \le 1$, the basis is called an *unconditional basis*. This is very similar to unconditional or absolute convergence of a numerical series [Don93b, You80, Mey90]. If the convergence depends on $m_k = 1$ for some $g(t)$, the basis is called a *conditional basis*.

An unconditional basis means all subsequences converge and all sequences of subsequences converge. It means convergence does not depend on the order of the terms in the summation or on the sign of the coefficients. This implies a very robust basis where the coefficients drop off rapidly for all members of the function class. That is indeed the case for wavelets which are unconditional bases for a very wide set of function classes [Dau92, Mey93, Gri93].

Unconditional bases have a special property that makes them near-optimal for signal processing in several situations. This property has to do with the geometry of the space of expansion coefficients of a class of functions in an unconditional basis. This is described in [Don93b].

The fundamental idea of bases or frames is representing a continuous function by a sequence of expansion coefficients. We have seen that the Parseval's theorem relates the L^2 norm of the function to the ℓ^2 norm of coefficients for orthogonal bases and tight frames (4.24). Different function spaces are characterized by different norms on the continuous function. If we have an unconditional basis for the function space, the norm of the function in the space not only can be related to some norm of the coefficients in the basis expansion, but the absolute values of the coefficients have the sufficient information to establish the relation. So there is no condition on the sign or phase information of the expansion coefficients if we only care about the norm of the function, thus *unconditional*.

For this tutorial discussion, it is sufficient to know that there are theoretical reasons why wavelets are an excellent expansion system for a wide set of signal processing problems. Being an unconditional basis also sets the stage for efficient and effective nonlinear processing of the wavelet transform of a signal for compression, denoising, and detection which are discussed in Chapter 10.

Chapter 5

The Scaling Function and Scaling Coefficients, Wavelet and Wavelet Coefficients

We will now look more closely at the basic scaling function and wavelet to see when they exist and what their properties are [DD87, Mal89b, Law90, Law91b, Law91a, LR91, Dau92]. Using the same approach that is used in the theory of differential equations, we will examine the properties of $\varphi(t)$ by considering the equation of which it is a solution. The basic recursion equation (2.13) that comes from the multiresolution formulation is

$$\varphi(t) = \sum_n h(n) \sqrt{2}\, \varphi(2t - n) \tag{5.1}$$

with $h(n)$ being the scaling coefficients and $\varphi(t)$ being the scaling function which satisfies this equation which is sometimes called the *refinement equation*, the *dilation equation*, or the *multiresolution analysis equation* (MRA).

In order to state the properties accurately, some care has to be taken in specifying just what classes of functions are being considered or are allowed. We will attempt to walk a fine line to present enough detail to be correct but not so much as to obscure the main ideas and results. A few of these ideas were presented in Section 2.1 and a few more will be given in the next section. A more complete discussion can be found in [VD95], in the introductions to [VK95, Wic95, Koo93], or in any book on function analysis.

5.1 Tools and Definitions

Signal Classes

There are three classes of signals that we will be using. The most basic is called $L^2(\mathbf{R})$ which contains all functions which have a finite, well-defined integral of the square: $f \in L^2 \Rightarrow \int |f(t)|^2 \, dt = E < \infty$. This class is important because it is a generalization of normal Euclidean geometry and because it gives a simple representation of the energy in a signal.

The next most basic class is $L^1(\mathbf{R})$, which requires a finite integral of the absolute value of the function: $f \in L^1 \Rightarrow \int |f(t)| \, dt = K < \infty$. This class is important because one may interchange infinite summations and integrations with *these* functions although not necessarily with L^2 functions. These classes of function spaces can be generalized to those with $int|f(t)|^p\,dt = K < \infty$ and designated L^p.

A more general class of signals than any L^p space contains what are called *distributions*. These are generalized functions which are not defined by their having "values" but by the value of an "inner product" with a normal function. An example of a distribution would be the Dirac delta function $\delta(t)$ where it is defined by the property: $f(T) = \int f(t)\,\delta(t-T)\,dt$.

Another detail to keep in mind is that the integrals used in these definitions are *Lebesque integrals* which are somewhat more general than the basic Riemann integral. The value of a Lebesque integral is not affected by values of the function over any countable set of values of its argument (or, more generally, a set of measure zero). A function defined as one on the rationals and zero on the irrationals would have a zero Lebesque integral. As a result of this, properties derived using measure theory and Lebesque integrals are sometime said to be true "almost everywhere," meaning they may not be true over a set of measure zero.

Many of these ideas of function spaces, distributions, Lebesque measure, etc. came out of the early study of Fourier series and transforms. It is interesting that they are also important in the theory of wavelets. As with Fourier theory, one can often ignore the signal space classes and can use distributions as if they were functions, but there are some cases where these ideas are crucial. For an introductory reading of this book or of the literature, one can usually skip over the signal space designation or assume Riemann integrals. However, when a contradiction or paradox seems to arise, its resolution will probably require these details.

Fourier Transforms

We will need the Fourier transform of $\varphi(t)$ which, if it exists, is defined to be

$$\Phi(\omega) = \int_{-\infty}^{\infty} \varphi(t)\,e^{-i\omega t}\,dt \qquad (5.2)$$

and the discrete-time Fourier transform (DTFT) [OS89] of $h(n)$ defined to be

$$H(\omega) = \sum_{n=-\infty}^{\infty} h(n)\,e^{-i\omega n} \qquad (5.3)$$

where $i = \sqrt{-1}$ and n is an integer ($n \in \mathbf{Z}$). If convolution with $h(n)$ is viewed as a digital filter, as defined in Section 3.1, then the DTFT of $h(n)$ is the filter's frequency response, [OS89, PB87] which is 2π periodic.

If $\Phi(\omega)$ exits, the defining recursive equation (5.1) becomes

$$\Phi(\omega) = \frac{1}{\sqrt{2}} H(\omega/2)\,\Phi(\omega/2) \qquad (5.4)$$

which after iteration becomes

$$\Phi(\omega) = \prod_{k=1}^{\infty} \left\{ \frac{1}{\sqrt{2}} H(\frac{\omega}{2^k}) \right\} \Phi(0). \qquad (5.5)$$

if $\sum_n h(n) = \sqrt{2}$ and $\Phi(0)$ is well defined. This may be a distribution or it may be a smooth function depending on $H(\omega)$ and, therefore, $h(n)$ [VD95, Dau92]. This makes sense only if $\Phi(0)$

is well defined. Although (5.1) and (5.5) are equivalent term-by-term, the requirement of $\Phi(0)$ being well defined and the nature of the limits in the appropriate function spaces may make one preferable over the other. Notice how the zeros of $H(\omega)$ determine the zeros of $\Phi(\omega)$.

Refinement and Transition Matrices

There are two matrices that are particularly important to determining the properties of wavelet systems. The first is the *refinement matrix M*, which is obtained from the basic recursion equation (5.1) by evaluating $\varphi(t)$ at integers [MP89, DL91, DL92, SN96, Str96a]. This looks like a convolution matrix with the even (or odd) rows removed. Two particular submatrices that are used later in Section 5.10 to evaluate $\varphi(t)$ on the dyadic rationals are illustrated for $N = 6$ by

$$\sqrt{2} \begin{bmatrix} h_0 & 0 & 0 & 0 & 0 & 0 \\ h_2 & h_1 & h_0 & 0 & 0 & 0 \\ h_4 & h_3 & h_2 & h_1 & h_0 & 0 \\ 0 & h_5 & h_4 & h_3 & h_2 & h_1 \\ 0 & 0 & 0 & h_5 & h_4 & h_3 \\ 0 & 0 & 0 & 0 & 0 & h_5 \end{bmatrix} \begin{bmatrix} \varphi_0 \\ \varphi_1 \\ \varphi_2 \\ \varphi_3 \\ \varphi_4 \\ \varphi_5 \end{bmatrix} = \begin{bmatrix} \varphi_0 \\ \varphi_1 \\ \varphi_2 \\ \varphi_3 \\ \varphi_4 \\ \varphi_5 \end{bmatrix} \tag{5.6}$$

which we write in matrix form as

$$\mathbf{M_0} \, \underline{\varphi} = \underline{\varphi} \tag{5.7}$$

with $\mathbf{M_0}$ being the 6×6 matrix of the $h(n)$ and $\underline{\varphi}$ being 6×1 vectors of integer samples of $\varphi(t)$. In other words, the vector $\underline{\varphi}$ with entries $\varphi(k)$ is the eigenvector of $\mathbf{M_0}$ for an eigenvalue of unity. The second submatrix is a shifted version illustrated by

$$\sqrt{2} \begin{bmatrix} h_1 & h_0 & 0 & 0 & 0 & 0 \\ h_3 & h_2 & h_1 & h_0 & 0 & 0 \\ h_5 & h_4 & h_3 & h_2 & h_1 & h_0 \\ 0 & 0 & h_5 & h_4 & h_3 & h_2 \\ 0 & 0 & 0 & 0 & h_5 & h_4 \\ 0 & 0 & 0 & 0 & 0 & 0 \end{bmatrix} \begin{bmatrix} \varphi_0 \\ \varphi_1 \\ \varphi_2 \\ \varphi_3 \\ \varphi_4 \\ \varphi_5 \end{bmatrix} = \begin{bmatrix} \varphi_{1/2} \\ \varphi_{3/2} \\ \varphi_{5/2} \\ \varphi_{7/2} \\ \varphi_{9/2} \\ \varphi_{11/2} \end{bmatrix} \tag{5.8}$$

with the matrix being denoted $\mathbf{M_1}$. The general refinement matrix $\mathbf{M}$ is the infinite matrix of which $\mathbf{M_0}$ and $\mathbf{M_1}$ are partitions. If the matrix $\mathbf{H}$ is the convolution matrix for $h(n)$, we can denote the $\mathbf{M}$ matrix by $[\downarrow 2]\mathbf{H}$ to indicate the down-sampled convolution matrix $\mathbf{H}$. Clearly, for $\varphi(t)$ to be defined on the dyadic rationals, $\mathbf{M_0}$ must have a unity eigenvalue.

A third, less obvious but perhaps more important, matrix is called the *transition matrix* $\mathbf{T}$ and it is built up from the autocorrelation matrix of $h(n)$. The transition matrix is constructed by

$$\mathbf{T} = [\downarrow 2]\mathbf{HH^T}. \tag{5.9}$$

This matrix (sometimes called the Lawton matrix) was used by Lawton (who originally called it the Wavelet-Galerkin matrix) [Law91a] to derive necessary and sufficient conditions for an orthogonal wavelet basis. As we will see later in this chapter, its eigenvalues are also important in determining the properties of $\varphi(t)$ and the associated wavelet system.

5.2   Necessary Conditions

Theorem 1 *If $\varphi(t) \in L^1$ is a solution to the basic recursion equation (5.1) and if $\int \varphi(t)\,dt \neq 0$, then*

$$\sum_n h(n) = \sqrt{2}. \tag{5.10}$$

The proof of this theorem requires only an interchange in the order of a summation and integration (allowed in L^1) but no assumption of orthogonality of the basis functions or any other properties of $\varphi(t)$ other than a nonzero integral. The proof of this theorem and several of the others stated here are contained in Appendix A.

   This theorem shows that, unlike linear constant coefficient differential equations, not just any set of coefficients will support a solution. The coefficients must satisfy the linear equation (5.10). This is the weakest condition on the $h(n)$.

Theorem 2 *If $\varphi(t)$ is an L^1 solution to the basic recursion equation (5.1) with $\int \varphi(t)\,dt = 1$, and*

$$\sum_\ell \varphi(t - \ell) = \sum_\ell \varphi(\ell) = 1 \tag{5.11}$$

with $\Phi(\pi + 2\pi k) \neq 0$ for some k, then

$$\sum_n h(2n) = \sum_n h(2n+1) \tag{5.12}$$

where (5.11) may have to be a distributional sum. Conversely, if (5.12) is satisfied, then (5.11) is true.

Equation (5.12) is called the *fundamental condition,* and it is weaker than requiring orthogonality but stronger than (5.10). It is simply a result of requiring the equations resulting from evaluating (5.1) on the integers be consistent. Equation (5.11) is called a partitioning of unity (or the Strang condition or the Shoenberg condition).

   A similar theorem by Cavaretta, Dahman and Micchelli [CDM91] and by Jia [Jia95] states that if $\varphi \in L^p$ and the integer translates of $\varphi(t)$ form a Riesz basis for the space they span, then $\sum_n h(2n) = \sum_n h(2n+1)$.

Theorem 3 *If $\varphi(t)$ is an $L^2 \cap L^1$ solution to (5.1) and if integer translates of $\varphi(t)$ are orthogonal as defined by*

$$\int \varphi(t)\,\varphi(t-k)\,dt = E\,\delta(k) = \begin{cases} E & \text{if } k = 0 \\ 0 & \text{otherwise,} \end{cases} \tag{5.13}$$

then

$$\sum_n h(n)\,h(n-2k) = \delta(k) = \begin{cases} 1 & \text{if } k = 0 \\ 0 & \text{otherwise.} \end{cases} \tag{5.14}$$

Notice that this does not depend on a particular normalization of $\varphi(t)$.

   If $\varphi(t)$ is normalized by dividing by the square root of its energy $\sqrt{E}$, then integer translates of $\varphi(t)$ are orthonormal defined by

$$\int \varphi(t)\,\varphi(t-k)\,dt = \delta(k) = \begin{cases} 1 & \text{if } k = 0 \\ 0 & \text{otherwise.} \end{cases} \tag{5.15}$$

This theorem shows that in order for the solutions of (5.1) to be orthogonal under integer translation, it is necessary that the coefficients of the recursive equation be orthogonal themselves after decimating or downsampling by two. If $\varphi(t)$ and/or $h(n)$ are complex functions, complex conjugation must be used in (5.13), (5.14), and (5.15).

Coefficients $h(n)$ that satisfy (5.14) are called a *quadrature mirror filter* (QMF) or conjugate mirror filter (CMF), and the condition (5.14) is called the *quadratic condition* for obvious reasons.

Corollary 1 *Under the assumptions of Theorem 3, the norm of $h(n)$ is automatically unity.*

$$\sum_n |h(n)|^2 = 1 \tag{5.16}$$

Not only must the sum of $h(n)$ equal $\sqrt{2}$, but for orthogonality of the solution, the sum of the squares of $h(n)$ must be one, both independent of any normalization of $\varphi(t)$. This unity normalization of $h(n)$ is the result of the $\sqrt{2}$ term in (5.1).

Corollary 2 *Under the assumptions of Theorem 3,*

$$\sum_n h(2n) = \sum_n h(2n+1) = \frac{1}{\sqrt{2}} \tag{5.17}$$

This result is derived in the Appendix by showing that not only must the sum of $h(n)$ equal $\sqrt{2}$, but for orthogonality of the solution, the individual sums of the even and odd terms in $h(n)$ must be $1/\sqrt{2}$, independent of any normalization of $\varphi(t)$. Although stated here as necessary for orthogonality, the results hold under weaker non-orthogonal conditions as is stated in Theorem 2.

Theorem 4 *If $\varphi(t)$ has compact support on $0 \leq t \leq N-1$ and if $\varphi(t-k)$ are linearly independent, then $h(n)$ also has compact support over $0 \leq n \leq N-1$:*

$$h(n) = 0 \ \ for \ \ n < 0 \ and \ n > N-1 \tag{5.18}$$

Thus N is the length of the $h(n)$ sequence.

If the translates are not independent (or some equivalent restriction), one can have $h(n)$ with infinite support while $\varphi(t)$ has finite support [Ron92].

These theorems state that if $\varphi(t)$ has compact support and is orthogonal over integer translates, $\frac{N}{2}$ bilinear or quadratic equations (5.14) must be satisfied in addition to the one linear equation (5.10). The support or length of $h(n)$ is N, which must be an even number. The number of degrees of freedom in choosing these N coefficients is then $\frac{N}{2} - 1$. This freedom will be used in the design of a wavelet system developed in Chapter 6 and elsewhere.

5.3   Frequency Domain Necessary Conditions

We turn next to frequency domain versions of the necessary conditions for the existence of $\varphi(t)$. Some care must be taken in specifying the space of functions that the Fourier transform operates on and the space that the transform resides in. We do not go into those details in this book but the reader can consult [VD95].

Theorem 5 *If $\varphi(t)$ is a L^1 solution of the basic recursion equation (5.1), then the following equivalent conditions must be true:*

$$\sum_n h(n) = H(0) = \sqrt{2} \qquad\qquad (5.19)$$

This follows directly from (5.3) and states that the basic existence requirement (5.10) is equivalent to requiring that the FIR filter's frequency response at DC ($\omega = 0$) be $\sqrt{2}$.

Theorem 6 *For $h(n) \in \ell^1$, then*

$$\sum_n h(2n) = \sum_n h(2n+1) \quad \text{if and only if} \quad H(\pi) = 0 \qquad (5.20)$$

which says the frequency response of the FIR filter with impulse response $h(n)$ is zero at the so-called Nyquist frequency ($\omega = \pi$). This follows from (5.4) and (7.7), and supports the fact that $h(n)$ is a lowpass digital filter. This is also equivalent to the $\mathbf{M}$ and $\mathbf{T}$ matrices having a unity eigenvalue.

Theorem 7 *If $\varphi(t)$ is a solution to (5.1) in $L^2 \cap L^1$ and $\Phi(\omega)$ is a solution of (5.4) such that $\Phi(0) \neq 0$, then*

$$\int \varphi(t)\,\varphi(t-k)\,dt = \delta(k) \quad \text{if and only if} \quad \sum_\ell |\Phi(\omega + 2\pi\ell)|^2 = 1 \qquad (5.21)$$

This is a frequency domain equivalent to the time domain definition of orthogonality of the scaling function [Mal89b, Mal89c, Dau92]. It allows applying the orthonormal conditions to frequency domain arguments. It also gives insight into just what time domain orthogonality requires in the frequency domain.

Theorem 8 *For any $h(n) \in \ell^1$,*

$$\sum_n h(n)\,h(n-2k) = \delta(k) \quad \text{if and only if} \quad |H(\omega)|^2 + |H(\omega+\pi)|^2 = 2 \qquad (5.22)$$

This theorem [Law90, Hei93, Dau92] gives equivalent time and frequency domain conditions on the scaling coefficients and states that the orthogonality requirement (5.14) is equivalent to the FIR filter with $h(n)$ as coefficients being what is called a Quadrature Mirror Filter (QMF) [SB86a]. Note that (5.22), (5.19), and (5.20) require $|H(\pi/2)| = 1$ and that the filter is a "half band" filter.

5.4   Sufficient Conditions

The above are necessary conditions for $\varphi(t)$ to exist and the following are sufficient. There are many forms these could and do take but we present the following as examples and give references for more detail [DD87, Mal89b, Law90, Law91b, Law91a, LR91, Dau92, LLS97b, LLS97a, Law97].

Theorem 9 *If* $\sum_n h(n) = \sqrt{2}$ *and* $h(n)$ *has finite support or decays fast enough so that* $\sum |h(n)|(1 + |n|)^\epsilon < \infty$ *for some* $\epsilon > 0$, *then a unique (within a scalar multiple)* $\varphi(t)$ *(perhaps a distribution) exists that satisfies (5.1) and whose distributional Fourier transform satisfies (5.5).*

This [DD87, Law90, Law] can be obtained in the frequency domain by considering the convergence of (5.5). It has recently been obtained using a much more powerful approach in the time domain by Lawton [Law97].

Because this theorem uses the weakest possible condition, the results are weak. The scaling function obtained from only requiring $\sum_n h(n) = \sqrt{2}$ may be so poorly behaved as to be impossible to calculate or use. The worst cases will not support a multiresolution analysis or provide a useful expansion system.

Theorem 10 *If* $\sum_n h(2n) = \sum_n h(2n + 1) = 1/\sqrt{2}$ *and* $h(n)$ *has finite support or decays fast enough so that* $\sum |h(n)|(1 + |n|)^\epsilon < \infty$ *for some* $\epsilon > 0$, *then a* $\varphi(t)$ *(perhaps a distribution) that satisfies (5.1) exists, is unique, and is well-defined on the dyadic rationals. In addition, the distributional sum*

$$\sum_k \varphi(t - k) = 1 \tag{5.23}$$

holds.

This condition, called the fundamental condition [SN96, LLS97b], gives a slightly tighter result than Theorem 9. While the scaling function still may be a distribution not in L^1 or L^2, it is better behaved than required by Theorem 9 in being defined on the dense set of dyadic rationals. This theorem is equivalent to requiring $H(\pi) = 0$ which from the product formula (5.5) gives a better behaved $\Phi(\omega)$. It also guarantees a unity eigenvalue for $\mathbf{M}$ and $\mathbf{T}$ but not that other eigenvalues do not exist with magnitudes larger than one.

The next several theorems use the transition matrix $\mathbf{T}$ defined in (5.9) which is a down-sampled autocorrelation matrix.

Theorem 11 *If the transition matrix* $\mathbf{T}$ *has eigenvalues on or in the unit circle of the complex plane and if any on the unit circle are multiple, they have a complete set of eigenvectors, then* $\varphi(t) \in L^2$.

If T has unity magnitude eigenvalues, the successive approximation algorithm (cascade algorithm) (5.71) converges weakly to $\varphi(t) \in L^2$ [Law].

Theorem 12 *If the transition matrix* T *has a simple unity eigenvalue with all other eigenvalues having magnitude less than one, then* $\varphi(t) \in L^2$.

Here the successive approximation algorithm (cascade algorithm) converges strongly to $\varphi(t) \in L^2$. This is developed in [SN96].

If in addition to requiring (5.10), we require the quadratic coefficient conditions (5.14), a tighter result occurs which gives $\varphi(t) \in L^2(\mathbf{R})$ and a multiresolution tight frame system.

Theorem 13 *(Lawton) If $h(n)$ has finite support or decays fast enough and if $\sum_n h(n) = \sqrt{2}$ and if $\sum_n h(n) h(n - 2k) = \delta(k)$, then $\varphi(t) \in L^2(\mathbf{R})$ exists, and generates a wavelet system that is a tight frame in L^2.*

This important result from Lawton [Law90, Law91b] gives the sufficient conditions for $\varphi(t)$ to exist and generate wavelet tight frames. The proof uses an iteration of the basic recursion equation (5.1) as a successive approximation similar to Picard's method for differential equations. Indeed, this method is used to calculate $\varphi(t)$ in Section 5.10. It is interesting to note that the scaling function may be very rough, even "fractal" in nature. This may be desirable if the signal being analyzed is also rough.

Although this theorem guarantees that $\varphi(t)$ generates a tight frame, in most practical situations, the resulting system is an orthonormal basis [Law91b]. The conditions in the following theorems are generally satisfied.

Theorem 14 *(Lawton) If $h(n)$ has compact support, $\sum_n h(n) = \sqrt{2}$, and $\sum_n h(n) h(n - 2k) = \delta(k)$, then $\varphi(t - k)$ forms an orthogonal set if and only if the transition matrix T has a simple unity eigenvalue.*

This powerful result allows a simple evaluation of $h(n)$ to see if it can support a wavelet expansion system [Law90, Law91b, Law91a]. An equivalent result using the frequency response of the FIR digital filter formed from $h(n)$ was given by Cohen.

Theorem 15 *(Cohen) If $H(\omega)$ is the DTFT of $h(n)$ with compact support and $\sum_n h(n) = \sqrt{2}$ with $\sum_n h(n) h(n - 2k) = \delta(k)$, and if $H(\omega) \neq 0$ for $-\pi/3 \leq \omega \leq \pi/3$, then the $\varphi(t - k)$ satisfying (5.1) generate an orthonormal basis in L^2.*

A slightly weaker version of this frequency domain sufficient condition is easier to prove [Mal89b, Mal89c] and to extend to the M-band case for the case of no zeros allowed in $-\pi/2 \leq \omega \leq \pi/2$ [Dau92]. There are other sufficient conditions that, together with those in Theorem 13, will guarantee an orthonormal basis. Daubechies' vanishing moments will guarantee an orthogonal basis.

Theorems 5, 6, and 15 show that $h(n)$ has the characteristics of a lowpass FIR digital filter. We will later see that the FIR filter made up of the wavelet coefficients is a high pass filter and the filter bank view developed in Chapter 3 and Section 7.2 further explains this view.

Theorem 16 *If $h(n)$ has finite support and if $\varphi(t) \in L^1$, then $\varphi(t)$ has finite support [Law].*

If $\varphi(t)$ is not restricted to L^1, it may have infinite support even if $h(n)$ has finite support.

These theorems give a good picture of the relationship between the recursive equation coefficients $h(n)$ and the scaling function $\varphi(t)$ as a solution of (5.1). More properties and characteristics are presented in Section 5.8.

Wavelet System Design

One of the main purposes for presenting the rather theoretical results of this chapter is to set up the conditions for designing wavelet systems. One approach is to require the minimum sufficient conditions as constraints in an optimization or approximation, then use the remaining degrees of freedom to choose $h(n)$ that will give the best signal representation, decomposition, or compression. In some cases, the sufficient conditions are overly restrictive and it is worthwhile to use the necessary conditions and then check the design to see if it is satisfactory. In many cases, wavelet systems are designed by a frequency domain design of $H(\omega)$ using digital filter design techniques with wavelet based constraints.

5.5   The Wavelet

Although this chapter is primarily about the scaling function, some basic wavelet properties are included here.

Theorem 17 *If the scaling coefficients $h(n)$ satisfy the conditions for existence and orthogonality of the scaling function and the wavelet is defined by (2.24), then the integer translates of this wavelet span $\mathcal{W}_0$, the orthogonal compliment of $\mathcal{V}_0$, both being in $\mathcal{V}_1$, i.e., the wavelet is orthogonal to the scaling function at the same scale,*

$$\int \varphi(t-n)\,\psi(t-m)\,dt = 0, \tag{5.24}$$

if and only if the coefficients $h_1(n)$ are given by

$$h_1(n) = \pm(-1)^n\,h(N-n) \tag{5.25}$$

where N is an arbitrary odd integer chosen to conveniently position $h_1(n)$.

An outline proof is in Appendix A.

Theorem 18 *If the scaling coefficients $h(n)$ satisfy the conditions for existence and orthogonality of the scaling function and the wavelet is defined by (2.24), then the integer translates of this wavelet span $\mathcal{W}_0$, the orthogonal compliment of $\mathcal{V}_0$, both being in $\mathcal{V}_1$; i.e., the wavelet is orthogonal to the scaling function at the same scale. If*

$$\int \varphi(t-n)\,\psi(t-m)\,dt = 0 \tag{5.26}$$

then

$$\sum_n h(n)\,h_1(n-2k) = 0 \tag{5.27}$$

which is derived in Appendix A, equation (12.39).

The translation orthogonality and scaling function-wavelet orthogonality conditions in (5.14) and (5.27) can be combined to give

$$\sum_n h_\ell(n)\,h_m(n-2k) = \delta(k)\,\delta(\ell-m) \tag{5.28}$$

if $h_0(n)$ is defined as $h(n)$.

Theorem 19 *If $h(n)$ satisfies the linear and quadratic admissibility conditions of (5.10) and (5.14), then*

$$\sum_n h_1(n) = H_1(0) = 0, \tag{5.29}$$

$$|H_1(\omega)| = |H(\omega + \pi)|, \tag{5.30}$$

$$|H(\omega)|^2 + |H_1(\omega)|^2 = 2, \tag{5.31}$$

and

$$\int \psi(t)\, dt = 0. \tag{5.32}$$

The wavelet is usually scaled so that its norm is unity.

The results in this section have not included the effects of integer shifts of the scaling function or wavelet coefficients $h(n)$ or $h_1(n)$. In a particular situation, these sequences may be shifted to make the corresponding FIR filter causal.

5.6   Alternate Normalizations

An alternate normalization of the scaling coefficients is used by some authors. In some ways, it is a cleaner form than that used here, but it does not state the basic recursion as a normalized expansion, and it does not result in a unity norm for $h(n)$. The alternate normalization uses the basic multiresolution recursive equation with no $\sqrt{2}$

$$\varphi(t) = \sum_n h(n)\, \varphi(2t - n). \tag{5.33}$$

Some of the relationships and results using this normalization are:

$$
\begin{aligned}
&\sum_n h(n) = 2 \\
&\sum_n |h(n)|^2 = 2 \\
&\sum_n h(n)\, h(h - 2k) = 2\, \delta(k) \\
&\sum_n h(2n) = \sum_n h(2n + 1) = 1 \\
&H(0) = 2 \\
&|H(\omega)|^2 + |H(\omega + \pi)|^2 = 4
\end{aligned}
\tag{5.34}
$$

A still different normalization occasionally used has a factor of 2 in (5.33) rather than $\sqrt{2}$ or unity, giving $\sum_n h(n) = 1$. Other obvious modifications of the results in other places in this book can be worked out. Take care in using scaling coefficients $h(n)$ from the literature as some must be multiplied or divided by $\sqrt{2}$ to be consistent with this book.

5.7   Example Scaling Functions and Wavelets

Several of the modern wavelets had never been seen or described before the 1980's. This section looks at some of the most common wavelet systems.

Haar Wavelets

The oldest and most basic of the wavelet systems that has most of our desired properties is constructed from the Haar basis functions. If one chooses a length $N = 2$ scaling coefficient set, after satisfying the necessary conditions in (5.10) and (5.14), there are no remaining degrees or freedom. The unique (within normalization) coefficients are

$$h(n) = \left\{ \frac{1}{\sqrt{2}}, \ \frac{1}{\sqrt{2}} \right\} \tag{5.35}$$

and the resulting normalized scaling function is

$$\varphi(t) = \begin{cases} 1 & \text{for } 0 < t < 1 \\ 0 & \text{otherwise.} \end{cases} \tag{5.36}$$

The wavelet is, therefore,

$$\psi(t) = \begin{cases} 1 & \text{for } 0 < t < 1/2 \\ -1 & \text{for } 1/2 < t < 1 \\ 0 & \text{otherwise.} \end{cases} \tag{5.37}$$

Their satisfying the multiresolution equation (2.13) is illustrated in Figure 2.2. Haar showed that translates and scalings of these functions form an orthonormal basis for $L^2(\mathcal{R})$. We can easily see that the Haar functions are also a compact support orthonormal wavelet system that satisfy Daubechies' conditions. Although they are as regular as can be achieved for $N = 2$, they are not even continuous. The orthogonality and nesting of spanned subspaces are easily seen because the translates have no overlap in the time domain. It is instructive to apply the various properties of Sections 5 and 5.8 to these functions and see how they are satisfied. They are illustrated in the example in Figures 2.11 through 2.15.

Sinc Wavelets

The next best known (perhaps the best known) basis set is that formed by the sinc functions. The sinc functions are usually presented in the context of the Shannon sampling theorem, but we can look at translates of the sinc function as an orthonormal set of basis functions (or, in some cases, a tight frame). They, likewise, usually form a orthonormal wavelet system satisfying the various required conditions of a multiresolution system.

The sinc function is defined as

$$\text{sinc}(t) = \frac{\sin(t)}{t} \tag{5.38}$$

where $\text{sinc}(0) = 1$. This is a very versatile and useful function because its Fourier transform is a simple rectangle function and the Fourier transform of a rectangle function is a sinc function. In order to be a scaling function, the sinc must satisfy (2.13) as

$$\text{sinc}(Kt) = \sum_n h(n) \, \text{sinc}(K2t - Kn) \tag{5.39}$$

for the appropriate scaling coefficients $h(n)$ and some K. If we construct the scaling function from the generalized sampling function as presented in (4.35), the sinc function becomes

$$\text{sinc}(Kt) = \sum_n \text{sinc}(KTn) \, \text{sinc}(\frac{\pi}{RT}t - \frac{\pi}{R}n). \tag{5.40}$$

In order for these two equations to be true, the sampling period must be $T = 1/2$ and the parameter

$$K = \frac{\pi}{R} \tag{5.41}$$

which gives the scaling coefficients as

$$h(n) = \operatorname{sinc}(\frac{\pi}{2R}n). \tag{5.42}$$

We see that $\varphi(t) = \operatorname{sinc}(Kt)$ is a scaling function with infinite support and its corresponding scaling coefficients are samples of a sinc function. It $R = 1$, then $K = \pi$ and the scaling function generates an orthogonal wavelet system. For $R > 1$, the wavelet system is a tight frame, the expansion set is not orthogonal or a basis, and R is the amount of redundancy in the system as discussed in Chapter 5.

For the orthogonal sinc scaling function, the wavelet is simply expressed by

$$\psi(t) = 2\,\varphi(2t) - \varphi(t). \tag{5.43}$$

The sinc scaling function and wavelet do not have compact support, but they do illustrate an infinitely differentiable set of functions that result from an infinitely long $h(n)$. The orthogonality and multiresolution characteristics of the orthogonal sinc basis is best seen in the frequency domain where there is no overlap of the spectra. Indeed, the Haar and sinc systems are Fourier duals of each other. The sinc generating scaling function and wavelet are shown in Figure 5.1.

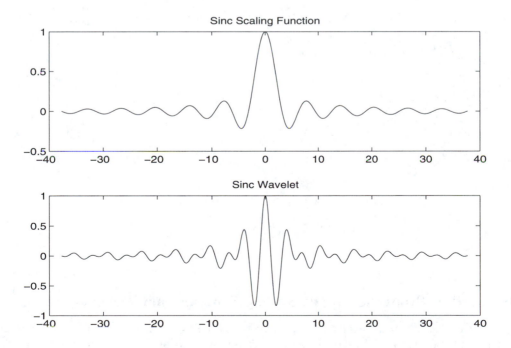

Figure 5.1. Sinc Scaling Function and Wavelet

Spline and Battle-Lemarié Wavelet Systems

The triangle scaling function illustrated in Figure 2.2 is a special case of a more general family of spline scaling functions. The scaling coefficient system $h(n) = \{\frac{1}{2\sqrt{2}}, \frac{1}{\sqrt{2}}, \frac{1}{2\sqrt{2}}, 0\}$ gives rise to the piecewise linear, continuous triangle scaling function. This function is a first-order spline, being a concatenation of two first order polynomials to be continuous at the junctions or "knots". A quadratic spline is generated from $h = \{1/4, 3/4, 3/4, 1/4\}/\sqrt{2}$ as three sections of second order polynomials connected to give continuous first order derivatives at the junctions. The cubic spline is generated from $h(n) = \{1/16.1/4, 3/8, 1/4, 1/16\}/\sqrt{2}$. This is generalized to an arbitrary Nth order spline with continuous $(N-1)$th order derivatives and with compact support of $N+1$. These functions have excellent mathematical properties, but they are not orthogonal over integer translation. If orthogonalized, their support becomes infinite (but rapidly decaying) and they generate the "Battle-Lemarié wavelet system" [Dau92, SN96, Chu92a, Chu92b]. Figure 5.2 illustrates the first-order spline scaling function which is the triangle function along with the second-, third-, and fourth-order spline scaling functions.

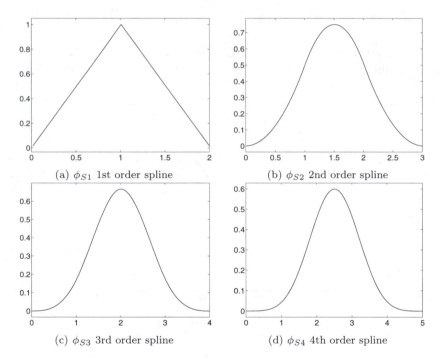

(a) ϕ_{S1} 1st order spline   (b) ϕ_{S2} 2nd order spline

(c) ϕ_{S3} 3rd order spline   (d) ϕ_{S4} 4th order spline

Figure 5.2. Spline Scaling Functions

5.8   Further Properties of the Scaling Function and Wavelet

The scaling function and wavelet have some remarkable properties that should be examined in order to understand wavelet analysis and to gain some intuition for these systems. Likewise, the scaling and wavelet coefficients have important properties that should be considered.

We now look further at the properties of the scaling function and the wavelet in terms of the basic defining equations and restrictions. We also consider the relationship of the scaling function and wavelet to the equation coefficients. A multiplicity or rank of two is used here but the more general multiplicity-M case is easily derived from these (See Section 7.2 and Appendix B). Derivations or proofs for some of these properties are included in Appendix B.

The basic recursive equation for the scaling function, defined in (5.1) as

$$\varphi(t) = \sum_n h(n) \sqrt{2}\, \varphi(2t - n), \tag{5.44}$$

is homogeneous, so its solution is unique only within a normalization factor. In most cases, both the scaling function and wavelet are normalized to unit energy or unit norm. In the properties discussed here, we normalize the energy as $E = \int |\varphi(t)|^2\, dt = 1$. Other normalizations can easily be used if desired.

General Properties not Requiring Orthogonality

There are several properties that are simply a result of the multiresolution equation (5.44) and, therefore, hold for orthogonal and biorthogonal systems.

Property 1 *The normalization of $\varphi(t)$ is arbitrary and is given in (5.13) as E. Here we usually set $E = 1$ so that the basis functions are orthonormal and coefficients can easily be calculated with inner products.*

$$\int |\varphi(t)|^2\, dt = E = 1 \tag{5.45}$$

Property 2 *Not only can the scaling function be written as a weighted sum of functions in the next higher scale space as stated in the basic recursion equation (5.44), but it can also be expressed in higher resolution spaces:*

$$\varphi(t) = \sum_n h^{(j)}(n)\, 2^{j/2}\, \varphi(2^j t - n) \tag{5.46}$$

where $h^{(1)}(n) = h(n)$ and for $j \geq 1$

$$h^{(j+1)}(n) = \sum_k h^{(j)}(k)\, h^{(j)}(n - 2k). \tag{5.47}$$

Property 3 *A formula for the sum of dyadic samples of $\varphi(t)$*

$$\sum_k \varphi\left(\frac{k}{2^J}\right) = 2^J \tag{5.48}$$

Property 4 *A "partition of unity" follows from (5.48) for $J = 0$*

$$\sum_m \varphi(m) = 1 \tag{5.49}$$

Property 5 *A generalized partition of unity exists if $\varphi(t)$ is continuous*

$$\sum_m \varphi(t - m) = 1 \tag{5.50}$$

Property 6 *A frequency domain statement of the basic recursion equation (5.44)*

$$\Phi(\omega) = \frac{1}{\sqrt{2}} H(\omega/2) \Phi(\omega/2) \tag{5.51}$$

Property 7 *Successive approximations in the frequency domain is often easier to analyze than the time domain version in (5.44). The convergence properties of this infinite product are very important.*

$$\Phi(\omega) = \prod_{k=1}^{\infty} \left\{ \frac{1}{\sqrt{2}} H(\frac{\omega}{2^k}) \right\} \Phi(0) \tag{5.52}$$

This formula is derived in (5.74).

Properties that Depend on Orthogonality

The following properties depend on the orthogonality of the scaling and wavelet functions.

Property 8 *The square of the integral of $\varphi(t)$ is equal to the integral of the square of $\varphi(t)$, or $A_0^2 = E$.*

$$\left[\int \varphi(t) \, dt \right]^2 = \int \varphi(t)^2 \, dt \tag{5.53}$$

Property 9 *The integral of the wavelet is necessarily zero*

$$\int \psi(t) \, dt = 0 \tag{5.54}$$

The norm of the wavelet is usually normalized to one such that $\int |\psi(t)|^2 \, dt = 1$.

Property 10 *Not only are integer translates of the wavelet orthogonal; different scales are also orthogonal.*

$$\int 2^{j/2} \psi(2^j t - k) \, 2^{i/2} \psi(2^i t - \ell) \, dt = \delta(k - \ell) \, \delta(j - i) \tag{5.55}$$

where the norm of $\psi(t)$ is one.

Property 11 *The scaling function and wavelet are orthogonal over both scale and translation.*

$$\int 2^{j/2} \psi(2^j t - k) \, 2^{i/2} \varphi(2^i t - \ell) \, dt = 0 \tag{5.56}$$

for all integer i, j, k, ℓ where $j \leq i$.

Property 12 *A frequency domain statement of the orthogonality requirements in (5.13) . It also is a statement of equivalent energy measures in the time and frequency domains as in Parseval's theorem, which is true with an orthogonal basis set.*

$$\sum_k |\Phi(\omega + 2\pi k)|^2 = \int |\Phi(\omega)|^2 \, d\omega = \int |\varphi(t)|^2 \, dt = 1 \tag{5.57}$$

Property 13 *The scaling coefficients can be calculated from the orthogonal or tight frame scaling functions by*

$$h(n) = \sqrt{2} \int \varphi(t)\varphi(2t - n)\, dt. \tag{5.58}$$

Property 14 *The wavelet coefficients can be calculated from the orthogonal or tight frame scaling functions by*

$$h_1(n) = \sqrt{2} \int \psi(t)\, \varphi(2t - n)\, dt. \tag{5.59}$$

Derivations of some of these properties can be found in Appendix B. Properties in equations (5.1), (5.10), (5.14), (5.53), (5.51), (5.52), and (5.57) are independent of any normalization of $\varphi(t)$. Normalization affects the others. Those in equations (5.1), (5.10), (5.48), (5.49), (5.51), (5.52), and (5.57) do not require orthogonality of integer translates of $\varphi(t)$. Those in (5.14), (5.16), (5.17), (5.22), (5.20), (5.53), (5.58) require orthogonality. No properties require compact support. Many of the derivations interchange order of summations or of summation and integration. Conditions for those interchanges must be met.

5.9 Parameterization of the Scaling Coefficients

The case where $\varphi(t)$ and $h(n)$ have compact support is very important. It aids in the time localization properties of the DWT and often reduces the computational requirements of calculating the DWT. If $h(n)$ has compact support, then the filters described in Chapter 3 are simple FIR filters. We have stated that N, the length of the sequence $h(n)$, must be even and $h(n)$ must satisfy the linear constraint of (5.10) and the $\frac{N}{2}$ bilinear constraints of (5.14). This leaves $\frac{N}{2} - 1$ degrees of freedom in choosing $h(n)$ that will still guarantee the existence of $\varphi(t)$ and a set of essentially orthogonal basis functions generated from $\varphi(t)$.

Length-2 Scaling Coefficient Vector

For a length-2 $h(n)$, there are no degrees of freedom left after satisfying the required conditions in (5.10) and (5.14). These requirements are

$$h(0) + h(1) = \sqrt{2} \tag{5.60}$$

and

$$h^2(0) + h^2(1) = 1 \tag{5.61}$$

which are uniquely satisfied by

$$h_{D2} = \{h(0), h(1)\} = \left\{ \frac{1}{\sqrt{2}}, \frac{1}{\sqrt{2}} \right\}. \tag{5.62}$$

These are the Haar scaling functions coefficients which are also the length-2 Daubechies coefficients [Dau92] used as an example in Chapter 2 and discussed later in this book.

Length-4 Scaling Coefficient Vector

For the length-4 coefficient sequence, there is one degree of freedom or one parameter that gives all the coefficients that satisfy the required conditions:

$$h(0) + h(1) + h(2) + h(3) = \sqrt{2}, \tag{5.63}$$

$$h^2(0) + h^2(1) + h^2(2) + h^2(3) = 1 \tag{5.64}$$

and

$$h(0)\,h(2) + h(1)\,h(3) = 0 \tag{5.65}$$

Letting the parameter be the angle α, the coefficients become

$$
\begin{aligned}
h(0) &= (1 - \cos(\alpha) + \sin(\alpha))/(2\sqrt{2}) \\
h(1) &= (1 + \cos(\alpha) + \sin(\alpha))/(2\sqrt{2}) \\
h(2) &= (1 + \cos(\alpha) - \sin(\alpha))/(2\sqrt{2}) \\
h(3) &= (1 - \cos(\alpha) - \sin(\alpha))/(2\sqrt{2}).
\end{aligned}
\tag{5.66}
$$

These equations also give the length-2 Haar coefficients (5.62) for $\alpha = 0, \pi/2, 3\pi/2$ and a degenerate condition for $\alpha = \pi$. We get the Daubechies coefficients (discussed later in this book) for $\alpha = \pi/3$. These Daubechies-4 coefficients have a particularly clean form,

$$h_{D4} = \left\{ \frac{1+\sqrt{3}}{4\sqrt{2}}, \frac{3+\sqrt{3}}{4\sqrt{2}}, \frac{3-\sqrt{3}}{4\sqrt{2}}, \frac{1-\sqrt{3}}{4\sqrt{2}} \right\} \tag{5.67}$$

Length-6 Scaling Coefficient Vector

For a length-6 coefficient sequence $h(n)$, the two parameters are defined as α and β and the resulting coefficients are

$$
\begin{aligned}
h(0) &= [(1 + \cos(\alpha) + \sin(\alpha))(1 - \cos(\beta) - \sin(\beta)) + 2\sin(\beta)\,\cos(\alpha)]/(4\sqrt{2}) \\
h(1) &= [(1 - \cos(\alpha) + \sin(\alpha))(1 + \cos(\beta) - \sin(\beta)) - 2\sin(\beta)\,\cos(\alpha)]/(4\sqrt{2}) \\
h(2) &= [1 + \cos(\alpha - \beta) + \sin(\alpha - \beta)]/(2\sqrt{2}) \\
h(3) &= [1 + \cos(\alpha - \beta) - \sin(\alpha - \beta)]/(2\sqrt{2}) \\
h(4) &= 1/\sqrt{2} - h(0) - h(2) \\
h(5) &= 1/\sqrt{2} - h(1) - h(3)
\end{aligned}
\tag{5.68}
$$

Here the Haar coefficients are generated for any $\alpha = \beta$ and the length-4 coefficients (5.66) result if $\beta = 0$ with α being the free parameter. The length-4 Daubechies coefficients are calculated for $\alpha = \pi/3$ and $\beta = 0$. The length-6 Daubechies coefficients result from $\alpha = 1.35980373244182$ and $\beta = -0.78210638474440$.

The inverse of these formulas which will give α and β from the allowed $h(n)$ are

$$\alpha = \arctan\left(\frac{2(h(0)^2 + h(1)^2) - 1 + (h(2) + h(3))/\sqrt{2}}{2\,(h(1)\,h(2) - h(0)\,h(3)) + \sqrt{2}(h(0) - h(1))} \right) \tag{5.69}$$

$$\beta = \alpha - \arctan\left(\frac{h(2) - h(3)}{h(2) + h(3) - 1/\sqrt{2}} \right) \tag{5.70}$$

As α and β range over $-\pi$ to π all possible $h(n)$ are generated. This allows informative experimentation to better see what these compactly supported wavelets look like. This parameterization is implemented in the MATLAB programs in Appendix C and in the Aware, Inc. software, UltraWave [The89].

Since the scaling functions and wavelets are used with integer translations, the location of their support is not important, only the size of the support. Some authors shift $h(n)$, $h_1(n)$, $\varphi(t)$, and $\psi(t)$ to be approximately centered around the origin. This is achieved by having the initial nonzero scaling coefficient start at $n = -\frac{N}{2} + 1$ rather than zero. We prefer to have the origin at $n = t = 0$.

MATLAB programs that calculate $h(n)$ for $N = 2, 4, 6$ are furnished in Appendix C. They calculate $h(n)$ from α and β according to (5.62), (5.66), and (5.68). They also work backwards to calculate α and β from allowable $h(n)$ using (5.70). A program is also included that calculates the Daubechies coefficients for any length using the spectral factorization techniques in [Dau92] and Chapter 6 of this book.

Longer $h(n)$ sequences are more difficult to parameterize but can be done with the techniques of Pollen [Polar] and Wells [Wel93] or the lattice factorization by Viadyanathan [Vai92] developed in Chapter 8. Selesnick derived explicit formulas for $N = 8$ using the symbolic software system, Maple, and set up the formulation for longer lengths [Sel97]. It is over the space of these independent parameters that one can find optimal wavelets for a particular problem or class of signals [Dau92, GOB94].

5.10 Calculating the Basic Scaling Function and Wavelet

Although one never explicitly uses the scaling function or wavelet (one uses the scaling and wavelet coefficients) in most practical applications, it is enlightening to consider methods to calculate $\varphi(t)$ and $\psi(t)$. There are two approaches that we will discuss. The first is a form of successive approximations that is used theoretically to prove existence and uniqueness of $\varphi(t)$ and can also be used to actually calculate them. This can be done in the time domain to find $\varphi(t)$ or in the frequency domain to find the Fourier transform of $\varphi(t)$ which is denoted $\Phi(\omega)$. The second method solves for the exact values of $\varphi(t)$ on the integers by solving a set of simultaneous equations. From these values, it is possible to then exactly calculate values at the half integers, then at the quarter integers and so on, giving values of $\varphi(t)$ on what are called the dyadic rationals.

Successive Approximations or the Cascade Algorithm

In order to solve the basic recursion equation (5.1), we propose an iterative algorithm that will generate successive approximations to $\varphi(t)$. If the algorithm converges to a fixed point, then that fixed point is a solution to (5.1). The iterations are defined by

$$\varphi^{(k+1)}(t) \;=\; \sum_{n=0}^{N-1} h(n)\,\sqrt{2}\,\varphi^{(k)}(2t - n) \tag{5.71}$$

for the kth iteration where an initial $\varphi^{(0)}(t)$ must be given. Because this can be viewed as applying the same operation over and over to the output of the previous application, it is sometimes called the *cascade algorithm*.

Using definitions (5.2) and (5.3), the frequency domain form becomes

$$\Phi^{(k+1)}(\omega) = \frac{1}{\sqrt{2}} H\left(\frac{\omega}{2}\right) \Phi^{(k)}\left(\frac{\omega}{2}\right) \tag{5.72}$$

and the limit can be written as an infinite product in the form

$$\Phi^{(\infty)}(\omega) = \left[\prod_{k=1}^{\infty}\left[\frac{1}{\sqrt{2}} H\left(\frac{\omega}{2^k}\right)\right]\right] \Phi^{(\infty)}(0). \tag{5.73}$$

If this limit exists, the Fourier transform of the scaling function is

$$\Phi(\omega) = \left[\prod_{k=1}^{\infty}\left[\frac{1}{\sqrt{2}} H\left(\frac{\omega}{2^k}\right)\right]\right] \Phi(0). \tag{5.74}$$

The limit does not depend on the shape of the initial $\varphi^{(0)}(t)$, but only on $\Phi^{(k)}(0) = \int \varphi^{(k)}(t)\,dt = A_0$, which is invariant over the iterations. This only makes sense if the limit of $\Phi(\omega)$ is well-defined as when it is continuous at $\omega = 0$.

The MATLAB program in Appendix C implements the algorithm in (5.71) which converges reliably to $\varphi(t)$, even when it is very discontinuous. From this scaling function, the wavelet can be generated from (2.24). It is interesting to try this algorithm, plotting the function at each iteration, on both admissible $h(n)$ that satisfy (5.10) and (5.14) and on inadmissible $h(n)$. The calculation of a scaling function for $N = 4$ is shown at each iteration in Figure 5.3.

Because of the iterative form of this algorithm, applying the same process over and over, it is sometimes called the *cascade algorithm* [SN96, Str96a].

Iterating the Filter Bank

An interesting method for calculating the scaling function also uses an iterative procedure which consists of the stages of the filter structure of Chapter 3 which calculates wavelet expansions coefficients (DWT values) at one scale from those at another. A scaling function, wavelet expansion of a scaling function itself would be a single nonzero coefficient at the scale of $j = 1$. Passing this single coefficient through the synthesis filter structure of Figure 3.7 and equation (3.20) would result in a fine scale output that for large j would essentially be samples of the scaling function.

Successive approximations in the frequency domain

The Fourier transform of the scaling function defined in (5.2) is an important tool for studying and developing wavelet theory. It could be approximately calculated by taking the DFT of the samples of $\varphi(t)$ but a more direct approach is available using the infinite product in (5.74). From this formulation we can see how the zeros of $H(\omega)$ determine the zeros of $\Phi(\omega)$. The existence conditions in Theorem 5 require $H(\pi) = 0$ or, more generally, $H(\omega) = 0$ for $\omega = (2k+1)\pi$. Equation (5.74) gives the relation of these zeros of $H(\omega)$ to the zeros of $\Phi(\omega)$. For the index $k = 1$, $H(\omega/2) = 0$ at $\omega = 2(2k+1)\pi$. For $k = 2$, $H(\omega/4) = 0$ at $\omega = 4(2k+1)\pi$, $H(\omega/8) = 0$

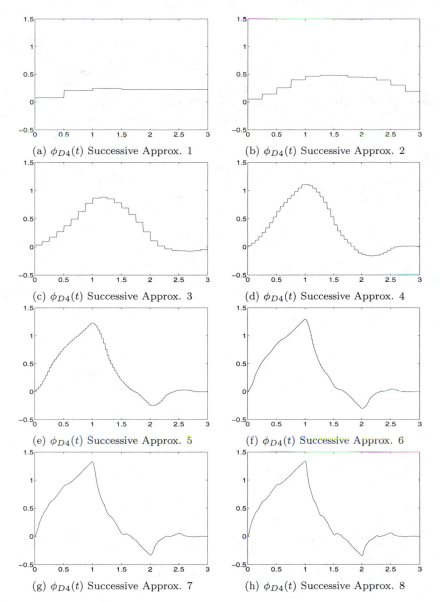

(a) $\phi_{D4}(t)$ Successive Approx. 1

(b) $\phi_{D4}(t)$ Successive Approx. 2

(c) $\phi_{D4}(t)$ Successive Approx. 3

(d) $\phi_{D4}(t)$ Successive Approx. 4

(e) $\phi_{D4}(t)$ Successive Approx. 5

(f) $\phi_{D4}(t)$ Successive Approx. 6

(g) $\phi_{D4}(t)$ Successive Approx. 7

(h) $\phi_{D4}(t)$ Successive Approx. 8

Figure 5.3. Iterations of the Successive Approximations for φ_{D4}

at $\omega = 8(2k+1)\pi$, etc. Because (5.74) is a product of stretched versions of $H(\omega)$, these zeros of $H(\omega/2^j)$ are the zeros of the Fourier transform of $\varphi(t)$. Recall from Theorem 15 that $H(\omega)$ has no zeros in $-\pi/3 < \omega < \pi/3$. All of this gives a picture of the shape of $\Phi(\omega)$ and the location of its zeros. From an asymptotic analysis of $\Phi(\omega)$ as $\omega \to \infty$, one can study the smoothness of $\varphi(t)$.

A MATLAB program that calculates $\Phi(\omega)$ using this frequency domain successive approximations approach suggested by (5.74) is given in Appendix C. Studying this program gives further insight into the structure of $\Phi(\omega)$. Rather than starting the calculations given in (5.74) for the index $j = 1$, they are started for the largest $j = J$ and worked backwards. If we calculate a length-N DFT consistent with $j = J$ using the FFT, then the samples of $H(\omega/2^j)$ for $j = J - 1$ are simply every other sample of the case for $j = J$. The next stage for $j = J - 2$ is done likewise and if the original N is chosen a power of two, the process in continued down to $j = 1$ without calculating any more FFTs. This results in a very efficient algorithm. The details are in the program itself.

This algorithm is so efficient, using it plus an inverse FFT might be a good way to calculate $\varphi(t)$ itself. Examples of the algorithm are illustrated in Figure 5.4 where the transform is plotted for each step of the iteration.

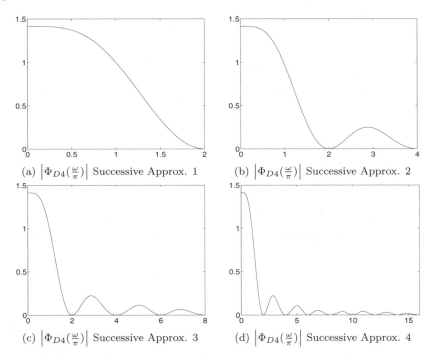

(a) $\left|\Phi_{D4}(\frac{\omega}{\pi})\right|$ Successive Approx. 1   (b) $\left|\Phi_{D4}(\frac{\omega}{\pi})\right|$ Successive Approx. 2

(c) $\left|\Phi_{D4}(\frac{\omega}{\pi})\right|$ Successive Approx. 3   (d) $\left|\Phi_{D4}(\frac{\omega}{\pi})\right|$ Successive Approx. 4

Figure 5.4. Iterations of the Successive Approximations for $\Phi(\omega)$

The Dyadic Expansion of the Scaling Function

The next method for evaluating the scaling function uses a completely different approach. It starts by calculating the values of the scaling function at integer values of t, which can be done exactly (within our ability to solve simultaneous linear equations). Consider the basic recursion equation (5.1) for integer values of $t = k$

$$\varphi(k) = \sum_n h(n) \sqrt{2}\, \varphi(2k - n), \tag{5.75}$$

and assume $h(n) \neq 0$ for $0 \leq n \leq N - 1$.

This is the refinement matrix illustrated in (5.6) for $N = 6$ which we write in matrix form as

$$\mathbf{M_0}\, \underline{\varphi} = \underline{\varphi}. \tag{5.76}$$

In other words, the vector of $\varphi(k)$ is the eigenvector of $\mathbf{M_0}$ for an eigenvalue of unity. The simple sum of $\sum_n h(n) = \sqrt{2}$ in (5.10) does not guarantee that M_0 always has such an eigenvalue, but $\sum_n h(2n) = \sum_n h(2n+1)$ in (5.12) does guarantee a unity eigenvalue. This means that if (5.12) is not satisfied, $\varphi(t)$ is not defined on the dyadic rationals and is, therefore, probably not a very nice signal.

Our problem is to now find that eigenvector. Note from (5.6) that $\varphi(0) = \varphi(N-1) = 0$ or $h(0) = h(N-1) = 1/\sqrt{2}$. For the Haar wavelet system, the second is true but for longer systems, this would mean all the other $h(n)$ would have to be zero because of (5.10) and that is not only not interesting, it produces a very poorly behaved $\varphi(t)$. Therefore, the scaling function with $N > 2$ and compact support will always be zero on the extremes of the support. This means that we can look for the eigenvector of the smaller 4 by 4 matrix obtained by eliminating the first and last rows and columns of M_0.

From (5.76) we form $[M_0 - I]\underline{\varphi} = \underline{0}$ which shows that $[M_0 - I]$ is singular, meaning its rows are not independent. We remove the last row and assume the remaining rows are now independent. If that is not true, we remove another row. We next replace that row with a row of ones in order to implement the normalizing equation

$$\sum_k \varphi(k) \;=\; 1 \tag{5.77}$$

This augmented matrix, $[M_0 - I]$ with a row replaced by a row of ones, when multiplied by $\underline{\varphi}$ gives a vector of all zeros except for a one in the position of the replaced row. This equation should not be singular and is solved for $\underline{\varphi}$ which gives $\varphi(k)$, the scaling function evaluated at the integers.

From these values of $\varphi(t)$ on the integers, we can find the values at the half integers using the recursive equation (5.1) or a modified form

$$\varphi(k/2) = \sum_n h(n)\,\sqrt{2}\,\varphi(k-n) \tag{5.78}$$

This is illustrated with the matrix equation (5.8) as

$$\mathbf{M_1}\underline{\varphi} = \underline{\varphi_2} \tag{5.79}$$

Here, the first and last columns and last row are not needed (because $\varphi_0 = \varphi_5 = \varphi_{11/2} = 0$) and can be eliminated to save some arithmetic.

The procedure described here can be repeated to find a matrix that when multiplied by a vector of the scaling function evaluated at the odd integers divided by k will give the values at the odd integers divided by $2k$. This modified matrix corresponds to convolving the samples of $\varphi(t)$ by an up-sampled $h(n)$. Again, convolution combined with up- and down-sampling is the basis of wavelet calculations. It is also the basis of digital filter bank theory. Figure 5.5 illustrates the dyadic expansion calculation of a Daubechies scaling function for $N = 4$ at each iteration of this method.

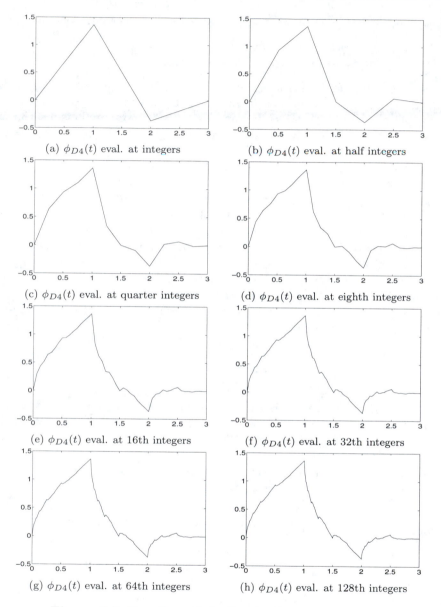

(a) $\phi_{D4}(t)$ eval. at integers

(b) $\phi_{D4}(t)$ eval. at half integers

(c) $\phi_{D4}(t)$ eval. at quarter integers

(d) $\phi_{D4}(t)$ eval. at eighth integers

(e) $\phi_{D4}(t)$ eval. at 16th integers

(f) $\phi_{D4}(t)$ eval. at 32th integers

(g) $\phi_{D4}(t)$ eval. at 64th integers

(h) $\phi_{D4}(t)$ eval. at 128th integers

Figure 5.5. Iterations of the Dyadic Expansion for ϕ_{D4}

Not only does this dyadic expansion give an explicit method for finding the exact values of $\varphi(t)$ of the dyadic rationals ($t = k/2^j$), but it shows how the eigenvalues of $\mathbf{M}$ say something about the $\varphi(t)$. Clearly, if $\varphi(t)$ is continuous, it says everything.

MATLAB programs are included in Appendix C to implement the successive approximation and dyadic expansion approaches to evaluating the scaling function from the scaling coefficients. They were used to generate the figures in this section. It is very illuminating to experiment with different $h(n)$ and observe the effects on $\varphi(t)$ and $\psi(t)$.

Chapter 6

Regularity, Moments, and Wavelet System Design

We now look at a particular way to use the remaining $\frac{N}{2} - 1$ degrees of freedom to design the N values of $h(n)$ after satisfying (5.10) and (5.14), which insure the existence and orthogonality (or property of being a tight frame) of the scaling function and wavelets [Dau92, Dau90, Dau93].

One of the interesting characteristics of the scaling functions and wavelets is that while satisfying (5.10) and (5.14) will guarantee the existence of an integrable scaling function, it may be extraordinarily irregular, even fractal in nature. This may be an advantage in analyzing rough or fractal signals but it is likely to be a disadvantage for most signals and images.

We will see in this section that the number of vanishing moments of $h_1(n)$ and $\psi(t)$ are related to the smoothness or differentiability of $\varphi(t)$ and $\psi(t)$. Unfortunately, smoothness is difficult to determine directly because, unlike with differential equations, the defining recursion equation (5.1) does not involve derivatives.

We also see that the representation and approximation of polynomials are related to the number of vanishing or minimized wavelet moments. Since polynomials are often a good model for certain signals and images, this property is both interesting and important.

The number of zero scaling function moments is related to the "goodness" of the approximation of high-resolution scaling coefficients by samples of the signal. They also affect the symmetry and concentration of the scaling function and wavelets.

This section will consider the basic 2-band or multiplier-2 case defined in (2.13). The more general M-band or multiplier-M case is discussed in Section 7.2.

6.1 K-Regular Scaling Filters

Here we start by defining a *unitary scaling filter* to be an FIR filter with coefficients $h(n)$ from the basic recursive equation (5.1) satisfying the admissibility conditions from (5.10) and orthogonality conditions from (5.14) as

$$\sum_n h(n) = \sqrt{2} \quad \text{and} \quad \sum_k h(k)\,h(k+2m) = \delta(m). \tag{6.1}$$

The term "scaling filter" comes from Mallat's algorithm, and the relation to filter banks discussed in Chapter 3. The term "unitary" comes from the orthogonality conditions expressed in filter bank language, which is explained in Chapter 8.

A unitary scaling filter is said to be K-*regular* if its z-transform has K zeros at $z = e^{i\pi}$. This looks like

$$H(z) = \left(\frac{1 + z^{-1}}{2} \right)^K Q(z) \tag{6.2}$$

where $H(z) = \sum_n h(n) z^{-n}$ is the z-transform of the scaling coefficients $h(n)$ and $Q(z)$ has no poles or zeros at $z = e^{i\pi}$. Note that we are presenting a definition of regularity of $h(n)$, not of the scaling function $\varphi(t)$ or wavelet $\psi(t)$. They are related but not the same. Note also from (5.20) that any unitary scaling filter is at least $K = 1$ regular.

The length of the scaling filter is N which means $H(z)$ is an $N-1$ degree polynomial. Since the multiple zero at $z = -1$ is order K, the polynomial $Q(z)$ is degree $N-1-K$. The existence of $\varphi(t)$ requires the zeroth moment be $\sqrt{2}$ which is the result of the linear condition in (6.1). Satisfying the conditions for orthogonality requires $N/2$ conditions which are the quadratic equations in (6.1). This means the degree of regularity is limited by

$$1 \leq K \leq \frac{N}{2}. \tag{6.3}$$

Daubechies used the degrees of freedom to obtain maximum regularity for a given N, or to obtain the minimum N for a given regularity. Others have allowed a smaller regularity and used the resulting extra degrees of freedom for other design purposes.

Regularity is defined in terms of zeros of the transfer function or frequency response function of an FIR filter made up from the scaling coefficients. This is related to the fact that the differentiability of a function is tied to how fast its Fourier series coefficients drop off as the index goes to infinity or how fast the Fourier transform magnitude drops off as frequency goes to infinity. The relation of the Fourier transform of the scaling function to the frequency response of the FIR filter with coefficients $h(n)$ is given by the infinite product (5.74). From these connections, we reason that since $H(z)$ is lowpass and, if it has a high order zero at $z = -1$ (i.e., $\omega = \pi$), the Fourier transform of $\varphi(t)$ should drop off rapidly and, therefore, $\varphi(t)$ should be smooth. This turns out to be true.

We next define the kth moments of $\varphi(t)$ and $\psi(t)$ as

$$m(k) = \int t^k \varphi(t)\, dt \tag{6.4}$$

and

$$m_1(k) = \int t^k \psi(t)\, dt \tag{6.5}$$

and the discrete k^{th} moments of $h(n)$ and $h_1(n)$ as

$$\mu(k) = \sum_n n^k h(n) \tag{6.6}$$

and

$$\mu_1(k) = \sum_n n^k h_1(n). \tag{6.7}$$

The partial moments of $h(n)$ (moments of samples) are defined as

$$\nu(k, \ell) = \sum_n (2n + \ell)^k h(2n + \ell). \tag{6.8}$$

Note that $\mu(k) = \nu(k, 0) + \nu(k, 1)$.

From these equations and the basic recursion equation (5.1) we obtain [GB92b]

$$m(k) = \frac{1}{(2^k - 1)\sqrt{2}} \sum_{\ell=1}^{k} \binom{k}{\ell} \mu(\ell) \, m(k - \ell) \tag{6.9}$$

which can be derived by substituting (5.1) into (6.4), changing variables, and using (6.6). Similarly, we obtain

$$m_1(k) = \frac{1}{2^k \sqrt{2}} \sum_{\ell=0}^{k} \binom{k}{\ell} \mu_1(\ell) \, m(k - \ell). \tag{6.10}$$

These equations exactly calculate the moments defined by the integrals in (6.4) and (6.5) with simple finite convolutions of the discrete moments with the lower order continuous moments. A similar equation also holds for the multiplier-M case described in Section 7.2 [GB92b]. A Matlab program that calculates the continuous moments from the discrete moments using (6.9) and (6.10) is given in Appendix C.

6.2   Vanishing Wavelet Moments

Requiring the moments of $\psi(t)$ to be zero has several interesting consequences. The following three theorems show a variety of equivalent characteristics for the K-regular scaling filter, which relate both to our desire for smooth scaling functions and wavelets as well as polynomial representation.

Theorem 20 (Equivalent Characterizations of K-Regular Filters) *A unitary scaling filter is K-regular if and only if the following equivalent statements are true:*

1. *All moments of the wavelet filters are zero, $\mu_1(k) = 0$, for $k = 0, 1, \cdots, (K-1)$*

2. *All moments of the wavelets are zero, $m_1(k) = 0$, for $k = 0, 1, \cdots, (K-1)$*

3. *The partial moments of the scaling filter are equal for $k = 0, 1, \cdots, (K-1)$*

4. *The frequency response of the scaling filter has a zero of order K at $\omega = \pi$, i.e. (6.2).*

5. *The kth derivative of the magnitude-squared frequency response of the scaling filter is zero at $\omega = 0$ for $k = 1, 2, \cdots, 2K - 1$.*

6. *All polynomial sequences up to degree $(K-1)$ can be expressed as a linear combination of shifted scaling filters.*

7. *All polynomials of degree up to $(K-1)$ can be expressed as a linear combination of shifted scaling functions at any scale.*

This is a very powerful result [SHGB93, Hel95]. It not only ties the number of zero moments to the regularity but also to the degree of polynomials that can be exactly represented by a sum of weighted and shifted scaling functions.

Theorem 21 *If $\psi(t)$ is K-times differentiable and decays fast enough, then the first $K-1$ wavelet moments vanish [Dau92]; i.e.,*

$$\left| \frac{d^k}{dt^k} \psi(t) \right| < \infty, \quad 0 \le k \le K \tag{6.11}$$

implies

$$m_1(k) = 0. \quad 0 \le k \le K \tag{6.12}$$

Unfortunately, the converse of this theorem is not true. However, we can relate the differentiability of $\psi(t)$ to vanishing moments by

Theorem 22 *There exists a finite positive integer L such that if $m_1(k) = 0$ for $0 \le k \le K - 1$ then*

$$\left| \frac{d^P}{dt^P} \psi(t) \right| < \infty \tag{6.13}$$

for $L P > K$.

For example, a three-times differentiable $\psi(t)$ must have three vanishing moments, but three vanishing moments results in only one-time differentiability.

These theorems show the close relationship among the moments of $h_1(n)$, $\psi(t)$, the smoothness of $H(\omega)$ at $\omega = 0$ and π and to polynomial representation. It also states a loose relationship with the smoothness of $\varphi(t)$ and $\psi(t)$ themselves.

6.3   Daubechies' Method for Zero Wavelet Moment Design

Daubechies used the above relationships to show the following important result which constructs orthonormal wavelets with compact support with the maximum number of vanishing moments.

Theorem 23 *The discrete-time Fourier transform of $h(n)$ having K zeros at $\omega = \pi$ of the form*

$$H(\omega) = \left(\frac{1 + e^{i\omega}}{2} \right)^K \mathcal{L}(\omega) \tag{6.14}$$

satisfies

$$|H(\omega)|^2 + |H(\omega + \pi)|^2 = 2 \tag{6.15}$$

if and only if $L(\omega) = |\mathcal{L}(\omega)|^2$ can be written

$$L(\omega) = P(\sin^2(\omega/2)) \tag{6.16}$$

with $K \le N/2$ where

$$P(y) = \sum_{k=0}^{K-1} \binom{K-1+k}{k} y^k + y^K R\left(\frac{1}{2} - y\right) \tag{6.17}$$

and $R(y)$ is an odd polynomial chosen so that $P(y) \ge 0$ for $0 \le y \le 1$.

If $R = 0$, the length N is minimum for a given regularity $K = N/2$. If $N > 2K$, the second term containing R has terms with higher powers of y whose coefficients can be used for purposes other than regularity.

The proof and a discussion are found in Daubechies [Dau88a, Dau92]. Recall from (5.20) that $H(\omega)$ always has at least one zero at $\omega = \pi$ as a result of $h(n)$ satisfying the necessary conditions for $\varphi(t)$ to exist and have orthogonal integer translates. We are now placing restrictions on $h(n)$ to have as high an order zero at $\omega = \pi$ as possible. That accounts for the form of (6.14). Requiring orthogonality in (5.22) gives (6.15).

Because the frequency domain requirements in (6.15) are in terms of the square of the magnitudes of the frequency response, spectral factorization is used to determine $H(\omega)$ and therefore $h(n)$ from $|H(\omega)|^2$. Equation (6.14) becomes

$$|H(\omega)|^2 = \left| \frac{1 + e^{i\omega}}{2} \right|^{2K} |\mathcal{L}(\omega)|^2. \tag{6.18}$$

If we use the functional notation:

$$M(\omega) = |H(\omega)|^2 \quad \text{and} \quad L(\omega) = |\mathcal{L}(\omega)|^2 \tag{6.19}$$

then (6.18) becomes

$$M(\omega) \;=\; |\cos^2(\omega/2)|^K \, L(\omega). \tag{6.20}$$

Since $M(\omega)$ and $L(\omega)$ are even functions of ω they can be written as polynomials in $\cos(\omega)$ and, using $\cos(\omega) = 1 - 2\sin^2(\omega/2)$, equation (6.20) becomes

$$\widetilde{M}(\sin^2(\omega/2)) \;=\; |\cos^2(\omega/2)|^K \, P(\sin^2(\omega/2)) \tag{6.21}$$

which, after a change of variables of $y = \sin^2(\omega/2) = 1 - \cos^2(\omega/2)$, becomes

$$\widetilde{M}(y) = (1 - y)^K \, P(y) \tag{6.22}$$

where $P(y)$ is an $(N - K)$ order polynomial which must be positive since it will have to be factored to find $H(\omega)$ from (6.19). This now gives (6.14) in terms of new variables which are easier to use.

In order that this description supports an orthonormal wavelet basis, we now require that (6.22) satisfies (5.22)

$$|H(\omega)|^2 + |H(\omega + \pi)|^2 \;=\; 2 \tag{6.23}$$

which using (6.19) and (6.22) becomes

$$M(\omega) + M(\omega + \pi) \;=\; (1 - y)^K \, P(y) + y^K \, P(1 - y) \;=\; 2. \tag{6.24}$$

Equations of this form have an explicit solution found by using Bezout's theorem. The details are developed by Daubechies [Dau92]. If all the $(N/2 - 1)$ degrees of freedom are used to set wavelet moments to zero, we set $K = N/2$ and the solution to (6.24) is given by

$$P(y) = \sum_{k=0}^{K-1} \binom{K - 1 + k}{k} \, y^k \tag{6.25}$$

which gives a complete parameterization of Daubechies' maximum zero wavelet moment design. It also gives a very straightforward procedure for the calculation of the $h(n)$ that satisfy these conditions. Herrmann derived this expression for the design of Butterworth or maximally flat FIR digital filters [Her71].

If the regularity is $K < N/2$, $P(y)$ must be of higher degree and the form of the solution is

$$P(y) = \sum_{k=0}^{K-1} \binom{K-1+k}{k} y^k + y^K R(\frac{1}{2} - y) \tag{6.26}$$

where $R(y)$ is chosen to give the desired filter length N, to achieve some other desired property, and to give $P(y) \geq 0$.

The steps in calculating the actual values of $h(n)$ are to first choose the length N (or the desired regularity) for $h(n)$, then factor $|H(\omega)|^2$ where there will be freedom in choosing which roots to use for $H(\omega)$. The calculations are more easily carried out using the z-transform form of the transfer function and using convolution in the time domain rather than multiplication (raising to a power) in the frequency domain. That is done in the MATLAB program [hn,h1n] = daub(N) in Appendix C where the polynomial coefficients in (6.25) are calculated from the binomial coefficient formula. This polynomial is factored with the roots command in MATLAB and the roots are mapped from the polynomial variable y_ℓ to the variable z_ℓ in (6.2) using first $\cos(\omega) = 1 - 2 y_\ell$, then with $i \sin(\omega) = \sqrt{\cos^2(\omega) - 1}$ and $e^{i\omega} = \cos(\omega) \pm i \sin(\omega)$ we use $z = e^{i\omega}$. These changes of variables are used by Herrmann [Her71] and Daubechies [Dau92].

Examine the MATLAB program to see the details of just how this is carried out. The program uses the sort command to order the roots of $H(z) H(1/z)$ after which it chooses the $N-1$ smallest ones to give a minimum phase $H(z)$ factorization. You could choose a different set of $N-1$ roots in an effort to get a more linear phase or even maximum phase. This choice allows some variation in Daubechies wavelets of the same length. The M-band generalization of this is developed by Heller in [SHGB93, Hel95]. In [Dau92], Daubechies also considers an alternation of zeros inside and outside the unit circle which gives a more symmetric $h(n)$. A completely symmetric real $h(n)$ that has compact support and supports orthogonal wavelets is not possible; however, symmetry is possible for complex $h(n)$, biorthogonal systems, infinitely long $h(n)$, and multiwavelets. Use of this zero moment design approach will also assure the resulting wavelets system is an orthonormal basis.

If all the degrees of freedom are used to set moments to zero, one uses $K = N/2$ in (6.14) and the above procedure is followed. It is possible to explicitly set a particular pair of zeros somewhere other than at $\omega = \pi$. In that case, one would use $K = (N/2) - 2$ in (6.14). Other constraints are developed later in this chapter and in later chapters.

To illustrate some of the characteristics of a Daubechies wavelet system, Table 6.1 shows the scaling function and wavelet coefficients, $h(n)$ and $h_1(n)$, and the corresponding discrete scaling coefficient moments and wavelet coefficient moments for a length-8 Daubechies system. Note the $N/2 = 4$ zero moments of the wavelet coefficients and the zeroth scaling coefficient moment of $\mu(0) = \sqrt{2}$.

Table 6.1. Scaling Function and Wavelet Coefficients plus their Discrete Moments for Daubechies-8

n	$h(n)$	$h_1(n)$	$\mu(k)$	$\mu_1(k)$	k
0	0.23037781330890	0.01059740178507	1.414213	0	0
1	0.71484657055292	0.03288301166689	1.421840	0	1
2	0.63088076792986	-0.03084138183556	1.429509	0	2
3	-0.02798376941686	-0.18703481171909	0.359097	0	3
4	-0.18703481171909	0.02798376941686	-2.890773	12.549900	4
5	0.03084138183556	0.63088076792986	-3.453586	267.067254	5
6	0.03288301166689	-0.71484657055292	23.909120	3585.681937	6
7	-0.01059740178507	0.23037781330890			7

Table 6.2 gives the same information for the length-6, 4, and 2 Daubechies scaling coefficients, wavelet coefficients, scaling coefficient moments, and wavelet coefficient moments. Again notice how many discrete wavelet moments are zero.

Table 6.3 shows the continuous moments of the scaling function $\varphi(t)$ and wavelet $\psi(t)$ for the Daubechies systems with lengths six and four. The discrete moments are the moments of the coefficients defined by (6.6) and (6.7) with the continuous moments defined by (6.4) and (6.5) calculated using (6.9) and (6.10) with the programs listed in Appendix C.

Table 6.2. Daubechies Scaling Function and Wavelet Coefficients plus their Moments

	Daubechies $N = 6$				
n	$h(n)$	$h_1(n)$	$\mu(k)$	$\mu_1(k)$	k
0	0.33267055295008	-0.03522629188571	1.414213	0	0
1	0.80689150931109	-0.08544127388203	1.155979	0	1
2	0.45987750211849	0.13501102001025	0.944899	0	2
3	-0.13501102001025	0.45987750211849	-0.224341	3.354101	3
4	-0.08544127388203	-0.80689150931109	-2.627495	40.679682	4
5	0.03522629188571	0.33267055295008	5.305591	329.323717	5
	Daubechies $N = 4$				
n	$h(n)$	$h_1(n)$	$\mu(k)$	$\mu_1(k)$	k
0	0.48296291314453	0.12940952255126	1.414213	0	0
1	0.83651630373781	0.22414386804201	0.896575	0	1
2	0.22414386804201	-0.83651630373781	0.568406	1.224744	2
3	-0.12940952255126	0.48296291314453	-0.864390	6.572012	3
	Daubechies $N = 2$				
n	$h(n)$	$h_1(n)$	$\mu(k)$	$\mu_1(k)$	k
0	0.70710678118655	0.70710678118655	1.414213	0	0
1	0.70710678118655	-0.70710678118655	0.707107	0.707107	1

Table 6.3. Daubechies Scaling Function and Wavelet Continuous and Discrete Moments

k	$N=6$ $\mu(k)$	$\mu_1(k)$	$m(k)$	$m_1(k)$
0	1.4142135	0	1.0000000	0
1	1.1559780	0	0.8174012	0
2	0.9448992	0	0.6681447	0
3	-0.2243420	3.3541019	0.4454669	0.2964635
4	-2.6274948	40.6796819	0.1172263	2.2824642
5	5.3055914	329.3237168	-0.0466511	11.4461157

k	$N=4$ $\mu(k)$	$\mu_1(k)$	$m(k)$	$m_1(k)$
0	1.4142136	0	1.0000000	0
1	0.8965755	0	0.6343975	0
2	0.5684061	1.2247449	0.4019238	0.2165063
3	-0.8643899	6.5720121	0.1310915	0.7867785
4	-6.0593531	25.9598790	-0.3021933	2.0143421
5	-23.4373939	90.8156100	-1.0658728	4.4442798

These tables are very informative about the characteristics of wavelet systems in general as well as particularities of the Daubechies system. We see the $\mu(0) = \sqrt{2}$ of (6.1) and (5.10) that is necessary for the existence of a scaling function solution to (5.1) and the $\mu_1(0) = m_1(0) = 0$ of (5.32) and (5.29) that is necessary for the orthogonality of the basis functions. Ortho*normality* requires (2.25) which is seen in comparison of the $h(n)$ and $h_1(n)$, and it requires $m(0) = 1$ from (5.53) and (5.45). After those conditions are satisfied, there are $N/2 - 1$ degrees of freedom left which Daubechies uses to set wavelet moments $m_1(k)$ equal zero. For length-6 we have two zero wavelet moments and for length-4, one. For all longer Daubechies systems we have exactly $N/2 - 1$ zero wavelet moments in addition to the one $m_1(0) = 0$ for a total of $N/2$ zero wavelet moments. Note $m(2) = m(1)^2$ as will be explained in (6.32) and there exist relationships among some of the values of the even-ordered scaling function moments, which will be explained in (6.52) through (6.55).

As stated earlier, these systems have a maximum number of zero moments of the wavelets which results in a high degree of smoothness for the scaling and wavelet functions. Figures 6.1 and 6.2 show the Daubechies scaling functions and wavelets for $N = 4, 6, 8, 10, 12, 16, 20, 40$. The coefficients were generated by the techniques described in Section 5.9 and Chapter 6. The MATLAB programs are listed in Appendix C and values of $h(n)$ can be found in [Dau92] or generated by the programs. Note the increasing smoothness as N is increased. For $N = 2$, the scaling function is not continuous; for $N = 4$, it is continuous but not differentiable; for $N = 6$, it is barely differentiable once; for $N = 14$, it is twice differentiable, and similarly for longer $h(n)$. One can obtain any degree of differentiability for sufficiently long $h(n)$.

The Daubechies coefficients are obtained by maximizing the number of moments that are zero. This gives regular scaling functions and wavelets, but it is possible to use the degrees of

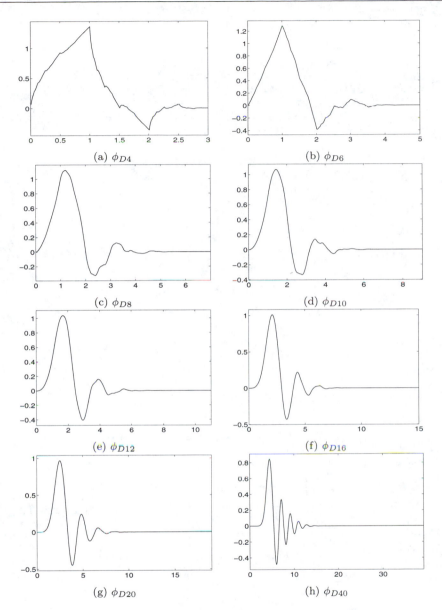

(a) ϕ_{D4}

(b) ϕ_{D6}

(c) ϕ_{D8}

(d) ϕ_{D10}

(e) ϕ_{D12}

(f) ϕ_{D16}

(g) ϕ_{D20}

(h) ϕ_{D40}

Figure 6.1. Daubechies Scaling Functions, $N = 4, 6, 8, \ldots, 40$

freedom to maximize the differentiability of $\varphi(t)$ rather than maximize the zero moments. This is not easily parameterized, and it gives only slightly greater smoothness than the Daubechies system [Dau92].

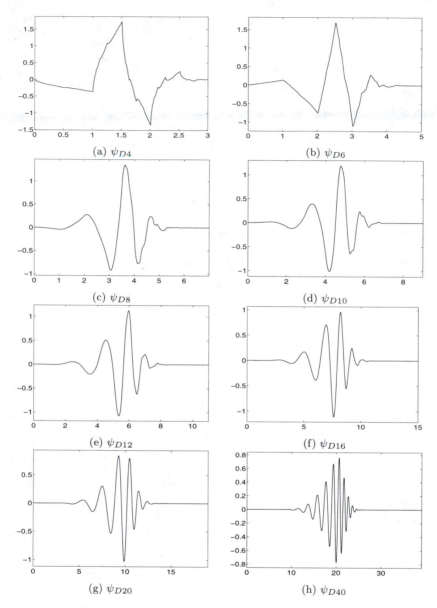

Figure 6.2. Daubechies Wavelets, $N = 4, 6, 8, \ldots, 40$

Examples of Daubechies scaling functions resulting from choosing different factors in the spectral factorization of $|H(\omega)|^2$ in (6.18) can be found in [Dau92].

6.4   Non-Maximal Regularity Wavelet Design

If the length of the scaling coefficient filter is longer than twice the desired regularity, i.e., $N > 2K$, then the parameterization of (6.26) should be used and the coefficients in the polynomial $R(y)$ must be determined. One interesting possibility is that of designing a system that has one or more zeros of $H(\omega)$ between $\omega = \pi/2$ and $\omega = \pi$ with the remaining zeros at π to contribute to the regularity. This will give better frequency separation in the filter bank in exchange for reduced regularity and lower degree polynomial representation.

If a zero of $H(\omega)$ is set at $\omega = \omega_0$, then the conditions of

$$M(\omega_0) = 0 \quad \text{and} \quad \frac{d\,M(\omega)}{d\,\omega}\bigg|_{\omega=\omega_0} = 0 \qquad (6.27)$$

are imposed with those in (6.26), giving a set of linear simultaneous equations that can be solved to find the scaling coefficients $h(n)$.

A powerful design system based on a Remez exchange algorithm allows the design of an orthogonal wavelet system that has a specified regularity and an optimal Chebyshev behavior in the stopband of $H(\omega)$. This and a variation that uses a constrained least square criterion [SLB97] is described in [LSOB94] and another Chebyshev design in [RD94].

An alternative approach is to design the wavelet system by setting an optimization criterion and using a general constrained optimization routine, such as that in the MATLAB optimization toolbox, to design the $h(n)$ with the existence and orthogonality conditions as constraints. This approach was used to generate many of the filters described in Table 6.7. Jun Tian used a Newton's method [Tia96, TW95] to design wavelet systems with zero moments.

6.5   Relation of Zero Wavelet Moments to Smoothness

We see from Theorems 21 and 22 that there is a relationship between zero wavelet moments and smoothness, but it is not tight. Although in practical application the degree of differentiability may not be the most important measure, it is an important theoretical question that should be addressed before application questions can be properly posed.

First we must define smoothness. From the mathematical point of view, smoothness is essentially the same as differentiability and there are at least two ways to pose that measure. The first is local (the Hölder measure) and the second is global (the Sobolev measure). Numerical algorithms for estimating the measures in the wavelet setting have been developed by Rioul [Rio92] and Heller and Wells [HW94, HW96a] for the Hölder and Sobolev measures respectively.

Definition 1 (Hölder continuity) *Let $\varphi : \mathbb{R} \to \mathbb{C}$ and let $0 < \alpha \le 1$. Then the function φ is Hölder continuous of order α if there exists a constant c such that*

$$|\varphi(x) - \varphi(y)| \le c\,|x - y|^{\alpha} \quad \text{for all } x, y \in \mathbb{R} \qquad (6.28)$$

Based on the above definition, one observes φ has to be a constant if $\alpha > 1$. Hence, this is not very useful for determining regularity of order $\alpha > 1$. However, using the above definition, Hölder regularity of any order $r > 0$ is defined as follows:

Definition 2 (Hölder regularity) *A function $\varphi : \mathbb{R} \to \mathbb{C}$ is regular of order $r = P + \alpha$ ($0 < \alpha \leq 1$) if $\varphi \in C^P$ and its Pth derivative is Hölder continuous of order α*

Definition 3 (Sobolev regularity) *Let $\varphi : \mathbb{R} \to \mathbb{C}$, then φ is said to belong to the Sobolev space of order s ($\varphi \in H^s$) if*

$$\int_{\mathbb{R}} (1 + |\omega|^2)^s \left| \hat{\Phi}(\omega) \right|^2 d\omega < \infty \tag{6.29}$$

Notice that, although Sobolev regularity does not give the explicit order of differentiability, it does yield lower and upper bounds on r, the Hölder regularity, and hence the differentiability of φ if $\varphi \in L^2$. This can be seen from the following inclusions:

$$H^{s+1/2} \subset C^r \subset H^s \tag{6.30}$$

A very interesting and important result by Volkmer [Vol92] and by Eirola [Eir92] gives an exact asymptotic formula for the Hölder regularity index (exponent) of the Daubechies scaling function.

Theorem 24 *The limit of the Hölder regularity index of a Daubechies scaling function as the length of the scaling filter goes to infinity is [Vol92]*

$$\lim_{N \to \infty} \frac{\alpha_N}{N} = \left[1 - \frac{\log(3)}{2 \log(2)} \right] = (0.2075 \cdots) \tag{6.31}$$

This result, which was also proven by A. Cohen and J. P. Conze, together with empirical calculations for shorter lengths, gives a good picture of the smoothness of Daubechies scaling functions. This is illustrated in Figure 6.3 where the Hölder index is plotted versus scaling filter length for both the maximally smooth case and the Daubechies case.

The question of the behavior of maximally smooth scaling functions was empirically addressed by Lang and Heller in [LH96]. They use an algorithm by Rioul to calculate the Hölder smoothness of scaling functions that have been designed to have maximum Hölder smoothness and the results are shown in Figure 6.3 together with the smoothness of the Daubechies scaling functions as functions of the length of the scaling filter. For the longer lengths, it is possible to design systems that give a scaling function over twice as smooth as with a Daubechies design. In most applications, however, the greater Hölder smoothness is probably not important.

Figure 6.4 shows the number of zero moments (zeros at $\omega = \pi$) as a function of the number of scaling function coefficients for both the maximally smooth and Daubechies designs.

One case from this figure is for $N = 26$ where the Daubechies smoothness is $S_H = 4.005$ and the maximum smoothness is $S_H = 5.06$. The maximally smooth scaling function has one more continuous derivative than the Daubechies scaling function.

Recent work by Heller and Wells [HW94, HW96a] gives a better connection of properties of the scaling coefficients and the smoothness of the scaling function and wavelets. This is done both for the scale factor or multiplicity of $M = 2$ and for general integer M.

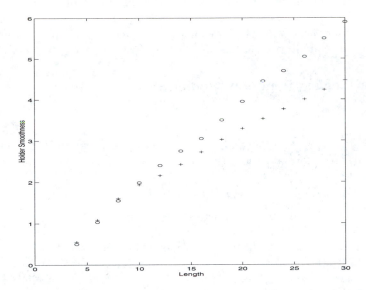

Figure 6.3. Hölder Smoothness versus Coefficient Length for Daubechies' (+) and Maximally Smooth (o) Wavelets.

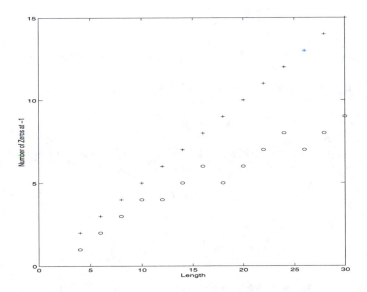

Figure 6.4. Number of Zeros at $\omega = \pi$ versus Coefficient Length for Daubechies' (+) and Maximally Smooth (o) Wavelets.

The usual definition of smoothness in terms of differentiability may not be the best measure for certain signal processing applications. If the signal is given as a sequence of numbers, not as a function of a continuous variable, what does smoothness mean? Perhaps the use of the *variation* of a signal may be a useful alternative [GB95c, OB96b, Ode96].

6.6 Vanishing Scaling Function Moments

While the moments of the wavelets give information about flatness of $H(\omega)$ and smoothness of $\psi(t)$, the moments of $\varphi(t)$ and $h(n)$ are measures of the "localization" and symmetry characteristics of the scaling function and, therefore, the wavelet transform. We know from (5.10) that $\sum_n h(n) = \sqrt{2}$ and, after normalization, that $\int \varphi(t)\, dt = 1$. Using (6.9), one can show [GB92b] that for $K \geq 2$, we have

$$m(2) = m^2(1). \tag{6.32}$$

This can be seen in Table 6.3. A generalization of this result has been developed by Johnson [JMNK96] and is given in (6.52) through (6.55).

A more general picture of the effects of zero moments can be seen by next considering two approximations. Indeed, this analysis gives a very important insight into the effects of zero moments. The mixture of zero scaling function moments with other specifications is addressed later in Section 6.9.

6.7 Approximation of Signals by Scaling Function Projection

The orthogonal projection of a signal $f(t)$ on the scaling function subspace $\mathcal{V}_j$ is given and denoted by

$$P^j\{f(t)\} = \sum_k \langle f(t), \varphi_{j,k}(t) \rangle\, \varphi_{j,k}(t) \tag{6.33}$$

which gives the component of $f(t)$ which is in $\mathcal{V}_j$ and which is the best least squares approximation to $f(t)$ in $\mathcal{V}_j$.

As given in (6.5), the ℓth moment of $\psi(t)$ is defined as

$$m_1(\ell) = \int t^\ell \psi(t)\, dt. \tag{6.34}$$

We can now state an important relation of the projection (6.33) as an approximation to $f(t)$ in terms of the number of zero wavelet moments and the scale.

Theorem 25 *If $m_1(\ell) = 0$ for $\ell = 0, 1, \cdots, L$ then the L^2 error is*

$$\epsilon_1 = \| f(t) - P^j\{f(t)\} \|_2 \leq C\, 2^{-j(L+1)}, \tag{6.35}$$

where C is a constant independent of j and L but dependent on $f(t)$ and the wavelet system [GLRT90, Uns96].

This states that at any given scale, the projection of the signal on the subspace at that scale approaches the function itself as the number of zero wavelet moments (and the length of the scaling filter) goes to infinity. It also states that for any given length, the projection goes to the function as the scale goes to infinity. These approximations converge exponentially fast. This projection is illustrated in Figure 6.5.

6.8 Approximation of Scaling Coefficients by Samples of the Signal

A second approximation involves using the samples of $f(t)$ as the inner product coefficients in the wavelet expansion of $f(t)$ in (6.33). We denote this sampling approximation by

$$S^j\{f(t)\} = \sum_k 2^{-j/2} f(k/2^j)\, \varphi_{j,k}(t) \tag{6.36}$$

and the scaling functior moment by

$$m(\ell) = \int t^\ell\, \varphi(t)\, dt \tag{6.37}$$

and can state [Tia96] the following

Theorem 26 *If $m(\ell) = 0$ for $\ell = 1, 2, \cdots, L$ then the L^2 error is*

$$\epsilon_2 = \|\, S^j\{f(t)\} - P^j\{f(t)\}\,\|_2 \le C_2\, 2^{-j(L+1)}, \tag{6.38}$$

where C_2 is a constant independent of j and L but dependent on $f(t)$ and the wavelet system.

This is a similar approximation or convergence result to the previous theorem but relates the projection of $f(t)$ on a j-scale subspace to the sampling approximation in that same subspace. These approximations are illustrated in Figure 6.5.

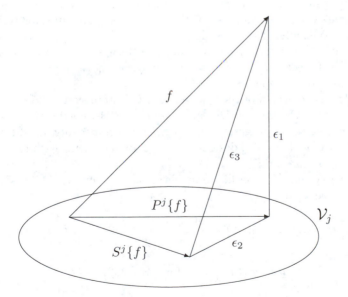

Figure 6.5. Approximation and Projection of $f(t)$ at a Finite Scale

This "vector space" illustration shows the nature and relationships of the two types of approximations. The use of samples as inner products is an approximation within the expansion subspace $\mathcal{V}_j$. The use of a finite expansion to represent a signal $f(t)$ is an approximation from L^2 onto the subspace $\mathcal{V}_j$. Theorems 25 and 26 show the nature of those approximations, which, for wavelets, is very good.

An illustration of the effects of these approximations on a signal is shown in Figure 6.6 where a signal with a very smooth component (a sinusoid) and a discontinuous component (a square wave) is expanded in a wavelet series using samples as the high resolution scaling function coefficients. Notice the effects of projecting onto lower and lower resolution scales.

If we consider a wavelet system where the same number of scaling function and wavelet moments are set zero and this number is as large as possible, then the following is true [WZ94, Tia96]:

Theorem 27 *If* $m(\ell) = m_1(\ell) = 0$ *for* $\ell = 1, 2, \cdots, L$ *and* $m_1(0) = 0$, *then the* L^2 *error is*

$$\epsilon_3 = \parallel f(t) - S^j\{f(t)\} \parallel_2 \leq C_3 \, 2^{-j(L+1)}, \tag{6.39}$$

where C_3 *is a constant independent of* j *and* L, *but dependent on* $f(t)$ *and the wavelet system.*

Here we see that for this wavelet system called a Coifman wavelet system, that using samples as the inner product expansion coefficients is an excellent approximation. This justifies that using samples of a signal as input to a filter bank gives a proper wavelet analysis. This approximation is also illustrated in Figure 6.5 and in [WB97].

6.9 Coiflets and Related Wavelet Systems

From the previous approximation theorems, we see that a combination of zero wavelet and zero scaling function moments used with samples of the signal may give superior results to wavelets with only zero wavelet moments. Not only does forcing zero scaling function moments give a better approximation of the expansion coefficients by samples, it causes the scaling function to be more symmetric. Indeed, that characteristic may be more important than the sample approximation in certain applications.

Daubechies considered the design of these wavelets which were suggested by Coifman [Dau92, Dau93, BCR91]. Gopinath [Gop90, GB92b] and Wells [WZ94, TW95] show how zero scaling function moments give a better approximation of high-resolution scaling coefficients by samples. Tian and Wells [Tia96, TW95] have also designed biorthogonal systems with mixed zero moments with very interesting properties.

The Coifman wavelet system (Daubechies named the basis functions "coiflets") is an orthonormal multiresolution wavelet system with

$$\int t^k \varphi(t) \, dt = m(k) = 0, \quad \text{for} \quad k = 1, 2, \cdots, L-1 \tag{6.40}$$

$$\int t^k \psi(t) \, dt = m_1(k) = 0, \quad \text{for} \quad k = 1, 2, \cdots, L-1. \tag{6.41}$$

This definition imposes the requirement that there be at least $L-1$ zero scaling function moments and at least $L-1$ wavelet moments in addition to the one zero moment of $m_1(0)$ required by orthogonality. This system is said to be of order or degree L and sometime has the additional requirement that the length of the scaling function filter $h(n)$, which is denoted N, is minimum [Dau92, Dau93].

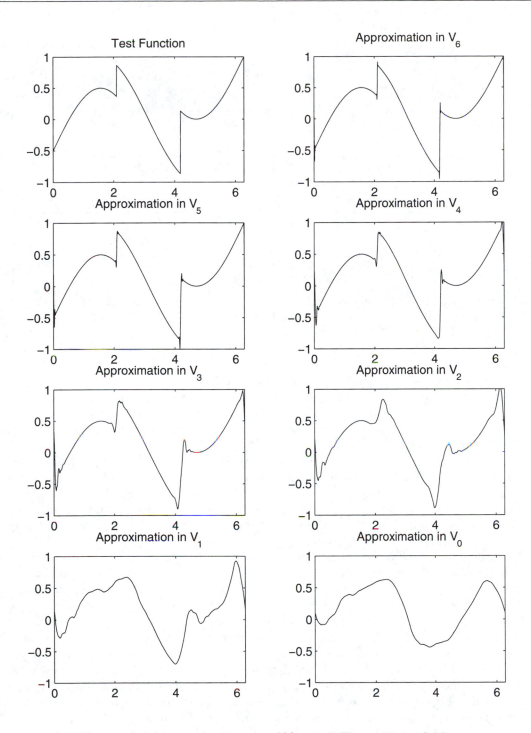

Figure 6.6. Approximations to $f(t)$ at a Different Finite Scales

The length-4 wavelet system has only one degree of freedom, so it cannot have both a scaling function moment and wavelet moment of zero (see Table 6.6). Tian [Tia96, TW95] has derived formulas for four length-6 coiflets. These are:

$$h = \left[\frac{-3+\sqrt{7}}{16\sqrt{2}}, \frac{1-\sqrt{7}}{16\sqrt{2}}, \frac{7-\sqrt{7}}{8\sqrt{2}}, \frac{7+\sqrt{7}}{8\sqrt{2}}, \frac{5+\sqrt{7}}{16\sqrt{2}}, \frac{1-\sqrt{7}}{16\sqrt{2}} \right], \tag{6.42}$$

or

$$h = \left[\frac{-3-\sqrt{7}}{16\sqrt{2}}, \frac{1+\sqrt{7}}{16\sqrt{2}}, \frac{7+\sqrt{7}}{8\sqrt{2}}, \frac{7-\sqrt{7}}{8\sqrt{2}}, \frac{5-\sqrt{7}}{16\sqrt{2}}, \frac{1+\sqrt{7}}{16\sqrt{2}} \right], \tag{6.43}$$

or

$$h = \left[\frac{-3+\sqrt{15}}{16\sqrt{2}}, \frac{1-\sqrt{15}}{16\sqrt{2}}, \frac{3-\sqrt{15}}{8\sqrt{2}}, \frac{3+\sqrt{15}}{8\sqrt{2}}, \frac{13+\sqrt{15}}{16\sqrt{2}}, \frac{9-\sqrt{15}}{16\sqrt{2}} \right], \tag{6.44}$$

or

$$h = \left[\frac{-3-\sqrt{15}}{16\sqrt{2}}, \frac{1+\sqrt{15}}{16\sqrt{2}}, \frac{3+\sqrt{15}}{8\sqrt{2}}, \frac{3-\sqrt{15}}{8\sqrt{2}}, \frac{13-\sqrt{15}}{16\sqrt{2}}, \frac{9+\sqrt{15}}{16\sqrt{2}} \right], \tag{6.45}$$

with the first formula (6.42) giving the same result as Daubechies [Dau92, Dau93] (corrected) and that of Odegard [BO96] and the third giving the same result as Wickerhauser [Wic95]. The results from (6.42) are included in Table 6.4 along with the discrete moments of the scaling function and wavelet, $\mu(k)$ and $\mu_1(k)$ for $k = 0, 1, 2, 3$. The design of a length-6 Coifman system specifies one zero scaling function moment and one zero wavelet moment (in addition to $\mu_1(0) = 0$), but we, in fact, obtain one extra zero scaling function moment. That is the result of $m(2) = m(1)^2$ from [GB92b]. In other words, we get one more zero scaling function moment than the two degrees of freedom would seem to indicate. This is true for all lengths $N = 6\ell$ for $\ell = 1, 2, 3, \cdots$ and is a result of the interaction between the scaling function moments and the wavelet moments described later.

The property of zero wavelet moments is shift invariant, but the zero scaling function moments are shift dependent [BCR91]. Therefore, a particular shift for the scaling function must be used. This shift is two for the length-6 example in Table 6.4, but is different for the solutions in (6.44) and (6.45). Compare this table to the corresponding one for Daubechies length-6 scaling functions and wavelets given in Table 6.2 where there are two zero discrete wavelet moments – just as many as the degrees of freedom in that design.

The scaling function from (6.42) is fairly symmetric, but not around its center and the other three designs in (6.43), (6.44), and (6.45) are not symmetric at all. The scaling function from (6.42) is also fairly smooth, and from (6.44) only slightly less so but the scaling function from (6.43) is very rough and from (6.45) seems to be fractal. Examination of the frequency response $H(\omega)$ and the zero location of the FIR filters $h(n)$ shows very similar frequency responses for (6.42) and (6.44) with (6.43) having a somewhat irregular but monotonic frequency response and (6.45) having a zero on the unit circle at $\omega = \pi/3$, i.e., not satisfying Cohen's condition [CS93] for an orthognal basis. It is also worth noticing that the design in (6.42) has the largest Hölder smoothness. These four designs, all satisfying the same necessary conditions, have very different characteristics. This tells us to be very careful in using zero moment methods to design wavelet systems. The designs are not unique and some are much better than others.

Table 6.4 contains the scaling function and wavelet coefficients for the length-6 and 12 designed by Daubechies and length-8 designed by Tian together with their discrete moments. We see the

extra zero scaling function moments for lengths 6 and 12 and also the extra zero for lengths 8
and 12 that occurs after a nonzero one.

The continuous moments can be calculated from the discrete moments and lower order continuous moments [BCR91, GB92b, SP93] using (6.9) and (6.10). An important relationship of
the discrete moments for a system with $K - 1$ zero wavelet moments is found by calculating
the derivatives of the magnitude squared of the discrete time Fourier transform of $h(n)$ which is
$H(\omega) = \sum_n h(n) e^{-i\omega n}$ and has $2K - 1$ zero derivatives of the magnitude squared at $\omega = 0$. This
gives [GB92b] the kth derivative for k even and $1 < k < 2K - 1$

$$\sum_{\ell=1}^{k} \binom{k}{\ell} (-1)^{\ell} \mu(\ell)\, \mu(k - \ell) = 0. \tag{6.46}$$

Solving for $\mu(k)$ in terms of lower order discrete moments and using $\mu(0) = \sqrt{2}$ gives for k even

$$\mu(k) = \frac{1}{2\sqrt{2}} \sum_{\ell=1}^{k-1} \binom{k}{\ell} (-1)^{\ell} \mu(\ell)\, \mu(k - \ell) \tag{6.47}$$

which allows calculating the even-order discrete scaling function moments in terms of the lower
odd-order discrete scaling function moments for $k = 2, 4, \cdots, 2K - 2$. For example:

$$\mu(2) = -\frac{1}{\sqrt{2}} \mu^2(1) \tag{6.48}$$

$$\mu(4) = \frac{1}{2\sqrt{2}} [8\mu(1)\, \mu(3) - 3\mu^4(1)] \tag{6.49}$$

$$\cdots$$

which can be seen from values in Table 6.2.

Johnson [JMNK96] noted from Beylkin [Bey92] and Unser [Uns96] that by using the moments
of the autocorrelation function of the scaling function, a relationship of the continuous scaling
function moments can be derived in the form

$$\sum_{\ell=0}^{k} \binom{k}{\ell} (-1)^{k-\ell} m(\ell)\, m(k - \ell) = 0 \tag{6.50}$$

where $0 < k < 2K$ if $K - 1$ wavelet moments are zero. Solving for $m(k)$ in terms of lower order
moments gives for k even

$$m(k) = \frac{1}{2} \sum_{\ell=1}^{k-1} \binom{k}{\ell} (-1)^{\ell} m(\ell)\, m(k - \ell) \tag{6.51}$$

which allows calculating the even-order scaling function moments in terms of the lower odd-order
scaling function moments for $k = 2, 4, \cdots, 2K - 2$. For example [JMNK96]:

$$m(2) = m^2(1) \tag{6.52}$$

$$m(4) = 4\, m(3)\, m(1) - 3\, m^4(1) \tag{6.53}$$

$$m(6) = 6\, m(5)\, m(1) + 10\, m^2(3) + 60\, m(3)\, m^3(1) + 45\, m^6(1) \tag{6.54}$$

$$m(8) = 8\, m(7)\, m(1) + 56\, m(5)\, m(3) - 168\, m(5)\, m^3(1)$$
$$+ 2520\, m(3)\, m^5(1) - 840\, m(3)\, m^2(1) - 1575\, m^8(1) \tag{6.55}$$

$$\cdots \qquad \cdots$$

Table 6.4. Coiflet Scaling Function and Wavelet Coefficients plus their Discrete Moments

	Length-$N = 6$,	Degree $L = 2$			
n	$h(n)$	$h_1(n)$	$\mu(k)$	$\mu_1(k)$	k
-2	-0.07273261951285	0.01565572813546	1.414213	0	0
-1	0.33789766245781	-0.07273261951285	0	0	1
0	0.85257202021226	-0.38486484686420	0	-1.163722	2
1	0.38486484686420	0.85257202021226	-0.375737	-3.866903	3
2	-0.07273261951285	-0.33789766245781	-2.872795	-10.267374	4
3	-0.01565572813546	-0.07273261951285			
	Length-$N = 8$,	Degree $L = 3$			
n	$h(n)$	$h_1(n)$	$\mu(k)$	$\mu_1(k)$	k
-4	0.04687500000000	0.01565572813546	1.414213	0	0
-3	-0.02116013576461	-0.07273261951285	0	0	1
-2	-0.14062500000000	-0.38486484686420	0	0	2
-1	0.43848040729385	1.38486484686420	-2.994111	0.187868	3
0	1.38486484686420	-0.43848040729385	0	11.976447	4
1	0.38486484686420	-0.14062500000000	-45.851020	-43.972332	5
2	-0.07273261951285	0.02116013576461	63.639610	271.348747	6
3	-0.01565572813546	0.04687500000000			
	Length-$N = 12$,	Degree $L = 4$			
n	$h(n)$	$h_1(n)$	$\mu(k)$	$\mu_1(k)$	k
-4	0.016387336463	0.000720549446	1.414213	0	0
-3	-0.041464936781	0.001823208870	0	0	1
-2	-0.067372554722	-0.005611434819	0	0	2
-1	0.386110066823	-0.023680171946	0	0	3
0	0.812723635449	0.059434418646	0	11.18525	4
1	0.417005184423	0.076488599078	-5.911352	175.86964	5
2	-0.076488599078	-0.417005184423	0	1795.33634	6
3	-0.059434418646	-0.812723635449	-586.341304	15230.54650	7
4	0.023680171946	-0.386110066823	3096.310009	117752.68833	8
5	0.005611434819	0.067372554722			
6	-0.001823208870	0.041464936781			
7	-0.000720549446	-0.016387336463			

if the wavelet moments are zero up to $k = K - 1$. Notice that setting $m(1) = m(3) = 0$ causes $m(2) = m(4) = m(6) = m(8) = 0$ if sufficient wavelet moments are zero. This explains the extra zero moments in Table 6.4. It also shows that the traditional specification of zero scaling function moments is redundant. In Table 6.4 $m(8)$ would be zero if more wavelet moments were zero.

To see the continuous scaling function and wavelet moments for these systems, Table 6.5 shows both the continuous and discrete moments for the length-6 and 8 coiflet systems. Notice the zero moment $m(4) = \mu(4) = 0$ for length-8. The length-14, 20, and 26 systems also have the "extra" zero scaling moment just after the first nonzero moment. This always occurs for length-$N = 6\ell + 2$ coiflets.

Table 6.5. Discrete and Continuous Moments for the Coiflet Systems

	$N = 6,$	$L = 2$			
k	$\mu(k)$	$\mu_1(k)$		$m(k)$	$m_1(k)$
0	1.4142135623	0		1.0000000000	0
1	0	0		0	0
2	0	-1.1637219122		0	-0.2057189138
3	-0.3757374752	-3.8669032118		-0.0379552166	-0.3417891854
4	-2.8727952940	-10.2673737288		-0.1354248688	-0.4537580992
5	-3.7573747525	-28.0624304008		-0.0857053279	-0.6103378310
	$N = 8,$	$L = 3$			
k	$\mu(k)$	$\mu_1(k)$		$m(k)$	$m_1(k)$
0	1.4142135623	0		1.0000000000	0
1	0	0		0	0
2	0	0		0	0
3	-2.9941117777	0.1878687376		-0.3024509630	0.0166054072
4	0	11.9764471108		0	0.5292891854
5	-45.8510203537	-43.9723329775		-1.0458570134	-0.9716604635

Figure 6.7 shows the length-6, 8, 10, and 12 coiflet scaling functions $\varphi(t)$ and wavelets $\psi(t)$. Notice their approximate symmetry and compare this to Daubechies' classical wavelet systems and her more symmetric ones achieved by using the different factorization mentioned in Section 6.3 and shown in [Dau92]. The difference between these systems and truly symmetric ones (which requires giving up orthogonality, realness, or finite support) is probably negligible in many applications.

Generalized Coifman Wavelet Systems

The preceding section shows that Coifman systems do not necessarily have an equal number of scaling function and wavelet moments equal to zero. Lengths $N = 6\ell + 2$ have equal number of zero scaling function and wavelet moments, but always have even-order "extra" zero scaling function moments located after the first nonzero one. Lengths $N = 6\ell$ always have an "extra" zero scaling function moment. Indeed, both will have several even-order "extra" zero moments for longer N as a result of the relationships illustrated in (6.52) through (6.55). Lengths $N = 6\ell - 2$ do not occur for the original definition of a Coifman If we generalize the investigation of the Coifman system to consider systems with approximately equal numbers of zero scaling function and wavelet moments, all lengths become possible and a larger class of coiflets are available.

Consider the general Coifman Wavelet System to be the same as stated in (6.40) and (6.41) but allow the number of zero scaling function and wavelet moments to differ by at most one [BO96, BO97]. That will include all the existing coiflets plus length-10, 16, 22, and $N = 6\ell - 2$. The length-10 was designed by Odegard [BO96] by setting the number of zero scaling functions to 3 and the number of zero wavelet moment to 2 rather than 2 and 2 for the length-8 or 3 and 3

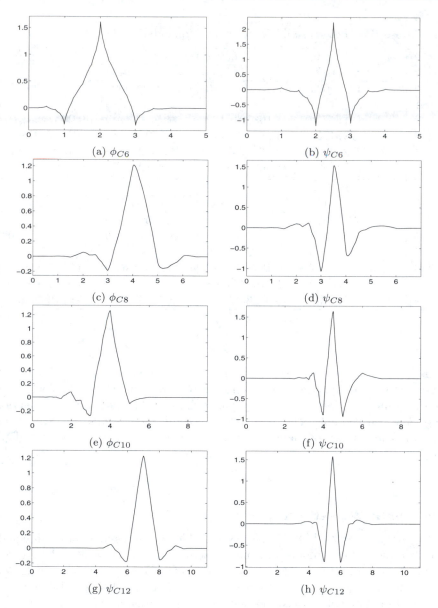

Figure 6.7. Length-6, 8, 10, and 12 Coiflet Scaling Functions and Wavelets

for the length-12 coiflets. The result in Table 6.6 shows that the length-10 design again gives one extra zero scaling function moment which is two more than the number of zero wavelet moments. This is an even-order moment predicted by (6.53) and results in a total number of zero moments between that for length-8 and length-12, as one would expect. A similar approach was used to design length-16, 22, and 28.

Table 6.6. Generalized Coiflet Scaling Function and Wavelet Coefficients plus their Discrete Moments

	Length-$N = 4$,	Degree $L = 1$			
n	$h(n)$	$h_1(n)$	$\mu(k)$	$\mu_1(k)$	k
-1	0.224143868042	0.129409522551	1.414213	0	0
0	0.836516303737	0.482962913144	0	-0.517638	1
1	0.482962913144	-0.836516303737	0.189468	0.189468	2
2	-0.129409522551	0.224143868042	-0.776457	0.827225	3
	Length-$N = 10$,	Degree $L = 3$			
n	$h(n)$	$h_1(n)$	$\mu(k)$	$\mu_1(k)$	k
-2	0.032128481856	0.000233764788	1.414213	0	0
-1	-0.075539271956	-0.000549618934	0	0	1
0	-0.096935064502	-0.013550370057	0	0	2
1	0.491549094027	0.033777338659	0	3.031570	3
2	0.805141083557	0.304413564385	0	24.674674	4
3	0.304413564385	-0.805141083557	-14.709025	138.980052	5
4	-0.033777338659	0.491549094027	64.986095	710.373341	6
5	-0.013550370057	0.096935064502			
6	0.000549618934	-0.075539271956			
7	0.000233764788	0.032128481856			

We have designed these "new" coiflet systems (e.g., $N = 10, 16, 22, 28$) by using the MATLAB optimization toolbox constrained optimization function. Wells and Tian [TW95] used Newton's method to design lengths $N = 6\ell + 2$ and $N = 6\ell$ coiflets up to length 30 [BO96]. Selesnick [SOB96] has used a filter design approach. Still another approach is given by Wei and Bovik [WB96].

Table 6.6 also shows the result of designing a length-4 system, using the one degree of freedom to ask for one zero scaling function moment rather than one zero wavelet moment as we did for the Daubechies system. For length-4, we do not get any "extra" zero moments because there are not enough zero wavelet moments. Here we see a direct trade-off between zero scaling function moments and wavelet moments. Adding these new lengths to our traditional coiflets gives Table 6.7.

The fourth and sixth columns in Table 6.7 contain the number of zero wavelet moments, excluding the $m_1(0) = 0$ which is zero because of orthogonality in all of these systems. The extra zero scaling function moments that occur after a nonzero moment for $N = 6\ell + 2$ are also excluded from the count. This table shows generalized coiflets for all even lengths. It shows the extra zero scaling function moments that are sometime achieved and how the total number of zero moments monotonically increases and how the "smoothness" as measured by the Hölder exponent [Rio93b, LH96, HW96a] increases with N and L.

When both scaling function and wavelet moments are set to zero, a larger number can be obtained than is expected from considering the degrees of freedom available. As stated earlier, of the N degrees of freedom available from the N coefficients, $h(n)$, one is used to insure existence of $\varphi(t)$ through the linear constraint (6.1), and $N/2$ are used to insure orthonormality through the quadratic constraints (6.1). This leaves $N/2-1$ degrees of freedom to achieve other characteristics.

Table 6.7. Moments for Various Length-N and Degree-L Generalized Coiflets, where (*) is the number of zero wavelet moments, excluding the $m_1(0) = 0$

N	L	$m = 0$ set	$m_1 = 0$ set*	$m = 0$ actual	$m_1 = 0$ actual*	Total zero moments	Hölder exponent
4	1	1	0	1	0	1	0.2075
6	2	1	1	2	1	3	1.0137
8	3	2	2	2	2	4	1.3887
10	3	3	2	4	2	6	1.0909
12	4	3	3	4	3	7	1.9294
14	5	4	4	4	4	8	1.7353
16	5	5	4	6	4	10	1.5558
18	6	5	5	6	5	11	2.1859
20	7	6	6	6	6	12	2.8531
22	7	7	6	8	6	14	2.5190
24	8	7	7	8	7	15	2.8300
26	9	8	8	8	8	16	3.4404
28	9	9	8	10	8	18	2.9734
30	10	9	9	10	9	19	3.4083

Daubechies used these to set the first $N/2 - 1$ wavelet moments to zero. From this approach, one would think that the coiflet system would allow $(N/2 - 1)/2$ wavelet moments to be set zero and the same number of scaling function moments. For the coiflets and generalized coiflets described in Table 6.7, one always obtains more than this. The structure of this problem allows more zero moments to be both set and achieved than the simple degrees of freedom would predict. In fact, the coiflets achieve approximately $2N/3$ total zero moments as compared with the number of degrees of freedom which is approximately $N/2$, and which is achieved by the Daubechies wavelet system.

As noted earlier and illustrated in Table 6.8, these generalized coiflets fall into three classes. Those with scaling filter lengths of $N = 6\ell + 2$ (due to Tian) have equal number of zero scaling function and wavelet moments, but always has "extra" zero scaling function moments located after the first nonzero one. Lengths $N = 6\ell$ (due to Daubechies) always have one more zero scaling function moment than zero wavelet moment and lengths $N = 6\ell - 2$ (new) always have two more zero scaling function moments than zero wavelet moments. These "extra" zero moments are predicted by (6.52) to (6.55), and there will be additional even-order zero moments for longer lengths. We have observed that within each of these classes, the Hölder exponent increases monotonically.

The lengths $N = 6\ell - 2$ were not found by earlier investigators because they have the same coiflet degree as the system that was just two shorter. However, they achieve two more zero scaling function moments than the shorter length system with the same degree.

Table 6.7 is just the beginning of a large collection of zero moment wavelet system designs with a wide variety of trade-offs that would be tailored to a particular application. In addition to the variety illustrated here, many (perhaps all) of these sets of specified zero moments have multiple solutions. This is certainly true for length-6 as illustrated in (6.42) through (6.45) and

Table 6.8. Number of Zero Moments for The Three Classes of Generalized Coiflets ($\ell = 1, 2, \cdots$), *excluding $\mu_1(0) = 0$, †excluding Non-Contiguous zeros

N Length	$m = 0$† achieved	$m_1 = 0$* achieved	Total zero moments
$N = 6\ell + 2$	$(N-2)/3$	$(N-2)/3$	$(2/3)(N-2)$
$N = 6\ell$	$N/3$	$(N-3)/3$	$(2/3)(N-3/2)$
$N = 6\ell - 2$	$(N+2)/3$	$(N-4)/3$	$(2/3)(N-1)$

for other lengths that we have found experimentally. The variety of solutions for each length can have different shifts, different Hölder exponents, and different degrees of being approximately symmetric.

The results of this chapter and section show the importance of moments to the characteristics of scaling functions and wavelets. It may not, however, be necessary or important to use the exact criteria of Daubechies or Coifman, but understanding the effects of zero moments *is* very important. It may be that setting a few scaling function moments and a few wavelets moments may be sufficient with the remaining degrees of freedom used for some other optimization, either in the frequency domain or in the time domain. As is noted in the next section, an alternative might be to minimize a larger number of various moments rather than to zero a few [OB96a].

Examples of the generalized Coiflet Systems are shown in Figure 6.7.

6.10   Minimization of Moments Rather than Zero Moments

Odegard has considered the case of minimization of a larger number of moments rather than setting $N/2 - 1$ equal to zero [OB97, OB96b, OB96a]. This results in some improvement in representing or approximating a larger class of signals at the expense of a better approximation of a smaller class. Indeed, Götze [GORB96] has shown that even in the designed zero moments wavelet systems, the implementation of the system in finite precision arithmetic results in nonzero moments and, in some cases, non-orthogonal systems.

Chapter 7

Generalizations of the Basic Multiresolution Wavelet System

Up to this point in the book, we have developed the basic two-band wavelet system in some detail, trying to provide insight and intuition into this new mathematical tool. We will now develop a variety of interesting and valuable generalizations and extensions to the basic system, but in much less detail. We hope the detail of the earlier part of the book can be transferred to these generalizations and, together with the references, will provide an introduction to the topics.

7.1 Tiling the Time–Frequency or Time–Scale Plane

A qualitative descriptive presentation of the decomposition of a signal using wavelet systems or wavelet transforms consists of partitioning the time–scale plane into *tiles* according to the indices k and j defined in (1.5). That is possible for orthogonal bases (or tight frames) because of Parseval's theorem. Indeed, it is Parseval's theorem that states that the signal energy can be partitioned on the time-scale plane. The shape and location of the tiles shows the logarithmic nature of the partitioning using basic wavelets and how the M-band systems or wavelet packets modify the basic picture. It also allows showing that the effects of time- or shift-varying wavelet systems, together with M-band and packets, can give an almost arbitrary partitioning of the plane.

The energy in a signal is given in terms of the DWT by Parseval's relation in (2.36) or (4.24). This shows the energy is a function of the translation index k and the scale index j.

$$\int |g(t)|^2 \, dt = \sum_{l=-\infty}^{\infty} |c(l)|^2 + \sum_{j=0}^{\infty} \sum_{k=-\infty}^{\infty} |d(j,k)|^2 \tag{7.1}$$

The wavelet transform allows analysis of a signal or parameterization of a signal that can locate energy in both the time and scale (or frequency) domain within the constraints of the uncertainty principle. The spectrogram used in speech analysis is an example of using the short-time Fourier transform to describe speech simultaneously in the time and frequency domains.

This graphical or visual description of the partitioning of energy in a signal using tiling depends on the *structure* of the system, not the *parameters* of the system. In other words, the tiling partitioning will depend on whether one uses $M = 2$ or $M = 3$, whether one uses wavelet packets or time-varying wavelets, or whether one uses over-complete frame systems. It does not depend on the particular coefficients $h(n)$ or $h_i(n)$, on the number of coefficients N, or the number of zero moments. One should remember that the tiling may look as if the indices j and k are continuous variables, but they are not. The energy is really a function of discrete variables in the DWT domain, and the boundaries of the tiles are symbolic of the partitioning. These tiling boundaries become more literal when the continuous wavelet transform (CWT) is used as described in Section 7.8, but even there it does not mean that the partitioned energy is literally confined to the tiles.

Nonstationary Signal Analysis

In many applications, one studies the decomposition of a signal in terms of basis functions. For example, stationary signals are decomposed into the Fourier basis using the Fourier transform. For nonstationary signals (i.e., signals whose frequency characteristics are time-varying like music, speech, images, etc.) the Fourier basis is ill-suited because of the poor time-localization. The classical solution to this problem is to use the short-time (or windowed) Fourier transform (STFT). However, the STFT has several problems, the most severe being the fixed time-frequency resolution of the basis functions. Wavelet techniques give a new class of (potentially signal dependent) bases that have desired time-frequency resolution properties. The "optimal" decomposition depends on the signal (or class of signals) studied. All classical time-frequency decompositions like the Discrete STFT (DSTFT), however, are signal independent. Each function in a basis can be considered *schematically* as a tile in the time-frequency plane, where most of its energy is concentrated. Orthonormality of the basis functions can be schematically captured by nonoverlapping tiles. With this assumption, the time-frequency tiles for the standard basis (i.e., delta basis) and the Fourier basis (i.e., sinusoidal basis) are shown in Figure 7.1.

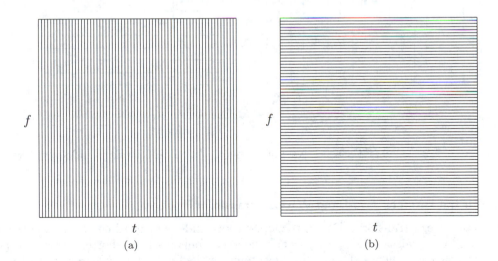

Figure 7.1. (a) Dirac Delta Function or Standard Time Domain Basis (b) Fourier or Standard Frequency Domain Basis

Tiling with the Discrete-Time Short-Time Fourier Transform

The DSTFT basis functions are of the form

$$w_{j,k}(t) = w(t - k\tau_0)e^{-\imath j\omega_0 t} \tag{7.2}$$

where $w(t)$ is a window function [Gab46]. If these functions form an orthogonal (orthonormal) basis, $x(t) = \sum_{j,k} \langle x, w_{j,k} \rangle\, w_{j,k}(t)$. The DSTFT coefficients, $\langle x, w_{j,k} \rangle$, estimate the presence of signal components centered at $(k\tau_0, j\omega_0)$ in the time-frequency plane, i.e., the DSTFT gives a uniform tiling of the time-frequency plane with the basis functions $\{w_{j,k}(t)\}$. If Δ_t and Δ_ω are time and frequency resolutions respectively of $w(t)$, then the uncertainty principle demands that $\Delta_t \Delta_\omega \leq 1/2$ [Dau90, Sie86]. Moreover, if the basis is orthonormal, the Balian-Low theorem implies either Δ_t or Δ_ω is infinite. Both Δ_t and Δ_ω can be controlled by the choice of $w(t)$, but for any particular choice, there will be signals for which either the time or frequency resolution is not adequate. Figure 7.1 shows the time-frequency tiles associated with the STFT basis for a narrow and wide window, illustrating the inherent time-frequency trade-offs associated with this basis. Notice that the tiling schematic holds for several choices of windows (i.e., each figure represents *all* DSTFT bases with the particular time-frequency resolution characteristic).

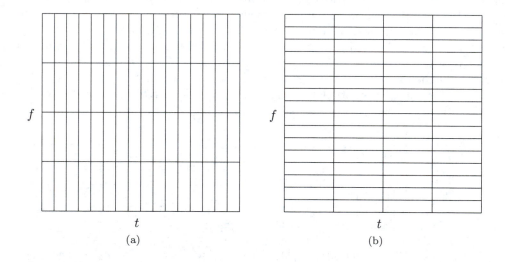

Figure 7.2. (a) STFT Basis - Narrow Window. (b) STFT Basis - Wide Window.

Tiling with the Discrete Two-Band Wavelet Transform

The discrete wavelet transform (DWT) is another signal-independent tiling of the time-frequency plane suited for signals where high frequency signal components have shorter duration than low frequency signal components. Time-frequency atoms for the DWT, $\{\psi_{j,k}(t)\} = \{2^{j/2}\psi(2^j t - k)\}$, are obtained by translates and scales of the wavelet function $\psi(t)$. One shrinks/stretches the wavelet to capture high-/low-frequency components of the signal. If these atoms form an orthonormal basis, then $x(t) = \sum_{j,k} \langle x, \psi_{j,k} \rangle\, \psi_{j,k}(t)$. The DWT coefficients, $\langle x, \psi_{j,k} \rangle$, are a measure of the energy of the signal components located at $(2^{-j}k, 2^j)$ in the time-frequency plane, giving yet another tiling of the time-frequency plane. As discussed in Chapters 3 and 8, the DWT (for compactly supported wavelets) can be efficiently computed using two-channel unitary FIR filter banks [Dau88a]. Figure 7.3 shows the corresponding tiling description which illustrates time-frequency resolution properties of a DWT basis. If you look along the frequency (or scale) axis at some particular time (translation), you can imagine seeing the frequency response of the filter bank as shown in Figure 3.5 with the logarithmic bandwidth of each channel. Indeed, each horizontal strip in the tiling of Figure 7.3 corresponds to each channel, which in turn corresponds to a scale j. The location of the tiles corresponding to each coefficient is shown in Figure 7.4. If at a particular scale, you imagine the translations along the k axis, you see the construction of the components of a signal at that scale. This makes it obvious that at lower resolutions (smaller j) the translations are large and at higher resolutions the translations are small.

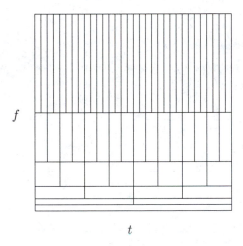

Figure 7.3. Two-band Wavelet Basis

The tiling of the time-frequency plane is a powerful graphical method for understanding the properties of the DWT and for analyzing signals. For example, if the signal being analyzed were a single wavelet itself, of the form

$$f(t) = \psi(4t - 2), \tag{7.3}$$

the DWT would have only one nonzero coefficient, $d_2(2)$. To see that the DWT is not time (or shift) invariant, imagine shifting $f(t)$ some noninteger amount and you see the DWT changes considerably. If the shift is some integer, the energy stays the same in each scale, but it "spreads out" along more values of k and spreads differently in each scale. If the shift is not an integer, the energy spreads in both j and k. There is no such thing as a "scale limited" signal corresponding to a band-limited (Fourier) signal if arbitrary shifting is allowed. For integer shifts, there is a corresponding concept [GOB94].

General Tiling

Notice that for general, nonstationary signal analysis, one desires methods for controlling the tiling of the time-frequency plane, not just using the two special cases above (their importance

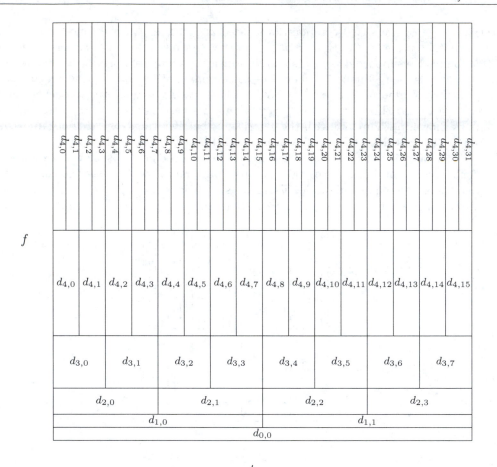

Figure 7.4. Relation of DWT Coefficients $d_{j,k}$ to Tiles

notwithstanding). An alternative way to obtain orthonormal wavelets $\psi(t)$ is using unitary FIR filter bank (FB) theory. That will be done with M-band DWTs, wavelet packets, and time-varying wavelet transforms addressed in Sections 7.2 and 7.3 and Chapter 8 respectively.

Remember that the tiles represent the relative size of the translations and scale change. They do not literally mean the partitioned energy is confined to the tiles. Representations with similar tilings can have very different characteristics.

7.2   Multiplicity-M (M-Band) Scaling Functions and Wavelets

While the use of a scale multiplier M of two in (5.1) or (7.4) fits many problems, coincides with the concept of an octave, gives a binary tree for the Mallat fast algorithm, and gives the constant-Q or logarithmic frequency bandwidths, the conditions given in Chapter 5 and Section 5.8 can be stated and proved in a more general setting where the basic scaling equation (5.1) uses a general scale factor or multiplicity of M [ZT92b, ZT92a, GB92c, SHGB93, Hel95, GOB92, Vai87b] rather than the specific doubling value of $M = 2$. Part of the motivation for a larger M comes from a desire to have a more flexible tiling of the time-scale plane than that resulting from the $M = 2$ wavelet or the short-time Fourier transform discussed in Section 7.1. It also comes from a desire for some regions of uniform band widths rather than the logarithmic spacing of the frequency responses illustrated in Figure 3.5. The motivation for larger M also comes from filter bank theory which is discussed in Chapter 8.

We pose the more general multiresolution formulation where (5.1) becomes

$$\varphi(x) = \sum_n h(n)\,\sqrt{M}\,\varphi(Mx - n). \tag{7.4}$$

In some cases, M may be allowed to be a rational number; however, in most cases it must be an integer, and in (5.1) it is required to be 2. In the frequency domain, this relationship becomes

$$\Phi(\omega) \;=\; \frac{1}{\sqrt{M}}\,H(\omega/M)\,\Phi(\omega/M) \tag{7.5}$$

and the limit after iteration is

$$\Phi(\omega) = \prod_{k=1}^{\infty} \left\{ \frac{1}{\sqrt{M}} H(\frac{\omega}{M^k}) \right\} \Phi(0) \tag{7.6}$$

assuming the product converges and $\Phi(0)$ is well defined. This is a generalization of (5.52) and is derived in (5.74).

Properties of M-Band Wavelet Systems

These theorems, relationships, and properties are generalizations of those given in Chapter 5 and Section 5.8 with some outline proofs or derivations given in the Appendix. For the multiplicity-M problem, if the support of the scaling function and wavelets and their respective coefficients is finite and the system is orthogonal or a tight frame, the length of the scaling function vector or filter $h(n)$ is a multiple of the multiplier M. This is $N = M\,G$, where Resnikoff and Wells [RW97] call M the rank of the system and G the genus.

The results of (5.10), (5.14), (5.16), and (5.17) become

Theorem 28 *If $\varphi(t)$ is an L^1 solution to (7.4) and $\int \varphi(t)\,dt \neq 0$, then*

$$\sum_n h(n) \;=\; \sqrt{M}. \tag{7.7}$$

This is a generalization of the basic multiplicity-2 result in (5.10) and does not depend on any particular normalization or orthogonality of $\varphi(t)$.

Theorem 29 *If integer translates of the solution to (7.4) are orthogonal, then*

$$\sum_n h(n + Mm)\,h(n) \;=\; \delta(m). \tag{7.8}$$

This is a generalization of (5.14) and also does not depend on any normalization. An interesting corollary of this theorem is

Corollary 3 *If integer translates of the solution to (7.4) are orthogonal, then*

$$\sum_n |h(n)|^2 = 1. \tag{7.9}$$

A second corollary to this theorem is

Corollary 4 *If integer translates of the solution to (7.4) are orthogonal, then*

$$\sum_n h(Mn + m) = 1/\sqrt{M}. \qquad m \in \mathbf{Z} \tag{7.10}$$

This is also true under weaker conditions than orthogonality as was discussed for the $M = 2$ case.

Using the Fourier transform, the following relations can be derived:

Theorem 30 *If $\varphi(t)$ is an L^1 solution to (7.4) and $\int \varphi(t)\, dt \neq 0$, then*

$$H(0) = \sqrt{M} \tag{7.11}$$

which is a frequency domain existence condition.

Theorem 31 *The integer translates of the solution to (7.4) are orthogonal if and only if*

$$\sum_\ell |\Phi(\omega + 2\pi\ell)|^2 = 1 \tag{7.12}$$

Theorem 32 *If $\varphi(t)$ is an L^1 solution to (7.4) and $\int \varphi(t)\, dt \neq 0$, then*

$$\sum_n h(n + Mm)\, h(n) = \delta(m) \tag{7.13}$$

if and only if

$$|H(\omega)|^2 + |H(\omega + 2\pi/M)|^2 + |H(\omega + 4\pi/M)|^2 + \cdots + |H(\omega + 2\pi(M-1)/M)|^2 = M. \tag{7.14}$$

This is a frequency domain orthogonality condition on $h(n)$.

Corollary 5

$$H(2\pi\,\ell/M) = 0, \quad for \;\; \ell = 1, 2, \cdots, M - 1 \tag{7.15}$$

which is a generalization of (5.20) stating where the zeros of $H(\omega)$, the frequency response of the scaling filter, are located. This is an interesting constraint on just where certain zeros of $H(z)$ must be located.

Theorem 33 *If* $\sum_n h(n) = \sqrt{M}$, *and* $h(n)$ *has finite support or decays fast enough, then a* $\varphi(t) \in L^2$ *that satisfies (7.4) exists and is unique.*

Theorem 34 *If* $\sum_n h(n) = \sqrt{M}$ *and if* $\sum_n h(n)\,h(n-Mk) = \delta(k)$, *then* $\varphi(t)$ *exists, is integrable, and generates a wavelet system that is a tight frame in* L^2.

These results are a significant generalization of the basic $M = 2$ wavelet system that we discussed in the earlier chapters. The definitions, properties, and generation of these more general scaling functions have the same form as for $M = 2$, but there is no longer a single wavelet associated with the scaling function. There are $M - 1$ wavelets. In addition to (7.4) we now have $M - 1$ wavelet equations, which we denote as

$$\psi_\ell(t) = \sum_n \sqrt{M}\, h_\ell(n)\, \varphi(Mt - n) \tag{7.16}$$

for

$$\ell = 1, 2, \cdots, M - 1.$$

Some authors use a notation $h_0(n)$ for $h(n)$ and $\varphi_0(t)$ for $\psi(t)$, so that $h_\ell(n)$ represents the coefficients for the scaling function and all the wavelets and $\varphi_\ell(t)$ represents the scaling function and all the wavelets.

Just as for the $M = 2$ case, the multiplicity-M scaling function and scaling coefficients are unique and are simply the solution of the basic recursive or refinement equation (7.4). However, the wavelets and wavelet coefficients are no longer unique or easy to design in general.

We now have the possibility of a more general and more flexible multiresolution expansion system with the M-band scaling function and wavelets. There are now $M - 1$ signal spaces spanned by the $M - 1$ wavelets at each scale j. They are denoted

$$\mathcal{W}_{\ell,j} = \operatorname*{Span}_{k}\{\psi_\ell(M^j t + k)\} \tag{7.17}$$

for $\ell = 1, 2, \cdots, M - 1$. For example with $M = 4$,

$$\mathcal{V}_1 = \mathcal{V}_0 \oplus \mathcal{W}_{1,0} \oplus \mathcal{W}_{2,0} \oplus \mathcal{W}_{3,0} \tag{7.18}$$

and

$$\mathcal{V}_2 = \mathcal{V}_1 \oplus \mathcal{W}_{1,1} \oplus \mathcal{W}_{2,1} \oplus \mathcal{W}_{3,1} \tag{7.19}$$

or

$$\mathcal{V}_2 = \mathcal{V}_0 \oplus \mathcal{W}_{1,0} \oplus \mathcal{W}_{2,0} \oplus \mathcal{W}_{3,0} \oplus \mathcal{W}_{1,1} \oplus \mathcal{W}_{2,1} \oplus \mathcal{W}_{3,1}. \tag{7.20}$$

In the limit as $j \to \infty$, we have

$$L^2 = \mathcal{V}_0 \oplus \mathcal{W}_{1,0} \oplus \mathcal{W}_{2,0} \oplus \mathcal{W}_{3,0} \oplus \mathcal{W}_{1,1} \oplus \mathcal{W}_{2,1} \oplus \mathcal{W}_{3,1} \oplus \cdots \oplus \mathcal{W}_{3,\infty}. \tag{7.21}$$

Our notation for $M = 2$ in Chapter 2 is $\mathcal{W}_{1,j} = \mathcal{W}_j$

This is illustrated pictorially in Figure 7.5 where we see the nested scaling function spaces $\mathcal{V}_j$ but each annular ring is now divided into $M - 1$ subspaces, each spanned by the $M - 1$ wavelets at that scale. Compare Figure 7.5 with Figure 2.3 for the classical $M = 2$ case.

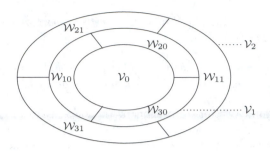

Figure 7.5. Vector Space Decomposition for a Four-Band Wavelet System, $\mathcal{W}_{\ell j}$

The expansion of a signal or function in terms of the M-band wavelets now involves a triple sum over ℓ, j, and k.

$$f(t) = \sum_k c(k)\, \varphi_k(t) + \sum_{k=-\infty}^{\infty} \sum_{j=0}^{\infty} \sum_{\ell=1}^{M-1} M^{j/2}\, d_{\ell,j}(k)\, \psi_\ell(M^j t - k) \tag{7.22}$$

where the expansion coefficients (DWT) are found by

$$c(k) = \int f(t)\, \varphi(t - k)\, dt \tag{7.23}$$

and

$$d_{\ell,j}(k) = \int f(t)\, M^{j/2}\, \psi_\ell(M^j t - k)\, dt. \tag{7.24}$$

We now have an M-band discrete wavelet transform.

Theorem 35 *If the scaling function $\varphi(t)$ satisfies the conditions for existence and orthogonality and the wavelets are defined by (7.16) and if the integer translates of these wavelets span $\mathcal{W}_{\ell,0}$ the orthogonal compliments of $\mathcal{V}_0$, all being in $\mathcal{V}_1$, i.e., the wavelets are orthogonal to the scaling function at the same scale; that is, if*

$$\int \varphi(t - n)\, \psi_\ell(t - m)\, dt = 0 \tag{7.25}$$

for $\ell = 1, 2, \cdots, M - 1$, then

$$\sum_n h(n)\, h_\ell(n - Mk) = 0 \tag{7.26}$$

for all integers k and for $\ell = 1, 2, \cdots, M - 1$.

Combining (7.8) and (7.26) and calling $h_0(n) = h(n)$ gives

$$\sum_n h_m(n)\, h_\ell(n - Mk) = \delta(k)\, \delta(m - \ell) \tag{7.27}$$

as necessary conditions on $h_\ell(n)$ for an orthogonal system.

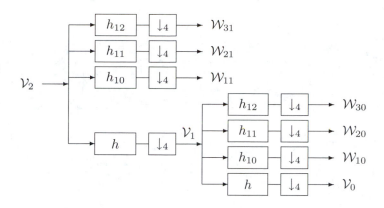

Figure 7.6. Filter Bank Structure for a Four-Band Wavelet System, $\mathcal{W}_{\ell j}$

Unlike the $M = 2$ case, for $M > 2$ there is no formula for $h_\ell(n)$ and there are many possible wavelets for a given scaling function.

Mallat's algorithm takes on a more complex form as shown in Figure 7.6. The advantage is a more flexible system that allows a mixture of linear and logarithmic tiling of the time–scale plane. A powerful tool that removes the ambiguity is choosing the wavelets by "modulated cosine" design.

Figure 7.7 shows the frequency response of the filter band, much as Figure 3.5 did for $M = 2$. Examples of scaling functions and wavelets are illustrated in Figure 7.8, and the tiling of the time-scale plane is shown in Figure 7.9. Figure 7.9 shows the time-frequency resolution characteristics of a four-band DWT basis. Notice how it is different from the Standard, Fourier, DSTFT and two-band DWT bases shown in earlier chapters. It gives a mixture of a logarithmic and linear frequency resolution.

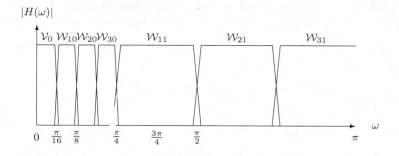

Figure 7.7. Frequency Responses for the Four-Band Filter Bank, $\mathcal{W}_{\ell j}$

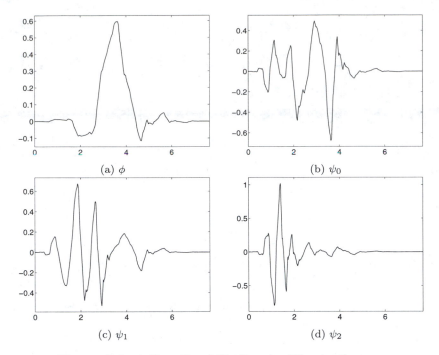

(a) ϕ

(b) ψ_0

(c) ψ_1

(d) ψ_2

Figure 7.8. A Four-Band Six-Regular Wavelet System

We next define the kth moments of $\psi_\ell(t)$ as

$$m_\ell(k) = \int t^k \, \psi_\ell(t) \, dt \qquad (7.28)$$

and the kth discrete moments of $h_\ell(n)$ as

$$\mu_\ell(k) = \sum_n n^k \, h_\ell(n). \qquad (7.29)$$

Theorem 36 (Equivalent Characterizations of K-Regular M-Band Filters) *A unitary scaling filter is K-regular if and only if the following equivalent statements are true:*

1. *All moments of the wavelet filters are zero, $\mu_\ell(k) = 0$, for $k = 0, 1, \cdots, (K - 1)$ and for $\ell = 1, 2, \cdots, (M - 1)$*

2. *All moments of the wavelets are zero, $m_\ell(k) = 0$, for $k = 0, 1, \cdots, (K - 1)$ and for $\ell = 1, 2, \cdots, (M - 1)$*

3. *The partial moments of the scaling filter are equal for $k = 0, 1, \cdots, (K - 1)$*

4. *The frequency response of the scaling filter has zeros of order K at the Mth roots of unity, $\omega = 2\pi \, \ell/M$ for $\ell = 1, 2, \cdots, M - 1$.*

5. *The magnitude-squared frequency response of the scaling filter is flat to order $2K$ at $\omega = 0$. This follows from (5.22).*

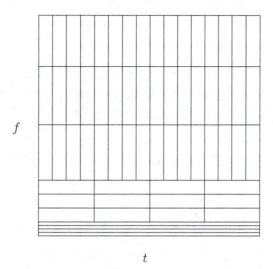

f

t

Figure 7.9. 4-band Wavelet Basis Tiling

6. *All polynomial sequences up to degree $(K-1)$ can be expressed as a linear combination of integer-shifted scaling filters.*

7. *All polynomials of degree up to $(K-1)$ can be expressed as a linear combination of integer-shifted scaling functions for all j.*

This powerful result [SHGB93, Hel95] is similar to the $M = 2$ case presented in Chapter 6. It not only ties the number of zero moments to the regularity but also to the degree of polynomials that can be exactly represented by a sum of weighted and shifted scaling functions. Note the location of the zeros of $H(z)$ are equally spaced around the unit circle, resulting in a narrower frequency response than for the half-band filters if $M = 2$. This is consistent with the requirements given in (7.14) and illustrated in Figure 7.7.

Sketches of some of the derivations in this section are given in the Appendix or are simple extensions of the $M = 2$ case. More details are given in [GB92c, SHGB93, Hel95].

M-Band Scaling Function Design

Calculating values of $\varphi(n)$ can be done by the same methods given in Section 5.10. However, the design of the scaling coefficients $h(n)$ parallels that for the two-band case but is somewhat more difficult [Hel95].

One special set of cases turns out to be a simple extension of the two-band system. If the multiplier $M = 2^m$, then the scaling function is simply a scaled version of the $M = 2$ case and a particular set of corresponding wavelets are those obtained by iterating the wavelet branches of the Mallat algorithm tree as is done for wavelet packets described in Section 7.3. For other values of M, especially odd values, the situation is more complex.

M-Band Wavelet Design and Cosine Modulated Methods

For $M > 2$ the wavelet coefficients $h_\ell(n)$ are not uniquely determined by the scaling coefficients, as was the case for $M = 2$. This is both a blessing and a curse. It gives us more flexibility in designing specific systems, but it complicates the design considerably. For small N and M, the designs can be done directly, but for longer lengths and/or for large M, direct design becomes impossible and something like the cosine modulated design of the wavelets from the scaling function as described in Chapter 8, is probably the best approach [KV92, Ngu92, NK92, Mal92, Mau92] [OGB94, GB95c, GB92a, GB95b, GB93, NH96, RT91, Ngu94, Ngu95a].

7.3   Wavelet Packets

The classical $M = 2$ wavelet system results in a logarithmic frequency resolution. The low frequencies have narrow bandwidths and the high frequencies have wide bandwidths, as illustrated in Figure 3.5. This is called "constant-Q" filtering and is appropriate for some applications but not all. The *wavelet packet* system was proposed by Ronald Coifman [Rus92, CW92] to allow a finer and adjustable resolution of frequencies at high frequencies. It also gives a rich structure that allows adaptation to particular signals or signal classes. The cost of this richer structure is a computational complexity of $O(N \log(N))$, similar to the FFT, in contrast to the classical wavelet transform which is $O(N)$.

Full Wavelet Packet Decomposition

In order to generate a basis system that would allow a higher resolution decomposition at high frequencies, we will iterate (split and down-sample) the highpass wavelet branch of the Mallat algorithm tree as well as the lowpass scaling function branch. Recall that for the discrete wavelet transform we repeatedly split, filter, and decimate the lowpass bands. The resulting three-scale analysis tree (three-stage filter bank) is shown in Figure 3.4. This type of tree results in a logarithmic splitting of the bandwidths and tiling of the time-scale plane, as shown in Figure 7.3.

If we split both the lowpass and highpass bands at all stages, the resulting filter bank structure is like a full binary tree as in Figure 7.10. It is this full tree that takes $O(N \log N)$ calculations and results in a completely evenly spaced frequency resolution. In fact, its structure is somewhat similar to the FFT algorithm. Notice the meaning of the subscripts on the signal spaces. The first integer subscript is the scale j of that space as illustrated in Figure 7.11. Each following subscript is a zero or one, depending the path taken through the filter bank illustrated in Figure 3.4. A "zero" indicates going through a lowpass filter (scaling function decomposition) and a "one" indicates going through a highpass filter (wavelet decomposition). This is different from the convention for the $M > 2$ case in Section 7.2.

Figure 7.11 pictorially shows the signal vector space decomposition for the scaling functions and wavelets. Figure 7.12 shows the frequency response of the packet filter bank much as Figure 3.5 did for $M = 2$ and Figure 7.7 for $M = 3$ wavelet systems.

Figure 7.14 shows the Haar wavelet packets with which we finish the example started in Section 2.8. This is an informative illustration that shows just what "packetizing" does to the regular wavelet system. It should be compared to the example at the end of Chapter 2. This is similar to the Walsh-Haddamar decomposition, and Figure 7.13 shows the full wavelet packet system generated from the Daubechies $\varphi_{D8'}$ scaling function. The "prime" indicates this is the Daubechies system with the spectral factorization chosen such that zeros are inside the unit circle

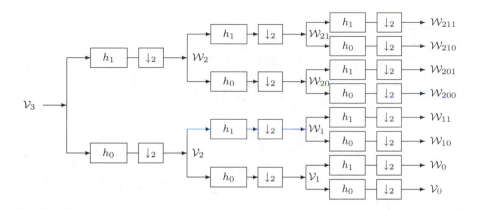

Figure 7.10. The full binary tree for the three-scale wavelet packet transform.

and some outside. This gives the maximum symmetry possible with a Daubechies system. Notice the three wavelets have increasing "frequency." They are somewhat like windowed sinusoids, hence the name, wavelet packet. Compare the wavelets with the $M = 2$ and $M = 4$ Daubechies wavelets.

Adaptive Wavelet Packet Systems

Normally we consider the outputs of each channel or band as the wavelet transform and from this have a nonredundant basis system. If, however, we consider the signals at the output of each band and at each stage or scale simultaneously, we have more outputs than inputs and clearly have a redundant system. From all of these outputs, we can choose an independent subset as a basis. This can be done in an adaptive way, depending on the signal characteristics according to some optimization criterion. One possibility is the regular wavelet decomposition shown in

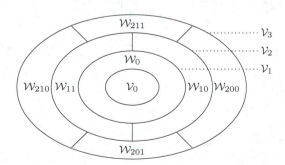

Figure 7.11. Vector Space Decomposition for a $M = 2$ Full Wavelet Packet System

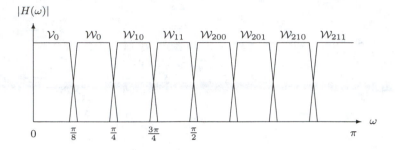

Figure 7.12. Frequency Responses for the Two-Band Wavelet Packet Filter Bank

Figure 3.3. Another is the full packet decomposition shown in Figure 7.10. Any pruning of this full tree would generate a valid packet basis system and would allow a very flexible tiling of the time-scale plane.

We can choose a set of basic vectors and form an orthonormal basis, such that some cost measure on the transformed coefficients is minimized. Moreover, when the cost is additive, the

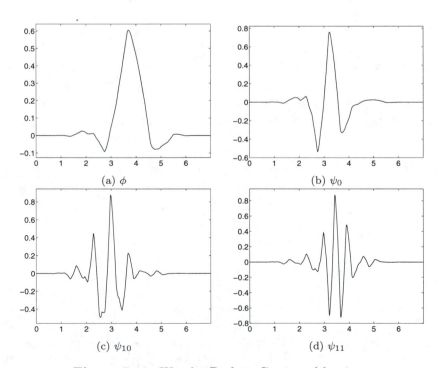

Figure 7.13. Wavelet Packets Generated by $\phi_{D8'}$

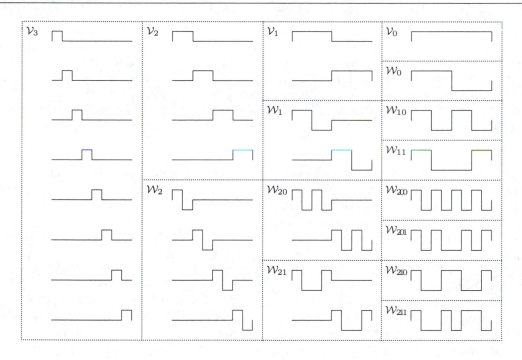

Figure 7.14. The Haar Wavelet Packet

best orthonormal wavelet packet transform can be found using a binary searching algorithm [CW92] in $O(N \log N)$ time.

Some examples of the resulting time-frequency tilings are shown in Figure 7.15. These plots demonstrate the frequency adaptation power of the wavelet packet transform.

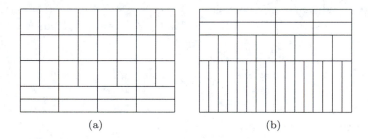

(a)   (b)

Figure 7.15. Examples of Time-Frequency Tilings of Different Three-Scale Orthonormal Wavelet Packet Transforms.

There are two approaches to using adaptive wavelet packets. One is to choose a particular decomposition (filter bank pruning) based on the characteristics of the class of signals to be processed, then to use the transform nonadaptively on the individual signals. The other is to adapt the decomposition for each individual signal. The first is a linear process over the class of signals. The second is not and will not obey superposition.

Let $P(J)$ denote the number of different J-scale orthonormal wavelet packet transforms. We can easily see that

$$P(J) = P(J-1)^2 + 1, \qquad\qquad P(1) = 1. \qquad\qquad (7.30)$$

So the number of possible choices grows dramatically as the scale increases. This is another reason for the wavelet packets to be a very powerful tool in practice. For example, the FBI standard for fingerprint image compression [BBH93, BBOH96] is based on wavelet packet transforms. The wavelet packets are successfully used for acoustic signal compression [Wic92]. In [RV93], a rate-distortion measure is used with the wavelet packet transform to improve image compression performance.

M-band DWTs give a flexible tiling of the time-frequency plane. They are associated with a particular tree-structured filter bank, where the lowpass channel at any depth is split into M bands. Combining the M-band and wavelet packet structure gives a rather arbitrary tree-structured filter bank, where all channels are split into sub-channels (using filter banks with a potentially different number of bands), and would give a very flexible signal decomposition. The wavelet analog of this is known as the wavelet packet decomposition [CW92]. For a given signal or class of signals, one can, for a fixed set of filters, obtain the best (in some sense) filter bank tree-topology. For a binary tree an efficient scheme using entropy as the criterion has been developed—the best wavelet packet basis algorithm [CW92, RV93].

7.4   Biorthogonal Wavelet Systems

Requiring the wavelet expansion system to be orthogonal across both translations and scale gives a clean, robust, and symmetric formulation with a Parseval's theorem. It also places strong limitations on the possibilities of the system. Requiring orthogonality uses up a large number of the degrees of freedom, results in complicated design equations, prevents linear phase analysis and synthesis filter banks, and prevents asymmetric analysis and synthesis systems. This section will develop the biorthogonal wavelet system using a nonorthogonal basis and dual basis to allow greater flexibility in achieving other goals at the expense of the energy partitioning property that Parseval's theorem states [CDF92, Wei95, Tia96, WTWB97, NH96, AAU96, Coh92, KT93, RG95, Swe96a]. Some researchers have considered "almost orthogonal" systems where there is some relaxation of the orthogonal constraints in order to improve other characteristics [OB95]. Indeed, many image compression schemes (including the fingerprint compression used by the FBI [BBH93, BBOH96]) use biorthogonal systems.

Two Channel Biorthogonal Filter Banks

In previous chapters for orthogonal wavelets, the analysis filters and synthesis filters are time reversal of each other; i.e., $\tilde{h}(n) = h(-n)$, $\tilde{g}(n) = g(-n)$. Here, for the biorthogonal case, we relax these restrictions. However, in order to perfectly reconstruct the input, these four filters still have to satisfy a set of relations.

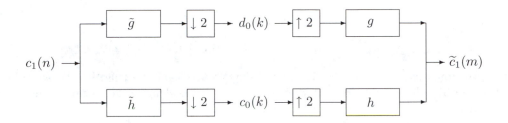

Figure 7.16. Two Channel Biorthogonal Filter Banks

Let $c_1(n), n \in \mathbf{Z}$ be the input to the filter banks in Figure 7.16, then the outputs of the analysis filter banks are

$$c_0(k) = \sum_n \tilde{h}(2k - n)c_1(n), \qquad d_0(k) = \sum_n \tilde{g}(2k - n)c_1(n). \qquad (7.31)$$

The output of the synthesis filter bank is

$$\tilde{c}_1(m) = \sum_k \left[h(2k - m)c_0(k) + g(2k - m)d_0(k) \right]. \qquad (7.32)$$

Substituting Equation (7.31) into (7.32) and interchanging the summations gives

$$\tilde{c}_1(m) = \sum_n \sum_k \left[h(2k - m)\tilde{h}(2k - n) + g(2k - m)\tilde{g}(2k - n) \right] c_1(n). \qquad (7.33)$$

For perfect reconstruction, i.e., $\tilde{c}_1(m) = c_1(m), \forall m \in \mathbf{Z}$, we need

$$\sum_k \left[h(2k - m)\tilde{h}(2k - n) + g(2k - m)\tilde{g}(2k - n) \right] = \delta(m - n). \qquad (7.34)$$

Fortunately, this condition can be greatly simplified. In order for it to hold, the four filters have to be related as [CDF92]

$$\tilde{g}(n) = (-1)^n h(1 - n), \qquad g(n) = (-1)^n \tilde{h}(1 - n), \qquad (7.35)$$

up to some constant factors. Thus they are cross-related by time reversal and flipping signs of every other element. Clearly, when $\tilde{h} = h$, we get the familiar relations between the scaling coefficients and the wavelet coefficients for orthogonal wavelets, $g(n) = (-1)^n h(1 - n)$. Substituting (7.35) back to (7.34), we get

$$\sum_n \tilde{h}(n)h(n + 2k) = \delta(k). \qquad (7.36)$$

In the orthogonal case, we have $\sum_n h(n)h(n + 2k) = \delta(k)$; i.e., $h(n)$ is orthogonal to even translations of itself. Here $\tilde{h}$ is orthogonal to h, thus the name *biorthogonal*.

Equation (7.36) is the key to the understanding of the biorthogonal filter banks. Let's assume $\tilde{h}(n)$ is nonzero when $\tilde{N}_1 \leq n \leq \tilde{N}_2$, and $h(n)$ is nonzero when $N_1 \leq n \leq N_2$. Equation (7.36) implies that [CDF92]

$$N_2 - \tilde{N}_1 = 2k + 1, \quad \tilde{N}_2 - N_1 = 2\tilde{k} + 1, \qquad k, \tilde{k} \in \mathbf{Z}. \tag{7.37}$$

In the orthogonal case, this reduces to the well-known fact that the length of h has to be even. Equations (7.37) also imply that the difference between the lengths of $\tilde{h}$ and h must be even. Thus their lengths must be both even or both odd.

Biorthogonal Wavelets

We now look at the scaling function and wavelet to see how removing orthogonality and introducing a dual basis changes their characteristics. We start again with the basic multiresolution definition of the scaling function and add to that a similar definition of a dual scaling function.

$$\phi(t) = \sum_n h(n)\sqrt{2}\phi(2t - n), \tag{7.38}$$

$$\tilde{\phi}(t) = \sum_n \tilde{h}(n)\sqrt{2}\tilde{\phi}(2t - n). \tag{7.39}$$

From Theorem 1 in Chapter 5, we know that for ϕ and $\tilde{\phi}$ to exist,

$$\sum_n h(n) = \sum_n \tilde{h}(n) = \sqrt{2}. \tag{7.40}$$

Continuing to parallel the construction of the orthogonal wavelets, we also define the wavelet and the dual wavelet as

$$\psi(t) = \sum_n g(n)\sqrt{2}\phi(2t - n) = \sum_n (-1)^n \tilde{h}(1 - n)\sqrt{2}\phi(2t - n), \tag{7.41}$$

$$\tilde{\psi}(t) = \sum_n \tilde{g}(n)\sqrt{2}\tilde{\phi}(2t - n) = \sum_n (-1)^n h(1 - n)\sqrt{2}\tilde{\phi}(2t - n). \tag{7.42}$$

Now that we have the scaling and wavelet functions and their duals, the question becomes whether we can expand and reconstruct arbitrary functions using them. The following theorem [CDF92] answers this important question.

Theorem 37 *For $\tilde{h}$ and h satisfying (7.36), suppose that for some $C, \epsilon > 0$,*

$$|\Phi(\omega)| \leq C(1 + \omega)^{-1/2-\epsilon}, \qquad |\tilde{\Phi}(\omega)| \leq C(1 + \omega)^{-1/2-\epsilon}. \tag{7.43}$$

If ϕ and $\tilde{\phi}$ defined above have sufficient decay in the frequency domain, then $\psi_{j,k} \stackrel{\text{def}}{=} 2^{j/2}\psi(2^j x - k)$, $j, k \in \mathbf{Z}$ constitute a frame in $L^2(\mathbf{R})$. Their dual frame is given by $\tilde{\psi}_{j,k} \stackrel{\text{def}}{=} 2^{j/2}\tilde{\psi}(2^j x - k)$, $j, k \in \mathbf{Z}$; for any $f \in L^2(\mathbf{R})$,

$$f = \sum_{j,k} \langle f, \psi_{j,k} \rangle \tilde{\psi}_{j,k} = \sum_{j,k} \langle f, \tilde{\psi}_{j,k} \rangle \psi_{j,k} \tag{7.44}$$

where the series converge strongly.

Moreover, the $\psi_{j,k}$ and $\tilde{\psi}_{j,k}$ constitute two Riesz bases, with

$$\left\langle \psi_{j,k}, \tilde{\psi}_{j',k'} \right\rangle = \delta(j - j')\delta(k - k') \tag{7.45}$$

if and only if

$$\int \phi(x)\, \tilde{\phi}(x - k)\, dx = \delta(k). \tag{7.46}$$

This theorem tells us that under some technical conditions, we can expand functions using the wavelets and reconstruct using their duals. The multiresolution formulations in Chapter 2 can be revised as

$$\cdots \subset \mathcal{V}_{-2} \subset \mathcal{V}_{-1} \subset \mathcal{V}_0 \subset \mathcal{V}_1 \subset \mathcal{V}_2 \subset \cdots \tag{7.47}$$

$$\cdots \subset \tilde{\mathcal{V}}_{-2} \subset \tilde{\mathcal{V}}_{-1} \subset \tilde{\mathcal{V}}_0 \subset \tilde{\mathcal{V}}_1 \subset \tilde{\mathcal{V}}_2 \subset \cdots \tag{7.48}$$

where

$$\mathcal{V}_j = \operatorname*{Span}_{k}\{\phi_{j.k}\}, \qquad \tilde{\mathcal{V}}_j = \operatorname*{Span}_{k}\{\tilde{\phi}_{j.k}\}. \tag{7.49}$$

If (7.46) holds, we have

$$\mathcal{V}_j \perp \widetilde{\mathcal{W}}_j, \qquad \tilde{\mathcal{V}}_j \perp \mathcal{W}_j, \tag{7.50}$$

where

$$\mathcal{W}_j = \operatorname*{Span}_{k}\{\psi_{j.k}\}, \qquad \widetilde{\mathcal{W}}_j = \operatorname*{Span}_{k}\{\tilde{\psi}_{j.k}\}. \tag{7.51}$$

Although $\mathcal{W}_j$ is not the orthogonal complement to $\mathcal{V}_j$ in $\mathcal{V}_{j+1}$ as before, the dual space $\widetilde{\mathcal{W}}_j$ plays the much needed role. Thus we have four sets of spaces that form two hierarchies to span $L^2(\mathbf{R})$.

In Section 5.8, we have a list of properties of the scaling function and wavelet that do not require orthogonality. The results for regularity and moments in Chapter 6 can also be generalized to the biorthogonal systems.

Comparisons of Orthogonal and Biorthogonal Wavelets

The biorthogonal wavelet systems generalize the classical orthogonal wavelet systems. They are more flexible and generally easy to design. The differences between the orthogonal and biorthogonal wavelet systems can be summarized as follows.

- The orthogonal wavelets filter and scaling filter must be of the same length, and the length must be even. This restriction has been greatly relaxed for biorthogonal systems.

- Symmetric wavelets and scaling functions are possible in the framework of biorthogonal wavelets. Actually, this is one of the main reasons to choose biorthogonal wavelets over the orthogonal ones.

- Parseval's theorem no longer holds in biorthogonal wavelet systems; i.e., the norm of the coefficients is not the same as the norm of the functions being spanned. This is one of the main disadvantages of using the biorthogonal systems. Many design efforts have been devoted to making the systems near orthogonal, so that the norms are close.

- In a biorthogonal system, if we switch the roles of the primary and the dual, the overall system is still sound. Thus we can choose the best arrangement for our application. For example, in image compression, we would like to use the smoother one of the pair to reconstruct the coded image to get better visual appearance.

- In statistical signal processing, white Gaussian noise remains white after orthogonal transforms. If the transforms are nonorthogonal, the noise becomes correlated or colored. Thus, when biorthogonal wavelets are used in estimation and detection, we might need to adjust the algorithm to better address the colored noise.

Example Families of Biorthogonal Systems

Because biorthogonal wavelet systems are very flexible, there are a wide variety of approaches to design different biorthogonal systems. The key is to design a pair of filters h and $\tilde{h}$ that satisfy (7.36) and (7.40) and have other desirable characteristics. Here we review several families of biorthogonal wavelets and discuss their properties and design methods.

Cohen-Daubechies-Feauveau Family of Biorthogonal Spline Wavelets

Splines have been widely used in approximation theory and numerical algorithms. Therefore, they may be desirable scaling functions, since they are symmetric, smooth, and have dyadic filter coefficients (see Section 5.7). However, if we use them as scaling functions in orthogonal wavelet systems, the wavelets have to have infinite support [Chu92a]. On the other hand, it is very easy to use splines in biorthogonal wavelet systems. Choose h to be a filter that can generate splines, then (7.36) and (7.40) are linear in the coefficients of $\tilde{h}$. Thus we only have to solve a set of linear equations to get $\tilde{h}$, and the resulting $\tilde{h}$ also have dyadic coefficients. In [CDF92], better methods are used to solve these equations indirectly.

The filter coefficients for some members of the Cohen-Daubechies-Feauveau family of biorthogonal spline wavelets are listed in Table 7.1. Note that they are symmetric. It has been shown that as the length increases, the regularity of ϕ and $\tilde{\phi}$ of this family also increases [CDF92].

$h/\sqrt{2}$	$\tilde{h}/\sqrt{2}$
$1/2, 1/2$	$-1/16, 1/16, 1/2, 1/16, -1/16$
$1/4, 1/2, 1/4$	$-1/8, 1/4, 3/4, 1/4, -1/8$
$1/8, 3/8, 3/8, 1/8$	$-5/512, 15/512, 19/512, -97/512, -13/256, 175/256, \cdots$

Table 7.1. Coefficients for Some Members of Cohen-Daubechies-Feauveau Family of Biorthogonal Spline Wavelets (For longer filters, we only list half of the coefficients)

Cohen-Daubechies-Feauveau Family of Biorthogonal Wavelets with Less Dissimilar Filter Length

The Cohen-Daubechies-Feauveau family of biorthogonal wavelets are perhaps the most widely used biorthogonal wavelets, since the scaling function and wavelet are symmetric and have similar lengths. A member of the family is used in the FBI fingerprint compression standard [BBH93, BBOH96]. The design method for this family is remarkably simple and elegant.

In the frequency domain, (7.36) can be written as

$$H(\omega)\tilde{H}^*(\omega) + H(\omega + \pi)\tilde{H}^*(\omega + \pi) \; = \; 2. \tag{7.52}$$

Recall from Chapter 6 that we have an explicit solution for $|H(\omega)|^2 = M(\omega)$ such that

$$M(\omega) + M(\omega + \pi) \; = \; 2, \tag{7.53}$$

and the resulting compactly supported orthogonal wavelet has the maximum number of zero moments possible for its length. In the orthogonal case, we get a scaling filter by factoring $M(\omega)$ as $H(\omega)H^*(\omega)$. Here in the biorthogonal case, we can factor the same $M(\omega)$ to get $H(\omega)$ and $\tilde{H}(\omega)$.

Factorizations that lead to symmetric h and $\tilde{h}$ with similar lengths have been found in [CDF92], and their coefficients are listed in Table 7.2. Plots of the scaling and wavelet functions, which are members of the family used in the FBI fingerprint compression standard, are in Figure 7.17.

$\tilde{h}$	h
0.85269867900889	0.78848561640637
0.37740285561283	0.41809227322204
-0.11062440441844	-0.04068941760920
-0.02384946501956	-0.06453888262876
0.03782845550726	

Table 7.2. Coefficients for One of the Cohen-Daubechies-Feauveau Family of Biorthogonal Wavelets that is Used in the FBI Fingerprint Compression Standard (We only list half of the coefficients)

Tian-Wells Family of Biorthogonal Coiflets

The coiflet system is a family of compactly supported orthogonal wavelets with zero moments of both the scaling functions and wavelets described in Section 6.9. Compared with Daubechies' wavelets with only zero wavelet moments, the coiflets are more symmetrical and may have better approximation properties when sampled data are used. However, finding the orthogonal coiflets involves solving a set of nonlinear equations. No closed form solutions have been found, and when the length increases, numerically solving these equations becomes less stable.

Tian and Wells [TW95, TWBOar] have constructed biorthogonal wavelet systems with both zero scaling function and wavelet moments. Closed form solutions for these biorthogonal coiflets have been found. They have approximation properties similar to the coiflets, and the filter coefficients are dyadic rationals as are the splines. The filter coefficients for these biorthogonal Coiflets are listed in Table 7.3. Some members of this family are also in the spline family described earlier.

Lifting Construction of Biorthogonal Systems

We have introduced several families of biorthogonal systems and their design methods. There is another method called a *lifting scheme*, which is very simple and general. It has a long history

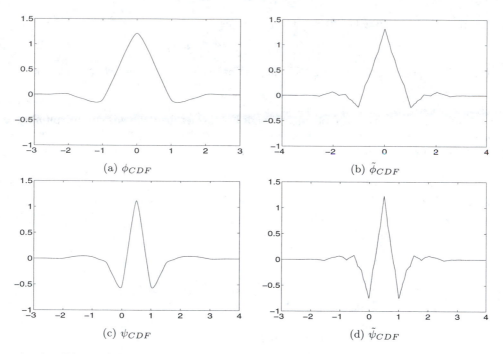

(a) ϕ_{CDF}

(b) $\tilde{\phi}_{CDF}$

(c) ψ_{CDF}

(d) $\tilde{\psi}_{CDF}$

Figure 7.17. Plots of Scaling Function and Wavelet and their Duals for one of the Cohen-Daubechies-Feauveau Family of Biorthogonal Wavelets that is Used in the FBI Fingerprint Compression Standard

[Mar92, Mar93, DM93, KS92, VH92, Donar], and has been systematically developed recently [Swe96a, Swe95]. The key idea is to build complicated biorthogonal systems using simple and invertible stages. The first stage does nothing but to separate even and odd samples, and it is easily invertible. The structure is shown in Figure 7.18, and is called the *lazy wavelet transform* in [Swe96a].

$\sqrt{2}h$	$\sqrt{2}\tilde{h}$
$1, 1$	$1, 1$
$1/2, 1, 1/2$	$-1/4, 1/2, 3/2, 1/2, -1/4$
$3/8, 1, 3/4, 0, -1/8$	$3/64, 0, -3/16, 3/8, 41/32, 3/4, -3/16, -1/8, 3/64$
$-1/16, 0, 9/16, 1, 9/16, 0, -1/16$	$-1/256, 0, 9/128, -1/16, -63/256, 9/16, 87/64, \cdots$

Table 7.3. Coefficients for some Members of the Biorthogonal Coiflets. For Longer Filters, We only List Half of the Coefficients.

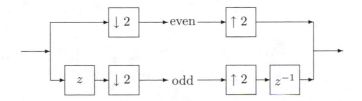

Figure 7.18. The Lazy Wavelet Transform

After splitting the data into two parts, we can predict one part from the other, and keep only the prediction error, as in Figure 7.19. We can reconstruct the data by recomputing the prediction and then add back the prediction. In Figure 7.19, s and t are prediction filters.

By concatenating simple stages, we can implement the forward and inverse wavelet transforms as in Figure 7.20. It is also called the *ladder structure*, and the reason for the name is clear from the figure. Clearly, the system is invertible, and thus biorthogonal. Moreover, it has been shown the orthogonal wavelet systems can also be implemented using lifting [DS96a]. The advantages of lifting are numerous:

- Lifting steps can be calculated inplace. As seen in Figure 7.20, the prediction outputs based on one channel of the data can be added to or subtracted from the data in other channels, and the results can be saved in the same place in the second channel. No auxiliary memory is needed.

- The predictors s and t do not have to be linear. Nonlinear operations like the medium filter or rounding can be used, and the system remains invertible. This allows a very simple generalization to nonlinear wavelet transform or nonlinear multiresolution analysis.

- The design of biorthogonal systems boils down to the design of the predictors. This may lead to simple approaches that do not relay on the Fourier transform [Swe96a], and can be generalized to irregular samples or manifolds.

- For biorthogonal systems, the lifting implementations require less numerical operations than direct implementations [DS96a]. For orthogonal cases, the lifting schemes have the computational complexity similar to the lattice factorizations, which is almost half of the direct implementation.

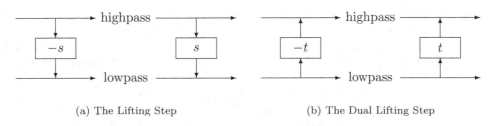

(a) The Lifting Step   (b) The Dual Lifting Step

Figure 7.19. The Lifting and Dual Lifting Step

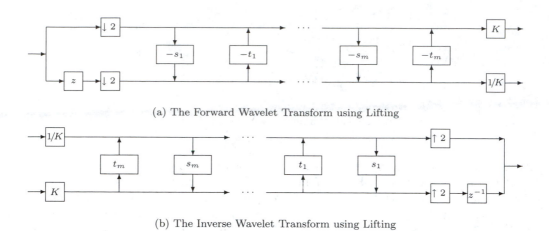

(a) The Forward Wavelet Transform using Lifting

(b) The Inverse Wavelet Transform using Lifting

Figure 7.20. Wavelet Transform using Lifting

7.5 Multiwavelets

In Chapter 2, we introduced the multiresolution analysis for the space of L^2 functions, where we have a set of nesting subspaces

$$\{0\} \subset \cdots \subset \mathcal{V}_{-2} \subset \mathcal{V}_{-1} \subset \mathcal{V}_0 \subset \mathcal{V}_1 \subset \mathcal{V}_2 \subset \cdots \subset L^2, \tag{7.54}$$

where each subspace is spanned by translations of scaled versions of a single scaling function φ; e.g.,

$$\mathcal{V}_j = \underset{k}{\text{Span}}\{2^{j/2}\varphi(2^j t - k)\}. \tag{7.55}$$

The direct difference between nesting subspaces are spanned by translations of a single *wavelet* at the corresponding scale; e.g.,

$$\mathcal{W}_j = \mathcal{V}_{j+1} \ominus \mathcal{V}_j = \underset{k}{\text{Span}}\{2^{j/2}\psi(2^j t - k)\}. \tag{7.56}$$

There are several limitations of this construction. For example, nontrivial orthogonal wavelets can not be symmetric. To avoid this problem, we generalized the basic construction, and introduced multiplicity-M (M-band) scaling functions and wavelets in Section 7.2, where the difference spaces are spanned by translations of $M-1$ wavelets. The scaling is in terms of the power of M; i.e.,

$$\varphi_{j,k}(t) = M^{j/2}\varphi(M^j t - k). \tag{7.57}$$

In general, there are more degrees of freedom to design the M-band wavelets. However, the nested $\mathcal{V}$ spaces are still spanned by translations of a single scaling function. It is the *multiwavelets* that removes the above restriction, thus allowing multiple scaling functions to span the nested $\mathcal{V}$ spaces [GLT93, GL94, Str96b]. Although it is possible to construct M-band multiwavelets, here we only present results on the two-band case, as most of the researches in the literature do.

Construction of Two-Band Multiwavelets

Assume that $\mathcal{V}_0$ is spanned by translations of R different scaling functions $\varphi_i(t)$, $i = 1, \ldots, R$. For a two-band system, we define the scaling and translation of these functions by

$$\varphi_{i,j,k}(t) = 2^{j/2}\varphi_i(2^j t - k). \tag{7.58}$$

The multiresolution formulation implies

$$\mathcal{V}_j = \underset{k}{\text{Span}}\{\varphi_{i,j,k}(t) : i = 1, \ldots, R\}. \tag{7.59}$$

We next construct a vector scaling function by

$$\Phi(t) = [\varphi_1(t), \ldots, \varphi_R(t)]^T. \tag{7.60}$$

Since $\mathcal{V}_0 \subset \mathcal{V}_1$, we have

$$\boxed{\Phi(t) = \sqrt{2}\sum_n H(n)\,\Phi(2t - n)} \tag{7.61}$$

where $H(k)$ is a $R \times R$ matrix for each $k \in \mathbf{Z}$. This is a matrix version of the scalar recursive equation (2.13). The first and simplest multiscaling functions probably appear in [Alp93], and they are shown in Figure 7.21.

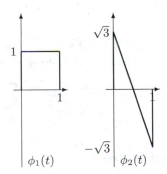

Figure 7.21. The Simplest Alpert Multiscaling Functions

The first scaling function $\varphi_1(t)$ is nothing but the Haar scaling function, and it is the sum of two time-compressed and shifted versions of itself, as shown in Figure 2.2(a). The second scaling function can be easily decomposed into linear combinations of time-compressed and shifted versions of the Haar scaling function and itself, as

$$\varphi_2(t) = \frac{\sqrt{3}}{2}\varphi_1(2t) + \frac{1}{2}\varphi_2(2t) - \frac{\sqrt{3}}{2}\varphi_1(2t - 1) + \frac{1}{2}\varphi_2(2t - 1). \tag{7.62}$$

This is shown in Figure 7.22

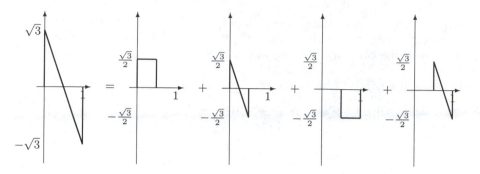

Figure 7.22. Multiwavelet Refinement Equation (7.62)

Putting the two scaling functions together, we have

$$\begin{bmatrix} \varphi_1(t) \\ \varphi_2(t) \end{bmatrix} = \begin{bmatrix} 1 & 0 \\ \sqrt{3}/2 & 1/2 \end{bmatrix} \begin{bmatrix} \varphi_1(2t) \\ \varphi_2(2t) \end{bmatrix} + \begin{bmatrix} 1 & 0 \\ -\sqrt{3}/2 & 1/2 \end{bmatrix} \begin{bmatrix} \varphi_1(2t-1) \\ \varphi_2(2t-1) \end{bmatrix}. \tag{7.63}$$

Further assume R wavelets span the difference spaces; i.e.,

$$\mathcal{W}_j = \mathcal{V}_{j+1} \ominus \mathcal{V}_j = \operatorname*{Span}_k \{\psi_{i,j,k}(t) : i = 1, \dots, R\}. \tag{7.64}$$

Since $\mathcal{W}_0 \subset \mathcal{V}_1$ for the stacked wavelets $\Psi(t)$ there must exist a sequence of $R \times R$ matrices $G(k)$, such that

$$\boxed{\Psi(t) = \sqrt{2} \sum_k G(k)\, \Phi(2t-k)} \tag{7.65}$$

These are vector versions of the two scale recursive equations (2.13) and (2.24).

We can also define the discrete-time Fourier transform of $H(k)$ and $G(k)$ as

$$\mathbf{H}(\omega) = \sum_k H(k) e^{i\omega k}, \qquad \mathbf{G}(\omega) = \sum_k G(k) e^{i\omega k}. \tag{7.66}$$

Properties of Multiwavelets

Approximation, Regularity and Smoothness

Recall from Chapter 6 that the key to regularity and smoothness is having enough number of zeros at π for $H(\omega)$. For multiwavelets, it has been shown that polynomials can be exactly reproduced by translates of $\Phi(t)$ if and only if $\mathbf{H}(\omega)$ can be factored in special form [Plo95c, Plo95a, Plo95b]. The factorization is used to study the regularity and convergence of refinable function vectors [CDP95], and to construct multi-scaling functions with approximation and symmetry [PS95]. Approximation and smoothness of multiple refinable functions are also studied in [HSS96, JRZ96a, JRZ97].

Support

In general, the finite length of $H(k)$ and $G(k)$ ensure the finite support of $\Phi(t)$ and $\Psi(t)$. However, there are no straightforward relations between the support length and the number of nonzero coefficients in $H(k)$ and $G(k)$. An explanation is the existence of nilpotent matrices [SW97]. A method to estimate the support is developed in [SW97].

Orthogonality

For these scaling functions and wavelets to be orthogonal to each other and orthogonal to their translations, we need [SS95]

$$\mathbf{H}(\omega)\mathbf{H}^\dagger(\omega) + \mathbf{H}(\omega + \pi)\mathbf{H}^\dagger(\omega + \pi) = I_R, \tag{7.67}$$

$$\mathbf{G}(\omega)\mathbf{G}^\dagger(\omega) + \mathbf{G}(\omega + \pi)\mathbf{G}^\dagger(\omega + \pi) = I_R, \tag{7.68}$$

$$\mathbf{H}(\omega)\mathbf{G}^\dagger(\omega) + \mathbf{H}(\omega + \pi)\mathbf{G}^\dagger(\omega + \pi) = 0_R, \tag{7.69}$$

where $\dagger$ denotes the complex conjugate transpose, I_R and 0_R are the $R \times R$ identity and zero matrix respectively. These are the matrix versions of (5.22) and (5.30). In the scalar case, (5.30) can be easily satisfied if we choose the wavelet filter by time-reversing the scaling filter and changing the signs of every other coefficients. However, for the matrix case here, since matrices do not commute in general, we cannot derive the $G(k)$'s from $H(k)$'s so straightforwardly. This presents some difficulty in finding the wavelets from the scaling functions; however, this also gives us flexibility to design different wavelets even if the scaling functions are fixed [SS95].

   The conditions in (7.67–7.69) are necessary but not sufficient. Generalization of Lawton's sufficient condition (Theorem 14 in Chapter 5) has been developed in [Plo97b, Plo97a, JRZ96b].

Implementation of Multiwavelet Transform

Let the expansion coefficients of multiscaling functions and multiwavelets be

$$c_{i,j}(k) = \langle f(t), \varphi_{i,j,k}(t) \rangle, \tag{7.70}$$

$$d_{i,j}(k) = \langle f(t), \psi_{i,j,k}(t) \rangle. \tag{7.71}$$

We create vectors by

$$C_j(k) = [c_{1,j}(k), \ldots, c_{R,j}(k)]^T, \tag{7.72}$$

$$D_j(k) = [d_{1,j}(k), \ldots, d_{R,j}(k)]^T. \tag{7.73}$$

For $f(t)$ in $\mathcal{V}_0$, it can be written as linear combinations of scaling functions and wavelets,

$$f(t) = \sum_k C_{j_0}(k)^T \Phi_{J_0,k}(t) + \sum_{j=j_0}^{\infty} \sum_k D_j(k)^T \Psi_{j,k}(t). \tag{7.74}$$

Using (7.61) and (7.65), we have

$$C_{j-1}(k) = \sqrt{2} \sum_n H(n) C_j(2k + n) \tag{7.75}$$

and

$$D_{j-1}(k) = \sqrt{2} \sum_n G(n) C_j(2k + n). \tag{7.76}$$

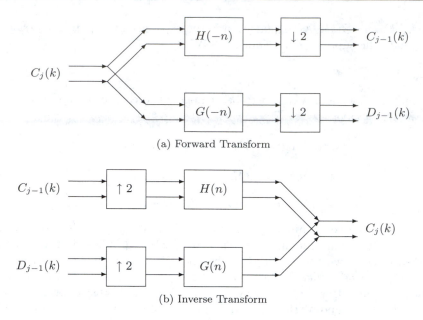

(a) Forward Transform

(b) Inverse Transform

Figure 7.23. Discrete Multiwavelet Transform

Moreover,

$$C_j(k) = \sqrt{2} \sum_k \left(H(k)^\dagger C_{j-1}(2k+n) + G(k)^\dagger D_{j-1}(2k+n) \right). \tag{7.77}$$

These are the vector forms of (3.9), (3.10), and (3.20). Thus the synthesis and analysis filter banks for multiwavelet transforms have similar structures as the scalar case. The difference is that the filter banks operate on blocks of R inputs and the filtering and rate-changing are all done in terms of blocks of inputs.

To start the multiwavelet transform, we need to get the scaling coefficients at high resolution. Recall that in the scalar case, the scaling functions are close to delta functions at very high resolution, so the samples of the function are used as the scaling coefficients. However, for multi-wavelets we need the expansion coefficients for R scaling functions. Simply using nearby samples as the scaling coefficients is a bad choice. Data samples need to be preprocessed (*prefiltered*) to produce reasonable values of the expansion coefficients for multi-scaling function at the highest scale. Prefilters have been designed based on interpolation [XGHS96], approximation [HR96], and orthogonal projection [VA96].

Examples

Because of the larger degree of freedom, many methods for constructing multiwavelets have been developed.

Geronimo-Hardin-Massopust Multiwavelets

A set of multiscaling filters based on fractal interpolation functions were developed in [GHM94], and the corresponding multiwavelets were constructed in [SS95]. As shown in Figure 7.24, they

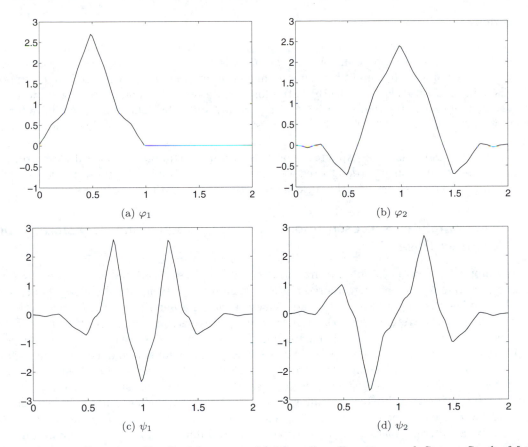

(a) φ_1

(b) φ_2

(c) ψ_1

(d) ψ_2

Figure 7.24.   Geronimo-Hardin-Massopust Multi-scaling Function and Strang-Strela Multi-wavelets

are both symmetrical and orthogonal—a combination which is impossible for two-band orthogonal scalar wavelets. They also have short support, and can exactly reproduce the hat function. These interesting properties make multiwavelet a promising expansion system.

Spline Multiwavelets

Spline bases have a maximal approximation order with respect to their length, however spline uniwavelets are only semiorthogonal [SS94]. A family of spline multiwavelets that are symmetric and orthogonal is developed in [RN96].

Other Constructions

Other types of multiwavelets are constructed using Hermite interpolating conditions [CL96], matrix spectral factorization [CNKS96], finite elements [SS94], and oblique projections [Ald96]. Similar to multiwavelets, vector-valued wavelets and vector filter banks are also developed [XS96].

Applications

Multiwavelets have been used in data compression [HSS*95, LLK96, SHS*96], noise reduction [DS96b, SHS*96], and solution of integral equations [CMX96]. Because multiwavelets are able to offer a combination of orthogonality, symmetry, higher order of approximation and short support, methods using multiwavelets frequently outperform those using the comparable scale wavelets. However, it is found that prefiltering is very important, and should be chosen carefully for the applications [DS96b, SHS*96, XGHS96]. Also, since discrete multiwavelet transforms operate on size-R blocks of data and generate blocks of wavelet coefficients, the correlation within each block of coefficients needs to be exploited. For image compression, predictions rules are proposed to exploit the correlation in order to reduce the bit rate [LLK96]. For noise reduction, joint thresholding coefficients within each block improve the performance [DS96b].

7.6   Overcomplete Representations, Frames, Redundant Transforms, and Adaptive Bases

In this chapter, we apply the ideas of frames and tight frames introduced in Chapter 4 as well as bases to obtain a more efficient representation of many interesting signal classes. It might be helpful to review the material on bases and frames in that chapter while reading this section.

Traditional basis systems such as Fourier, Gabor, wavelet, and wave packets are efficient representations for certain classes of signals, but there are many cases where a single system is not effective. For example, the Fourier basis is an efficient system for sinusoidal or smooth periodic signals, but poor for transient or chirp-like signals. Each system seems to be best for a rather well-defined but narrow class of signals. Recent research indicates that significant improvements in efficiency can be achieved by combining several basis systems. One can intuitively imagine removing Fourier components until the expansion coefficients quit dropping off rapidly, then switching to a different basis system to expand the residual and, after that expansion quits dropping off rapidly, switching to still another. Clearly, this is not a unique expansion because the order of expansion system used would give different results. This is because the total expansion system is a linear combination of the individual basis systems and is, therefore, not a basis itself but a frame. It is an overcomplete expansion system and a variety of criteria have been developed to use the freedom of the nonuniqueness of the expansion to advantage. The collection of basis systems from which a subset of expansion vectors is chosen is sometimes called a dictionary.

There are at least two descriptions of the problem. We may want a single expansion system to handle several different classes of signals, each of which are well-represented by a particular basis system or we may have a single class of signals, but the elements of that class are linear combinations of members of the well-represented classes. In either case, there are several criteria that have been identified as important [CD94, CD95a]:

- *Sparsity:* The expansion should have most of the important information in the smallest number of coefficients so that the others are small enough to be neglected or set equal to zero. This is important for compression and denoising.

- *Separation:* If the measurement consists of a linear combination of signals with different characteristics, the expansion coefficients should clearly separate those signals. If a single signal has several features of interest, the expansion should clearly separate those features. This is important for filtering and detection.

- *Superresolution:* The resolution of signals or characteristics of a signal should be much better than with a traditional basis system. This is likewise important for linear filtering, detection, and estimation.

- *Stability:* The expansions in terms of our new overcomplete systems should not be significantly changed by perturbations or noise. This is important in implementation and data measurement.

- *Speed:* The numerical calculation of the expansion coefficients in the new overcomplete system should be of order $O(N)$ or $O(N \log(N))$.

These criteria are often in conflict with each other, and various compromises will be made in the algorithms and problem formulations for an acceptable balance.

Overcomplete Representations

This section uses the material in Chapter 4 on bases and frames. One goal is to represent a signal using a "dictionary" of expansion functions that could include the Fourier basis, wavelet basis, Gabor basis, etc. We formulate a finite dimensional version of this problem as

$$y(n) = \sum_k \alpha_k \, x_k(n) \qquad n, k \in \mathbf{Z} \tag{7.78}$$

for $n = 0, 1, 2, \cdots, N - 1$ and $k = 0, 1, 2, \cdots, K - 1$. This can be written in matrix form as

$$\mathbf{y} = \mathbf{X} \, \alpha \tag{7.79}$$

where $\mathbf{y}$ is a $N \times 1$ vector with elements being the signal values $y(n)$, the matrix $\mathbf{X}$ is $N \times K$ the columns of which are made up of all the functions in the dictionary and α is a $K \times 1$ vector of the expansion coefficients α_k. The matrix operator has the basis signals $\mathbf{x_k}$ as its columns so that the matrix multiplication (7.79) is simply the signal expansion (7.78).

For a given signal representation problem, one has two decisions: what dictionary to use (i.e., choice of the $\mathbf{X}$) and how to represent the signal in terms of this dictionary (i.e., choice of α). Since the dictionary is overcomplete, there are several possible choices of α and typically one uses prior knowledge or one or more of the desired properties we saw earlier to calculate the α.

A Matrix Example

Consider a simple two-dimensional system with orthogonal basis vectors

$$\mathbf{x_1} = \begin{bmatrix} 1 \\ 0 \end{bmatrix} \quad \text{and} \quad \mathbf{x_2} = \begin{bmatrix} 0 \\ 1 \end{bmatrix} \tag{7.80}$$

which gives the matrix operator with $\mathbf{x_1}$ and $\mathbf{x_2}$ as columns

$$\mathbf{X} = \begin{bmatrix} 1 & 0 \\ 0 & 1 \end{bmatrix}. \tag{7.81}$$

Decomposition with this rather trivial operator gives a time-domain description in that the first expansion coefficient α_0 is simply the first value of the signal, $x(0)$, and the second coefficient is the second value of the signal. Using a different set of basis vectors might give the operator

$$\mathbf{X} = \begin{bmatrix} 0.7071 & 0.7071 \\ 0.7071 & -0.7071 \end{bmatrix} \tag{7.82}$$

which has the normalized basis vectors still orthogonal but now at a 45° angle from the basis vectors in (7.81). This decomposition is a sort of frequency domain expansion. The first column vector will simply be the constant signal, and its expansion coefficient $\alpha(0)$ will be the average of the signal. The coefficient of the second vector will calculate the difference in $y(0)$ and $y(1)$ and, therefore, be a measure of the change.

Notice that $\mathbf{y} = \begin{bmatrix} 1 \\ 0 \end{bmatrix}$ can be represented exactly with only one nonzero coefficient using (7.81) but will require two with (7.82), while for $\mathbf{y} = \begin{bmatrix} 1 \\ 1 \end{bmatrix}$ the opposite is true. This means the signals $\mathbf{y} = \begin{bmatrix} 1 \\ 0 \end{bmatrix}$ and $\mathbf{y} = \begin{bmatrix} 0 \\ 1 \end{bmatrix}$ can be represented sparsely by (7.81) while $\mathbf{y} = \begin{bmatrix} 1 \\ 1 \end{bmatrix}$ and $\mathbf{y} = \begin{bmatrix} 1 \\ -1 \end{bmatrix}$ can be represented sparsely by (7.82).

If we create an overcomplete expansion by a linear combination of the previous orthogonal basis systems, then it should be possible to have a sparse representation for all four of the previous signals. This is done by simply adding the columns of (7.82) to those of (7.81) to give

$$\mathbf{X} = \begin{bmatrix} 1 & 0 & 0.7071 & 0.7071 \\ 0 & 1 & 0.7071 & -0.7071 \end{bmatrix} \tag{7.83}$$

This is clearly overcomplete, having four expansion vectors in a two-dimensional system. Finding α_k requires solving a set of underdetermined equations, and the solution is not unique. For example, if the signal is given by

$$\mathbf{y} = \begin{bmatrix} 1 \\ 0 \end{bmatrix} \tag{7.84}$$

there are an infinity of solutions, several of which are listed in the following table.

Case	1	2	3	4	5	6	7
α_0	0.5000	1.0000	1.0000	1.0000	0	0	0
α_1	0.0000	0.0000	0	0	-1.0000	1.0000	0
α_2	0.3536	0	0.0000	0	1.4142	0	0.7071
α_3	0.3536	0	0	0.0000	0	1.4142	0.7071
$\|\alpha\|^2$	0.5000	1.0000	1.0000	1.0000	3.0000	3.0000	1.0000

Case 1 is the minimum norm solution of $\mathbf{y} = \mathbf{X}\alpha$ for α_k. It is calculated by a pseudo inverse with the MATLAB command a = pinv(X)*y . It is also the redundant DWT discussed in the next section and calculated by a = X'*y/2. Case 2 is the minimum norm solution, but for no more than two nonzero values of α_k. Case 2 can also be calculated by inverting the matrix (7.83) with columns 3 and 4 deleted. Case 3 is calculated the same way with columns 2 and 4 deleted, case 4 has columns 2 and 3 deleted, case 5 has 1 and 4 deleted, case 6 has 1 and 3 deleted, and case 7 has 1 and 2 deleted. Cases 3 through 7 are unique since the reduced matrix is square and nonsingular. The second term of α for case 1 is zero because the signal is orthogonal to that expansion vector. Notice that the norm of α is minimum for case 1 and is equal to the norm of $\mathbf{y}$ divided by the redundancy, here two. Also notice that the coefficients in cases 2, 3, and 4 are the same even though calculated by different methods.

Because $\mathbf{X}$ is not only a frame, but a tight frame with a redundancy of two, the energy (norm squared) of α is one-half the norm squared of $\mathbf{y}$. The other decompositions (not tight frame or basis) do not preserve the energy.

Next consider a two-dimensional signal that cannot be exactly represented by only one expansion vector. If the unity norm signal is given by

$$\mathbf{y} = \begin{bmatrix} 0.9806 \\ 0.1961 \end{bmatrix} \tag{7.85}$$

the expansion coefficients are listed next for the same cases described previously.

Case	1	2	3	4	5	6	7
α_0	0.4903	0.9806	0.7845	1.1767	0	0	0
α_1	0.0981	0.1961	0	0	-0.7845	1.1767	0
α_2	0.4160	0	0.2774	0	1.3868	0	0.8321
α_3	0.2774	0	0	-0.2774	0	1.3868	0.5547
$\|\alpha\|^2$	0.5000	1.0000	0.6923	1.4615	2.5385	3.3077	1.0000

Again, case 1 is the minimum norm solution; however, it has no zero components this time because there are no expansion vectors orthogonal to the signal. Since the signal lies between the 90° and 45° expansion vectors, it is case 3 which has the least two-vector energy representation.

There are an infinite variety of ways to construct the overcomplete frame matrix $\mathbf{X}$. The one in this example is a four-vector tight frame. Each vector is 45° degrees apart from nearby vectors. Thus they are evenly distributed in the 180° upper plane of the two dimensional space. The lower plane is covered by the negative of these frame vectors. A three-vector tight frame would have three columns, each 60° from each other in the two-dimension plane. A 36-vector tight frame would have 36 columns spaced 5° from each other. In that system, any signal vector would be very close to an expansion vector.

Still another alternative would be to construct a frame (not tight) with nonorthogonal rows. This would result in columns that are not evenly spaced but might better describe some particular class of signals. Indeed, one can imagine constructing a frame operator with closely spaced expansion vectors in the regions where signals are most likely to occur or where they have the most energy.

We next consider a particular modified tight frame constructed so as to give a shift-invariant DWT.

Shift-Invariant Redundant Wavelet Transforms and Nondecimated Filter Banks

One of the few flaws in the various wavelet basis decompositions and wavelet transforms is the fact the DWT is not translation-invariant. If you shift a signal, you would like the DWT coefficients to simply shift, but it does more than that. It significantly changes character.

Imagine a DWT of a signal that is a wavelet itself. For example, if the signal were

$$y(n) = \varphi(2^4 n - 10) \tag{7.86}$$

then the DWT would be

$$d_4(10) = 1 \quad \text{all other } d_j(k) = c(k) = 0. \tag{7.87}$$

In other words, the series expansion in the orthogonal wavelet basis would have only one nonzero coefficient.

If we shifted the signal to the right so that $y(n) = \varphi(2^4(n-1) - 10)$, there would be many nonzero coefficients because at this shift or translation, the signal is no longer orthogonal to most of the basis functions. The signal energy would be partitioned over many more coefficients and, therefore, because of Parseval's theorem, be smaller. This would degrade any denoising or compressions using thresholding schemes. The DWT described in Chapter 9 is periodic in that at each scale j the periodized DWT repeats itself after a shift of $n = 2^j$, but the period depends on the scale. This can also be seen from the filter bank calculation of the DWT where each scale goes through a different number of decimators and therefore has a different aliasing.

A method to create a linear, shift-invariant DWT is to construct a frame from the orthogonal DWT supplemented by shifted orthogonal DWTs using the ideas from the previous section. If you do this, the result is a frame and, because of the redundancy, is called the redundant DWT or RDWT.

The typical wavelet based signal processing framework consists of the following three simple steps, 1) wavelet transform; 2) point-by-point processing of the wavelet coefficients (e.g. thresholding for denoising, quantization for compression); 3) inverse wavelet transform. The diagram of the framework is shown in Figure 7.25. As mentioned before, the wavelet transform is not translation-invariant, so if we shift the signal, perform the above processing, and shift the output back, then the results are different for different shifts. Since the frame vectors of the RDWT consist of the shifted orthogonal DWT basis, if we replace the forward/inverse wavelet transform

Figure 7.25. The Typical Wavelet Transform Based Signal Processing Framework (Δ denotes the pointwise processing)

Figure 7.26. The Typical Redundant Wavelet Transform Based Signal Processing Framework (Δ denotes the pointwise processing)

in the above framework by the forward/inverse RDWT, then the result of the scheme in Figure 7.26 is the same as the average of all the processing results using DWTs with different shifts of the input data. This is one of the main reasons that RDWT-based signal processing tends to be more robust.

Still another view of this new transform can be had by looking at the Mallat-derived filter bank described in Chapters 5 and 8. The DWT filter banks illustrated in Figures 3.3 and 3.6 can be modified by removing the decimators between each stage to give the coefficients of the tight frame expansion (the RDWT) of the signal. We call this structure the undecimated filter bank. Notice that, without the decimation, the number of terms in the DWT is larger than N. However, since these are the expansion coefficients in our new overcomplete frame, that is consistent. Also, notice that this idea can be applied to M-band wavelets and wavelet packets in the same way.

These RDWTs are not precisely a tight frame because each scale has a different redundancy. However, except for this factor, the RDWT and undecimated filter have the same characteristics of a tight frame and, they support a form of Parseval's theorem or energy partitioning.

If we use this modified tight frame as a dictionary to choose a particular subset of expansion vectors as a new frame or basis, we can tailor the system to the signal or signal class. This is discussed in the next section on adaptive systems.

This idea of RDWT was suggested by Mallat [Mal91], Beylkin [Bey92], Shensa [She92], Dutilleux [Dut89], Nason [NS95], Guo [Guo94, Guo95], Coifman, and others. This redundancy comes at a price of the new RDWT having $O(N \log(N))$ arithmetic complexity rather than $O(N)$. Liang and Parks [LP94b, LP94a], Bao and Erdol [BE94, BE93], Marco and Weiss [MWJ94, MW94b, MW94a], Daubechies [Dau92], and others [PKC96] have used some form of averaging or "best basis" transform to obtain shift invariance.

Recent results indicate this nondecimated DWT, together with thresholding, may be the best denoising strategy [DJKP95b, DJKP95a, LGO*95, CD95b, LGOB95, GLOB95, LGO*96, Guo95]. The nondecimated DWT is shift invariant, is less affected by noise, quantization, and error, and has order $N \log(N)$ storage and arithmetic complexity. It combines with thresholding to give denoising and compression superior to the classical Donoho method for many examples. Further discussion of use of the RDWT can be found in Section 10.3.

Adaptive Construction of Frames and Bases

In the case of the redundant discrete wavelet transform just described, an overcomplete expansion system was constructed in such a way as to be a tight frame. This allowed a single linear shift-invariant system to describe a very wide set of signals, however, the description was adapted to the characteristics of the signal. Recent research has been quite successful in constructing expansion systems adaptively so as to give high sparsity and superresolution but at a cost of added computation and being nonlinear. This section will look at some of the recent results in this area [MZ93, DJ94a, CD94, CD95a].

While use of an adaptive paradigm results in a shift-invariant orthogonal transform, it is nonlinear. It has the property of $DWT\{a\,f(x)\} = a\,DWT\{f(x)\}$, but it does not satisfy superposition, i.e. $DWT\{\,f(x) + g(x)\} \neq DWT\{f(x)\} + DWT\{g(x)\}$. That can sometimes be a problem.

Since these finite dimensional overcomplete systems are a frame, a subset of the expansion vectors can be chosen to be a basis while keeping most of the desirable properties of the frame. This is described well by Chen and Donoho in [CD94, CD95a]. Several of these methods are outlined as follows:

- The method of frames (MOF) was first described by Daubechies [Dau88b, Dau90, Dau92] and uses the rather straightforward idea of solving the overcomplete frame (underdetermined set of equations) in (7.83) by minimizing the L^2 norm of α. Indeed, this is one of the classical definitions of solving the normal equations or use of a pseudo-inverse. That can easily be done in MATLAB by `a = pinv(X)*y`. This gives a frame solution, but it is usually not sparse.

- The best orthogonal basis method (BOB) was proposed by Coifman and Wickerhauser [CW92, DJ94a] to adaptively choose a best basis from a large collection. The method is fast (order $N \log N$) but not necessarily sparse.

- Mallat and Zhang [MZ93] proposed a sequential selection scheme called matching pursuit (MP) which builds a basis, vector by vector. The efficiency of the algorithm depends on the order in which vectors are added. If poor choices are made early, it takes many terms to correct them. Typically this method also does not give sparse representations.

- A method called basis pursuit (BP) was proposed by Chen and Donoho [CD94, CD95a] which solves (7.83) while minimizing the L^1 norm of α. This is done by linear programming and results in a globally optimal solution. It is similar in philosophy to the MOFs but uses an L^1 norm rather than an L^2 norm and uses linear programming to obtain the optimization. Using interior point methods, it is reasonably efficient and usually gives a fairly sparse solution.

- Krim et al. describe a best basis method in [KMDW95]. Tewfik et al. propose a method called optimal subset selection in [NAT96] and others are [BM95, Cro96].

All of these methods are very signal and problem dependent and, in some cases, can give much better results than the standard M-band or wavelet packet based methods.

7.7   Local Trigonometric Bases

In the material up to this point, all of the expansion systems have required the translation and scaling properties of (1.5) and the satisfaction of the multiresolution analysis assumption of (2.13). From this we have been able to generate orthogonal basis systems with the basis functions having compact support and, through generalization to M-band wavelets and wavelet packets, we have been able to allow a rather general tiling of the time-frequency or time-scale plane with flexible frequency resolution.

By giving up the multiresolution analysis (MRA) requirement, we will be able to create another basis system with a time-frequency tiling somewhat the dual of the wavelet or wavelet packet system. Much as we saw the multiresolution system dividing the frequency bands in a logarithmic spacing for the $M = 2$ systems and a linear spacing for the higher M case, and a rather general form for the wavelet packets, we will now develop the *local cosine* and *local sine* basis systems for a more flexible time segmenting of the time-frequency plane. Rather than modifying the MRA systems by creating the time-varying wavelet systems, we will abandon the MRA and build a basis directly.

What we are looking for is an expansion of a signal or function in the form

$$f(t) = \sum_{k,n} a_k(n)\, \chi_{k,n}(t), \tag{7.88}$$

where the functions $\chi_{j,k}(t)$ are of the form (for example)

$$\chi_{k,n}(t) = w_k(t) \cos(\alpha\pi(n+\beta)t + \gamma). \tag{7.89}$$

Here $w_k(t)$ is a window function giving localization to the basis function and α, β and γ are constants the choice of which we will get to shortly. k is a time index while n is a frequency index. By requiring orthogonality of these basis functions, the coefficients (the transform) are found by an inner product

$$a_k(n) = \langle f(t), \chi_{k,n}(t)\rangle = \int f(t)\chi_{k,n}(t)\, dt. \tag{7.90}$$

We will now examine how this can be achieved and what the properties of the expansion are.

Fundamentally, the wavelet packet system decomposes $L^2(\mathbb{R})$ into a direct sum of orthogonal spaces, each typically covering a certain frequency band and spanned by the translates of a particular element of the wavelet packet system. With wavelet packets time-frequency tiling with flexible frequency resolution is possible. However, the temporal resolution is determined by the frequency band associated with a particular element in the packet.

Local trigonometric bases [Wic95, AWW92] are duals of wavelet packets in the sense that these bases give flexible temporal resolution. In this case, $L^2(\mathbb{R})$ is decomposed into a direct sum of spaces each typically covering a particular time interval. The basis functions are all modulates of a fixed window function.

One could argue that an obvious approach is to partition the time axis into disjoint bins and use a Fourier series expansion in each temporal bin. However, since the basis functions are "rectangular-windowed" exponentials they are discontinuous at the bin boundaries and hence undesirable in the analysis of smooth signals. If one replaces the rectangular window with a "smooth" window, then, since products of smooth functions are smooth, one can generate smooth windowed exponential basis functions. For example, if the time axis is split uniformly, one is looking at basis functions of the form $\left\{w(t-k)e^{\iota 2\pi nt}\right\}, k, n \in \mathbf{Z}$ for some smooth window function $w(t)$. Unfortunately, orthonormality disallows the function $w(t)$ from being well-concentrated in time or in frequency - which is undesirable for time frequency analysis. More precisely, the Balian-Low theorem (see p.108 in [Dau92]) states that the Heisenberg product of g (the product of the time-spread and frequency-spread which is lower bounded by the Heisenberg uncertainty principle) is infinite. However, it turns out that windowed trigonometric bases (that use cosines and sines but not exponentials) can be orthonormal, and the window can have a finite Heisenberg product [DJJ]. That is the reason why we are looking for local trigonometric bases of the form given in (7.89).

Nonsmooth Local Trigonometric Bases

To construct local trigonometric bases we have to choose: (a) the window functions $w_k(t)$; and (b) the trigonometric functions (i.e., α, β and γ in Eq. 7.89). If we use the rectangular window (which we know is a bad choice), then it suffices to find a trigonometric basis for the interval that the window spans. Without loss of generality, we could consider the unit interval $(0, 1)$ and hence we are interested in trigonometric bases for $L^2((0, 1))$. It is easy to see that the following four sets of functions satisfy this requirement.

1. $\{\phi_n(t)\} = \{\sqrt{2}\cos(\pi(n + \frac{1}{2})t)\}$, $n \in \{0, 1, 2, \ldots\}$;

2. $\{\phi_n(t)\} = \{\sqrt{2}\sin(\pi(n + \frac{1}{2})t)\}$, $n \in \{0, 1, 2, \ldots\}$;

3. $\{\phi_n(t)\} = \{1, \sqrt{2}\cos(\pi n t)\}$, $n \in \{1, 2, \ldots\}$;

4. $\{\phi_n(t)\} = \{\sqrt{2}\sin(\pi n t)\}$, $n \in \{0, 1, 2, \ldots\}$.

Indeed, these orthonormal bases are obtained from the Fourier series on $(-2, 2)$ (the first two) and on $(-1, 1)$ (the last two) by appropriately imposing symmetries and hence are readily verified to be complete and orthonormal on $(0, 1)$. If we choose a set of nonoverlapping rectangular window functions $w_k(t)$ such that $\sum_k w_k(t) = 1$ for all $t \in \mathbb{R}$, and define $\chi_{k,n}(t) = w_k(t)\phi_n(t)$, then, $\{\chi_{k,n}(t)\}$ is a local trigonometric basis for $L^2(\mathbb{R})$, for each of the four choices of $phi_n(t)$ above.

Construction of Smooth Windows

We know how to construct orthonormal trigonometric bases for disjoint temporal bins or intervals. Now we need to construct smooth windows $w_k(t)$ that when applied to cosines and sines retain orthonormality. An outline of the process is as follows: A unitary operation is applied that "unfolds" the discontinuities of all the local basis functions at the boundaries of each temporal bin. Unfolding leads to overlapping (unfolded) basis functions. However, since unfolding is unitary, the resulting functions still form an orthonormal basis. The unfolding operator is parameterized by a function $r(t)$ that satisfies an algebraic constraint (which makes the operator unitary). The smoothness of the resulting basis functions depends on the smoothness of this underlying function $r(t)$.

The function $r(t)$, referred to as a *rising cutoff* function, satisfies the following conditions (see Fig. 7.27) :

$$|r(t)|^2 + |r(-t)|^2 = 1, \quad \text{for all } t \in \mathbb{R}; \qquad r(t) = \begin{cases} 0, & \text{if } t \leq -1 \\ 1, & \text{if } t \geq 1 \end{cases} \qquad (7.91)$$

$r(t)$ is called a rising cutoff function because it rises from 0 to 1 in the interval $[-1, 1]$ (note: it does not necessarily have to be monotone increasing). Multiplying a function by $r(t)$ would localize it to $[-1, \infty]$. Every real-valued function $r(t)$ satisfying (7.91) is of the form $r(t) = sin(\theta(t))$ where

$$\theta(t) + \theta(-t) = \frac{\pi}{2} \quad \text{for all } t \in \mathbb{R}; \qquad r(t) = \begin{cases} 0, & \text{if } t \leq -1. \\ \frac{\pi}{2}, & \text{if } t \geq 1. \end{cases} \qquad (7.92)$$

This ensures that $r(-t) = \sin(\theta(-t)) = \sin(\frac{\pi}{2} - \theta(t)) = \cos(\theta(t))$ and therefore $r^2(t) + r^2(-t) = 1$. One can easily construct arbitrarily smooth rising cutoff functions. We give one such recipe from [Wic95] (p.105) . Start with a function

$$r_{[0]}(t) = \begin{cases} 0, & \text{if } t \leq 1 \\ \sin\left(\frac{\pi}{4}(1+t)\right), & \text{if } -1 < t < 1 \\ 1, & \text{if } t \geq 1 \end{cases} \tag{7.93}$$

It is readily verified to be a rising cutoff function. Now recursively define $r_{[1]}(t), r_{[2]}(t), \ldots$ as follows:

$$r_{[n+1]}(t) = r_{[n]}(\sin(\frac{\pi}{2}t)). \tag{7.94}$$

Notice that $r_{[n]}(t)$ is a rising cutoff function for every n. Moreover, by induction on n it is easy to show that $r_{[n]}(t) \in C^{2^n-1}$ (it suffices to show that derivatives at $t = -1$ and $t = 1$ exist and are zero up to order $2^n - 1$).

Folding and Unfolding

Using a rising cutoff function $r(t)$ one can define the folding operator, U, and its inverse, the unfolding operator $U^\star$ as follows:

$$U(r)f(t) = \begin{cases} r(t)f(t) + r(-t)f(-t), & \text{if } t > 0 \\ r(-t)f(t) - r(t)f(-t), & \text{if } t < 0 \end{cases} \tag{7.95}$$

$$U^\star(r)f(t) = \begin{cases} r(t)f(t) - r(-t)f(-t), & \text{if } t > 0 \\ r(-t)f(t) + r(t)f(-t), & \text{if } t < 0 \end{cases} \tag{7.96}$$

Notice that $U^\star(r)U(r)f(t) = (|r(t)|^2 + |r(-t)|^2)f(t) = U(r)U^\star(r)f(t)$ and that $\|U(r)f\| = \|f\| = \|U^\star(r)f\|$ showing that $U(r)$ is a unitary operator on $L^2(\mathbb{R})$. Also these operators change $f(t)$ only in $[-1, 1]$ since $U(r)f(t) = f(t) = U^\star(r)f(t)$ for $t \leq -1$ and $t \geq 1$. The interval $[-1, 1]$ is the *action region* of the folding/unfolding operator. $U(r)$ is called a *folding operator* acting at zero because for smooth f, $U(r)f$ has a discontinuity at zero. By translation and dilation of $r(t)$ one can define $U(r, t_0, \epsilon)$ and $U^\star(r, t_0, \epsilon)$ that folds and unfolds respectively about $t = t_0$ with *action region* $[t_0 - \epsilon, t_0 + \epsilon]$ and *action radius* ϵ.

Notice (7.95) and (7.96) do not define the value $U(r)f(0)$ and $U^\star(r)f(0)$ because of the discontinuity that is potentially introduced. An elementary exercise in calculus divulges the nature of this discontinuity. If $f \in C^d(\mathbb{R})$, then $U(r)f \in C^d(\mathbb{R} \setminus \{0\})$. At $t = 0$, left and right derivatives exist with all even-order left-derivatives and all odd order right-derivatives (upto and including d) being zero. Conversely, given any function $f \in C^d(\mathbb{R} \setminus \{0\})$ which has a discontinuity of the above type, $U^\star(r)f$ has a unique extension across $t = 0$ (i.e., a choice of value for $(U^\star(r)f)(0)$) that is in $C^d(\mathbb{R})$. One can switch the signs in (7.95) and (7.96) to obtain another set of folding and unfolding operators. In this case, for $f \in C^d(\mathbb{R})$, $U(r)f$ will have its even-order right derivatives and odd-order left derivatives equal to zero. We will use U_+, $U_+^\star$ and U_-, $U_-^\star$, respectively to distinguish between the two types of folding/unfolding operators and call them positive and negative polarity folding/unfolding operators respectively.

So far we have seen that the folding operator is associated with a rising cutoff function, acts at a certain point, has a certain action region and radius and has a certain polarity. To get a qualitative idea of what these operators do, let us look at some examples.

First, consider a case where $f(t)$ is even- or-odd symmetric about the folding point on the action interval. Then, Uf corresponds to simply windowing f by an appropriate window function. Indeed, if $f(t) = f(-t)$ on $[-1,1]$,

$$U_+(r,0,1)f(t) = \begin{cases} (r(t) + r(-t))f(t), & \text{if } t > 0, \\ (r(-t) - r(t))f(t), & \text{if } t < 0, \end{cases} \tag{7.97}$$

and if $f(t) = -f(-t)$ on $[-1,1]$

$$U_+(r,0,1)f(t) = \begin{cases} (r(t) - r(-t))f(t), & \text{if } t > 0, \\ (r(-t) - r(t))f(t), & \text{if } t < 0. \end{cases} \tag{7.98}$$

Figure 7.27 shows a rising cutoff function and the action of the folding operators of both polarities on the constant function. Observe the nature of the discontinuity at $t = 0$ and the effect of polarity reversal.

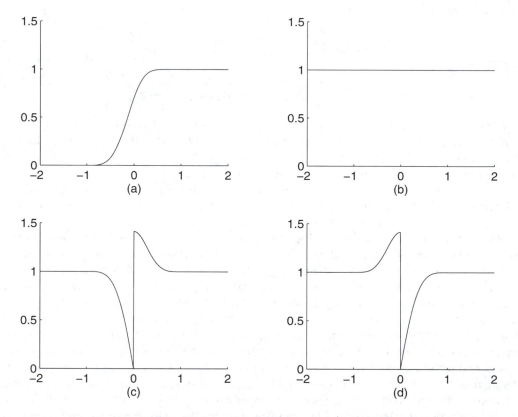

Figure 7.27. (a) The Rising Cutoff Function $r(t) = r_{[3]}(t)$; (b) $f(t) \equiv 1$; (c) $U_+(r_{[3]}, 0, 1)f(t)$; (d) $U_-(r_{[3]}, 0, 1)f(t)$

We saw that for signals with symmetry in the action region folding, corresponds to windowing. Next we look at signals that are supported to the right (or left) of the folding point and see what unfolding does to them. In this case, $U_+^\star(r)(f)$ is obtained by windowing the even (or odd) extension of $f(t)$ about the folding point. Indeed if $f(t) = 0, t < 0$

$$U_+^\star(r, 0, 1)f(t) = \begin{cases} r(t)f(t), & \text{if } t > 0, \\ r(t)f(-t), & \text{if } t < 0, \end{cases} \tag{7.99}$$

and if $f(t) = 0, t > 0$,

$$U_+^\star(r, 0, 1)f(t) = \begin{cases} --r(-t)f(-t), & \text{if } t > 0, \\ r(-t)f(t), & \text{if } t < 0, \end{cases} \tag{7.100}$$

Figure 7.28 shows the effect of the positive unfolding operator acting on cosine and sine functions supported on the right and left half-lines respectively. Observe that unfolding removes the discontinuities at $t = 0$. If the polarity is reversed, the effects on signals on the half-line are switched; the right half-line is associated with windowed odd extensions and left half-line with windowed even extensions.

Local Cosine and Sine Bases

Recall the four orthonormal trigonometric bases for $L^2((0,1))$ we described earlier.

1. $\{\phi_n(t)\} = \{\sqrt{2}\cos(\pi(n + \frac{1}{2})t)\}$, $n \in \{0, 1, 2, \ldots\}$;

2. $\{\phi_n(t)\} = \{\sqrt{2}\sin(\pi(n + \frac{1}{2})t)\}$, $n \in \{0, 1, 2, \ldots\}$;

3. $\{\phi_n(t)\} = \{1, \sqrt{2}\cos(\pi n t)\}$, $n \in \{1, 2, \ldots\}$;

4. $\{\phi_n(t)\} = \{\sqrt{2}\sin(\pi n t)\}$, $n \in \{0, 1, 2, \ldots\}$.

The bases functions have discontinuities at $t = 0$ and $t = 1$ because they are restrictions of the cosines and sines to the unit interval by rectangular windowing. The natural extensions of these basis functions to $t \in \mathbb{R}$ (i.e., unwindowed cosines and sines) are either even (say "+") or odd (say "-") symmetric (locally) about the endpoints $t = 0$ and $t = 1$. Indeed the basis functions for the four cases are $(+, -)$, $(-, +)$, $(+, +)$ and $(-, -)$ symmetric, respectively, at $(0, 1)$. From the preceding analysis, this means that unfolding these basis functions corresponds to windowing if the unfolding operator has the right polarity. Also observe that the basis functions are discontinuous at the endpoints. Moreover, depending on the symmetry at each endpoint all odd derivatives (for "+" symmetry) or even derivatives (for "−" symmetry) are zero. By choosing unfolding operators of appropriate polarity at the endpoints (with non overlapping action regions) for the four bases, we get smooth basis functions of compact support. For example, for $(+,-)$ symmetry, the basis function $U_+(r_0, 0, \epsilon_0)U_+(r_1, 1, \epsilon_1)\psi_n(t)$ is supported in $(-\epsilon_0, 1 + \epsilon_1)$ and is as many times continuously differentiable as r_0 and r_1 are.

Let $\{t_j\}$ be an ordered set of points in $\mathbb{R}$ defining a partition into disjoint intervals $I_j = [t_j, t_{j+1}]$. Now choose one of the four bases above for each interval such that at t_j the basis functions for I_{j-1} and that for I_j have opposite symmetries. We say the polarity at t_j is positive if the symmetry is $-)(+$ and negative if it is $+)(-$. At each t_j choose a smooth cutoff function

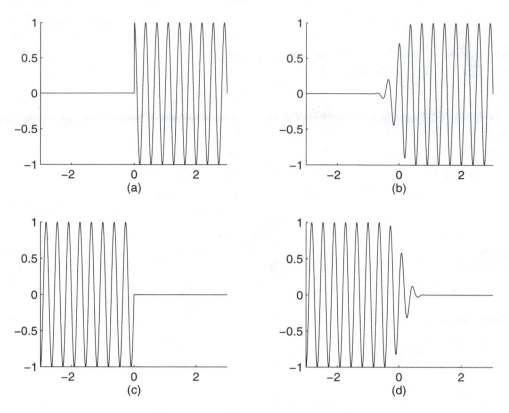

Figure 7.28. Folding Functions Supported on Half-Lines. (a) $f(t) = \cos(\frac{\pi}{2}11t)u(t)$ (($u(t)$ is the Unit Step or Heaviside Function) ; (b) $U_+(r_{[3]}, 0, 1)f(t)$; (c) $f(t) = \sin(\frac{\pi}{2}11t)u(-t)$; (d) $U_+(r_{[3]}, 0, 1)f(t)$

$r_j(t)$ and action radius ϵ_j so that the action intervals do not overlap. Let $p(j)$ be the polarity of t_j and define the unitary operator

$$U^\star = \prod_j U^\star_{p(j)}(r_j, t_j, \epsilon_j). \tag{7.101}$$

Let $\{\psi_n(t)\}$ denote all the basis functions for all the intervals put together. Then $\{\psi_n(t)\}$ forms a nonsmooth orthonormal basis for $L^2(\mathbb{R})$. Simultaneously $\{U^\star\psi_n(t)\}$ also forms a smooth orthonormal basis for $L^2(\mathbb{R})$. To find the expansion coefficients of a function $f(t)$ in this basis we use

$$\langle f, U^\star\psi_n \rangle = \langle Uf, \psi_n \rangle. \tag{7.102}$$

In other words, to compute the expansion coefficients of f in the new (smooth) basis, one merely folds f to Uf and finds its expansion coefficients with respect to the original basis. This allows one to exploit fast algorithms available for coefficient computation in the original basis.

So for an arbitrary choice of polarities at the end points t_j we have smooth local trigonometric bases. In particular by choosing the polarity to be positive for all t_j (consistent with the choice

of the first basis in all intervals) we get local cosine bases. If the polarity is negative for all
t_j (consistent with the choice of the second basis for all intervals), we get local sine bases.
Alternating choice of polarity (consistent with the alternating choice of the third and fourth
bases in the intervals) thus leads to alternating cosine/sine bases.

All these bases can be constructed in discrete time by sampling the cosines/sines basis functions [Wic95]. Local cosine bases in discrete time were constructed originally by Malvar and are
sometimes called lapped orthogonal transforms [Mal92]. In the discrete case, the efficient implementation of trigonometric transforms (using DCT I-IV and DST I-IV) can be utilized after
folding. In this case, expanding in local trigonometric bases corresponds to computing a DCT
after preprocesing (or folding) the signal.

For a sample basis function in each of the four bases, Figure 7.7 shows the corresponding
smooth basis function after unfolding. Observe that for local cosine and sine bases, the basis
functions are not linear phase; while the window is symmetric, the windowed functions are not.
However, for alternating sine/cosine bases the (unfolded) basis functions are linear phase. There
is a link between local sine (or cosine) bases and modulated filter banks that cannot have linear
phase filters (discussed in Chapter 8). So there is also a link between alternating cosine/sine
bases and linear-phase modulated filter banks (again see Chapter 8). This connection is further
explored in [Gop96a].

Local trigonometric bases have been applied to several signal processing problems. For example, they have been used in adaptive spectral analysis and in the segmentation of speech into
voiced and unvoiced regions [Wic95]. They are also used for image compression and are known
in the literature as lapped-orthogonal transforms [Mal92].

Signal Adaptive Local Trigonometric Bases

In the adaptive wavelet packet analysis described in Section 7.3, we considered a full filter bank
tree of decompositions and used some algorithm (best-basis algorithm, for instance) to prune the
tree to get the best tree topology (equivalently frequency partition) for a given signal. The idea
here is similar. We partition the time axis (or interval) into bins and successively refine each
partition into further bins, giving a tree of partitions for the time axis (or interval). If we use
smooth local trigonometric bases at each of the leaves of a full or pruned tree, we get a smooth
basis for all signals on the time axis (or interval). In adaptive local bases one grows a full tree
and prunes it based on some criterion to get the optimal set of temporal bins.

Figure 7.30 schematically shows a sample time-frequency tiling associated with a particular
local trigonometric basis. Observe that this is the dual of a wavelet packet tiling (see Figure 7.15)—in the sense that one can switch the time and frequency axes to go between the
two.

7.8   Discrete Multiresolution Analysis, the Discrete-Time Wavelet
Transform, and the Continuous Wavelet Transform

Up to this point, we have developed wavelet methods using the series wavelet expansion of
continuous-time signals called the discrete wavelet transform (DWT), even though it probably
should be called the continuous-time wavelet series. This wavelet expansion is analogous to the

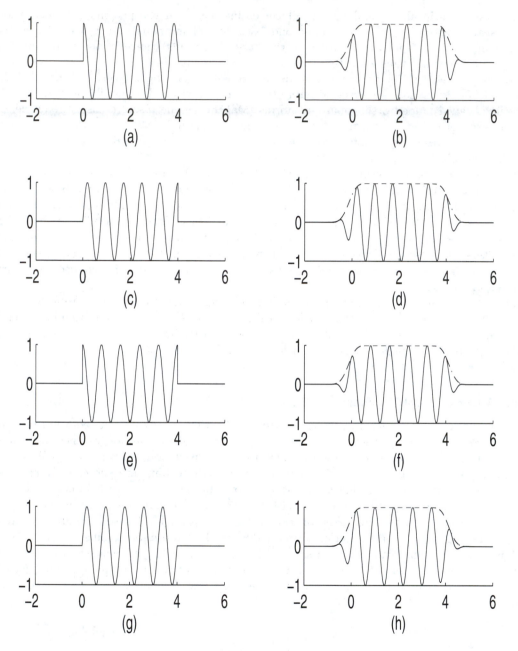

Figure 7.29. Trigonometric basis functions - before and after unfolding. (a) $f(t) = \cos(\frac{\pi}{4}(n + .5)t)u(t)u(4 - t)$ where $n = 10$; (b) $U_+(r_{[3]}, 0, 1)U_+(r_{[3]}, 4, 1)f(t)$; (c) $f(t) = \sin(\frac{\pi}{4}(n + .5)t)u(t)u(4-t)$ where $n = 10$; (d) $U_-(r_{[3]}, 0, 1)U_-(r_{[3]}, 4, 1)f(t)$; (e) $f(t) = \cos(\frac{\pi}{4}(n)t)u(t)u(4-t)$ where $n = 10$. (f) $U_+(r_{[3]}, 0, 1)U_-(r_{[3]}, 4, 1)f(t)$; (g) $f(t) = \sin(\frac{\pi}{4}(n)t)u(t)u(4 - t)$ where $n = 10$. (h) $U_-(r_{[3]}, 0, 1)U_+(r_{[3]}, 4, 1)f(t)$.

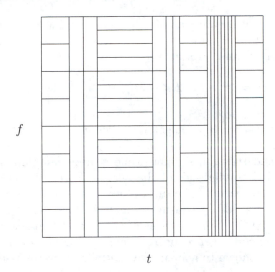

Figure 7.30. Local Basis

Fourier series in that both are series expansions that transform continuous-time signals into a
discrete sequence of coefficients. However, unlike the Fourier series, the DWT can be made
periodic or nonperiodic and, therefore, is more versatile and practically useful.

In this chapter we will develop a wavelet method for expanding discrete-time signals in a
series expansion since, in most practical situations, the signals are already in the form of discrete
samples. Indeed, we have already discussed when it is possible to use samples of the signal as
scaling function expansion coefficients in order to use the filter bank implementation of Mallat's
algorithm. We find there is an intimate connection between the DWT and DTWT, much as
there is between the Fourier series and the DFT. One expands signals with the FS but often
implements that with the DFT.

To further generalize the DWT, we will also briefly present the continuous wavelet transform
which, similar to the Fourier transform, transforms a function of continuous time to a representation with continuous scale and translation. In order to develop the characteristics of these
various wavelet representations, we will often call on analogies with corresponding Fourier representations. However, it is important to understand the differences between Fourier and wavelet
methods. Much of that difference is connected to the wavelet being concentrated in both time
and scale or frequency, to the periodic nature of the Fourier basis, and to the choice of wavelet
bases.

Discrete Multiresolution Analysis and the Discrete-Time Wavelet Transform

Parallel to the developments in early chapters on multiresolution analysis, we can define a discrete
multiresolution analysis (DMRA) for l_2, where the basis functions are discrete sequences [Rio93b,
Rio93a, VK95]. The expansion of a discrete-time signal in terms of discrete-time basis function
is expressed in a form parallel to (1.8) as

$$f(n) = \sum_{j,k} d_j(k)\,\psi(2^j n - k) \qquad (7.103)$$

where $\psi(m)$ is the basic expansion function of an integer variable m. If these expansion functions are an orthogonal basis (or form a tight frame), the expansion coefficients (discrete-time wavelet transform) are found from an inner product by

$$d_j(k) = \langle f(n), \psi(2^j n - k)\rangle = \sum_n f(n)\,\psi(2^j n - k) \tag{7.104}$$

If the expansion functions are not orthogonal or even independent but do span ℓ^2, a biorthogonal system or a frame can be formed such that a transform and inverse can be defined.

Because there is no underlying continuous-time scaling function or wavelet, many of the questions, properties, and characteristics of the analysis using the DWT in Chapters 1, 2, 6, etc. do not arise. In fact, because of the filter bank structure for calculating the DTWT, the design is often done using multirate frequency domain techniques, e.g., the work by Smith and Barnwell and associates [AS96]. The questions of zero wavelet moments posed by Daubechies, which are related to ideas of convergence for iterations of filter banks, and Coifman's zero scaling function moments that were shown to help approximate inner products by samples, seem to have no DTWT interpretation.

The connections between the DTWT and DWT are:

- If the starting sequences are the scaling coefficients for the continuous multiresolution analysis at very fine scale, then the discrete multiresolution analysis generates the same coefficients as does the continuous multiresolution analysis on dyadic rationals.

- When the number of scales is large, the basis sequences of the discrete multiresolution analysis converge in shape to the basis functions of the continuous multiresolution analysis.

The DTWT or DMRA is often described by a matrix operator. This is especially easy if the transform is made periodic, much as the Fourier series or DFT are. For the discrete time wavelet transform (DTWT), a matrix operator can give the relationship between a vector of inputs to give a vector of outputs. Several references on this approach are in [RW97, HRW92, KT94a, KT93, RGN94, RG95, KT94b, Kwo94].

Continuous Wavelet Transforms

The natural extension of the redundant DWT in Section 7.6 is to the continuous wavelet transform (CWT), which transforms a continuous-time signal into a wavelet transform that is a function of continuous shift or translation and a continuous scale. This transform is analogous to the Fourier transform, which is redundant, and results in a transform that is easier to interpret, is shift invariant, and is valuable for time-frequency/scale analysis. [HW89, GKM89, You80, RD92, Dau92, VK95, Rio91]

The definition of the CWT in terms of the wavelet $w(t)$ is given by

$$F(s, \tau) = s^{-1/2} \int f(t)\, w\left(\frac{t - \tau}{s}\right) dt \tag{7.105}$$

where the inverse transform is

$$f(t) = K \int \int \frac{1}{s^2}\, F(s, \tau)\, w\left(\frac{t - \tau}{s}\right) ds\, d\tau \tag{7.106}$$

with the normalizing constant given by

$$K = \int \frac{|W(\omega)|^2}{|\omega|} d\omega, \tag{7.107}$$

with $W(\omega)$ being the Fourier transform of the wavelet $w(t)$. In order for the wavelet to be *admissible* (for (7.106) to hold), $K < \infty$. In most cases, this simply requires that $W(0) = 0$ and that $W(\omega)$ go to zero ($W(\infty) = 0$) fast enough that $K < \infty$.

These admissibility conditions are satisfied by a very large set of functions and give very little insight into what basic wavelet functions should be used. In most cases, the wavelet $w(t)$ is chosen to give as good localization of the energy in both time and scale as possible for the class of signals of interest. It is also important to be able to calculate samples of the CWT as efficiently as possible, usually through the DWT and Mallat's filter banks or FFTs. This, and the interpretation of the CWT, is discussed in [Dau92, VK95, Gop90, GB90, P P89, GKM89, JB82, RD92, VLU97].

The use of the CWT is part of a more general time-frequency analysis that may or may not use wavelets [Coh89, Coh95, HB92, Boa92, LP89].

Analogies between Fourier Systems and Wavelet Systems

In order to better understand the wavelet transforms and expansions, we will look at the various forms of Fourier transforms and expansion. If we denote continuous time by CT, discrete time by DT, continuous frequency by CF, and discrete frequency by DF, the following table will show what the discrete Fourier transform (DFT), Fourier series (FS), discrete-time Fourier transform (DTFT), and Fourier transform take as time domain signals and produce as frequency domain transforms or series. For example, the Fourier series takes a continuous-time input signal and produces a sequence of discrete-frequency coefficients while the DTFT takes a discrete-time sequence of numbers as an input signal and produces a transform that is a function of continuous frequency.

	DT	CT
DF	DFT	FS
CF	DTFT	FT

Table 7.4. Continuous and Discrete Input and Output for Four Fourier Transforms

Because the basis functions of all four Fourier transforms are periodic, the transform of a periodic signal (CT or DT) is a function of discrete frequency. In other words, it is a sequence of series expansion coefficients. If the signal is infinitely long and not periodic, the transform is a function of continuous frequency and the inverse is an integral, not a sum.

Periodic in time ⇔ Discrete in frequency

Periodic in frequency ⇔ Discrete in time

A bit of thought and, perhaps, referring to appropriate materials on signal processing and Fourier methods will make this clear and show why so many properties of Fourier analysis are created by the periodic basis functions.

Also recall that in most cases, it is the Fourier transform, discrete-time Fourier transform, or Fourier series that is needed but it is the DFT that can be calculated by a digital computer and that is probably using the FFT algorithm. If the coefficients of a Fourier series drop off fast enough or, even better, are zero after some harmonic, the DFT of samples of the signal will give the Fourier series coefficients. If a discrete-time signal has a finite nonzero duration, the DFT of its values will be samples of its DTFT. From this, one sees the relation of samples of a signal to the signal and the relation of the various Fourier transforms.

Now, what is the case for the various wavelet transforms? Well, it is both similar and different. The table that relates the continuous and discrete variables is given by where DW indicates discrete values for scale and translation given by j and k, with CW denoting continuous values for scale and translation.

	DT	CT
DW	DTWT	DWT
CW	DTCWT	CWT

Table 7.5. Continuous and Discrete Input and Output for Four Wavelet Transforms

We have spent most this book developing the DWT, which is a series expansion of a continuous time signal. Because the wavelet basis functions are concentrated in time and not periodic, both the DTWT and DWT will represent infinitely long signals. In most practical cases, they are made periodic to facilitate efficient computation. Chapter 9 gives the details of how the transform is made periodic. The discrete-time, continuous wavelet transform (DTCWT) is seldom used and not discussed here.

The naming of the various transforms has not been consistent in the literature and this is complicated by the wavelet transforms having two transform variables, scale and translation. If we could rename all the transforms, it would be more consistent to use Fourier series (FS) or wavelet series (WS) for a series expansion that produced discrete expansion coefficients, Fourier transforms (FT) or wavelet transforms (WT) for integral expansions that produce functions of continuous frequency or scale or translation variable together with DT (discrete time) or CT (continuous time) to describe the input signal. However, in common usage, only the DTFT follows this format!

Common name	Consistent name	Time, C or D	Transform C or D	Input periodic	Output periodic
FS	CTFS	C	D	Yes	No
DFT	DTFS	D	D	Yes	Yes
DTFT	DTFT	D	C	No	Yes
FT	CTFT	C	C	No	No
DWT	CTWS	C	D	Y or N	Y or N
DTWT	DTWS	D	D	Y or N	Y or N
–	DTWT	D	C	N	N
CWT	CTWT	C	C	N	N

Table 7.6. Continuous and Discrete, Periodic and Nonperiodic Input and Output for Transforms

Recall that the difference between the DWT and DTWT is that the input to the DWT is a sequence of expansion coefficients or a sequence of inner products while the input to the DTWT is the signal itself, probably samples of a continuous-time signal. The Mallat algorithm or filter bank structure is exactly the same. The approximation is made better by zero moments of the scaling function (see Section 6.8) or by some sort of prefiltering of the samples to make them closer to the inner products [SN96].

As mentioned before, both the DWT and DTWT can be formulated as nonperiodic, on-going transforms for an exact expansion of infinite duration signals or they may be made periodic to handle finite-length or periodic signals. If they are made periodic (as in Chapter 9), then there is an aliasing that takes place in the transform. Indeed, the aliasing has a different period at the different scales which may make interpretation difficult. This does not harm the inverse transform which uses the wavelet information to "unalias" the scaling function coefficients. Most (but not all) DWT, DTWT, and matrix operators use a periodized form [SV93].

Chapter 8

Filter Banks and Transmultiplexers

8.1 Introduction

In this chapter, we develop the properties of wavelet systems in terms of the underlying filter banks associated with them. This is an expansion and elaboration of the material in Chapter 3, where many of the conditions and properties developed from a signal expansion point of view in Chapter 5 are now derived from the associated filter bank. The Mallat algorithm uses a special structure of filters and downsamplers/upsamplers to calculate and invert the discrete wavelet transform. Such filter structures have been studied for over three decades in digital signal processing in the context of the *filter bank* and *transmultiplexer* problems [SB86b, Vai87a, VNDS89, VD89, Vet87, VL89, Mal92, Vai92, RC83]. Filter bank theory, besides providing efficient computational schemes for wavelet analysis, also gives valuable insights into the construction of wavelet bases. Indeed, some of the finer aspects of wavelet theory emanates from filter bank theory.

The Filter Bank

A filter bank is a structure that decomposes a signal into a collection of subsignals. Depending on the application, these subsignals help emphasize specific aspects of the original signal or may be easier to work with than the original signal. We have *linear* or *non-linear* filter banks depending on whether or not the subsignals depend linearly on the original signal. Filter banks were originally studied in the context of signal compression where the subsignals were used to "represent" the original signal. The subsignals (called subband signals) are downsampled so that the data rates are the same in the subbands as in the original signal—though this is not essential. Key points to remember are that the subsignals convey salient features of the original signal and are sufficient to reconstruct the original signal.

Figure 8.1 shows a linear filter bank that is used in signal compression (subband coding). The *analysis* filters $\{h_i\}$ are used to filter the input signal $x(n)$. The filtered signals are downsampled to give the *subband* signals. Reconstruction of the original signal is achieved by upsampling, filtering and adding up the subband signals as shown in the right-hand part of Figure 8.1. The desire for perfect reconstruction (i.e., $y(n) = x(n)$) imposes a set of bilinear constraints (since all operations in Figure 8.1 are linear) on the analysis and synthesis filters. This also constrains the downsampling factor, M, to be at most the number of subband signals, say L. Filter bank design involves choosing filters $\{h_i\}$ and $\{g_i\}$ that satisfy perfect reconstruction and simultaneously give informative and useful subband signals. In subband speech coding, for example, a natural choice of desired frequency responses—motivated by the nonuniform sensitivity of the human ear to various frequency bands—for the analysis and synthesis filters is shown in Figure 8.2.

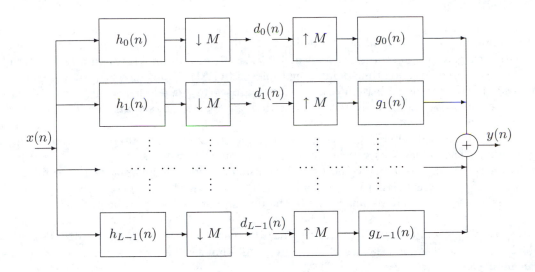

Figure 8.1. L-Band Filter Bank with Rate-Change Factor of M

In summary, the filter bank problem involves the design of the filters $h_i(n)$ and $g_i(n)$, with the following goals:

1. *Perfect Reconstruction* (i.e., $y(n) = x(n)$).

2. *Usefulness.* Clearly this depends on the application. For the subband coding application, the filter frequency responses might approximate the ideal responses in Figure 8.2. In other applications the filters may have to satisfy other constraints or approximate other frequency responses.

If the signals and filters are multidimensional in Figure 8.1, we have the multidimensional filter bank design problem.

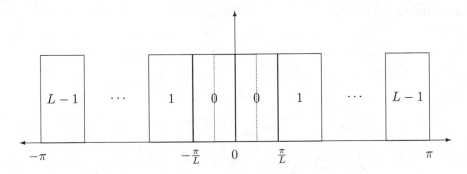

Figure 8.2. Ideal Frequency Responses in an L-band Filter Bank

Transmultiplexer

A transmultiplexer is a structure that combines a collection of signals into a single signal at a higher rate; i.e., it is the dual of a filter bank. If the combined signal depends linearly on the constituent signal, we have a *linear* transmultiplexer. Transmultiplexers were originally studied in the context of converting time-domain-multiplexed (TDM) signals into frequency domain multiplexed (FDM) signals with the goal of converting back to time-domain-multiplexed signals at some later point. A key point to remember is that the constituent signals should be recoverable from the combined signal. Figure 8.3 shows the structure of a transmultiplexer. The input signals $y_i(n)$ were upsampled, filtered, and combined (by a synthesis bank of filters) to give a composite signal $d(n)$. The signal $d(n)$ can be filtered (by an analysis bank of filters) and downsampled to give a set of signals $x_i(n)$. The goal in transmultiplexer design is a choice of filters that ensures perfect reconstruction (i.e., for all i, $x_i(n) = y_i(n)$). This imposes bilinear constraints on the synthesis and analysis filters. Also, the upsampling factor must be at least the number of constituent input signals, say L. Moreover, in classical TDM-FDM conversion the analysis and synthesis filters must approximate the ideal frequency responses in Figure 8.2. If the input signals, analysis filters and synthesis filters are multidimensional, we have a multidimensional transmultiplexer.

Perfect Reconstruction—A Closer Look

We now take a closer look at the set of bilinear constraints on the analysis and synthesis filters of a filter bank and/or transmultiplexer that ensures perfect reconstruction (PR). Assume that there are L analysis filters and L synthesis filters and that downsampling/upsampling is by some integer M. These constraints, broadly speaking, can be viewed in three useful ways, each applicable in specific situations.

1. Direct characterization - which is useful in wavelet theory (to characterize orthonormality and frame properties), in the study of a powerful class of filter banks (modulated filter banks), etc.

2. Matrix characterization - which is useful in the study of time-varying filter banks.

3. z-transform-domain (or polyphase-representation) characterization - which is useful in the design and implementation of (unitary) filter banks and wavelets.

Direct Characterization of PR

We will first consider the direct characterization of PR, which, for both filter banks and transmultiplexers, follows from an elementary superposition argument.

Theorem 38 *A filter bank is PR if and only if, for all integers n_1 and n_2,*

$$\sum_{i=0}^{L-1}\sum_{n} h_i(Mn + n_1)g_i(-Mn - n_2) = \delta(n_1 - n_2). \tag{8.1}$$

A transmultiplexer is PR if and only if, for all $i, j \in \{0, 1, \ldots, L-1\}$,

$$\sum_{n} h_i(n)g_j(-Ml - n) = \delta(l)\delta(i - j). \tag{8.2}$$

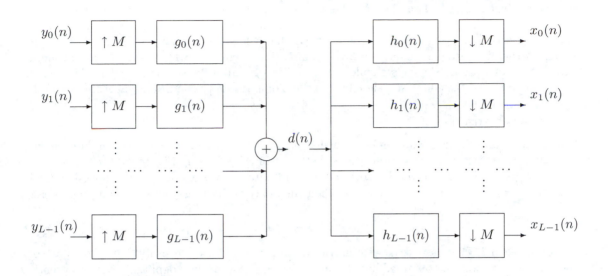

Figure 8.3. L-Band Transmultiplexer with Rate Change Factor of M

Moreover, if the number of channels is equal to the downsampling factor (i.e., $L = |M|$),(8.1) and (8.2) are equivalent.

Consider a PR filter bank. Since an arbitrary signal is a linear superposition of impulses, it suffices to consider the input signal, $x(n) = \delta(n - n_1)$, for arbitrary integer n_1. Then (see Figure 8.1) $d_i(n) = h_i(Mn - n_1)$ and therefore, $y(n_2) = \sum_i \sum_n g_i(n_2 - Mn)d_i(n)$. But by PR, $y(n_2) = \delta(n_2 - n_1)$. The filter bank PR property is precisely a statement of this fact:

$$y(n_2) = \sum_i \sum_n g_i(n_2 - Mn)d_i(n) = \sum_i \sum_n g_i(n_2 - Mn)h_i(Mn - n_1) = \delta(n_2 - n_1).$$

Consider a PR transmultiplexer. Once again because of linear superposition, it suffices to consider only the input signals $x_i(n) = \delta(n)\delta(i - j)$ for all i and j. Then, $d(n) = g_j(n)$ (see Figure 8.3), and $y_i(l) = \sum_n h_i(n)d(Ml - n)$. But by PR $y_i(l) = \delta(l)\delta(i - j)$. The transmultiplexer PR property is precisely a statement of this fact:

$$y_i(l) = \sum_n h_i(n)d(Ml - n) = \sum_n h_i(n)g_j(Ml - n) = \delta(l)\delta(i - j).$$

Remark: Strictly speaking, in the superposition argument proving (8.2), one has to consider the input signals $x_i(n) = \delta(n - n_1)\delta(i - j)$ for arbitrary n_1. One readily verifies that for all n_1 (8.2) has to be satisfied.

The equivalence of (8.1) and (8.2) when $L = M$ is not obvious from the direct characterization. However, the transform domain characterization that we shall see shortly will make this connection obvious. For a PR filter, bank the L channels should contain sufficient information to reconstruct the original signal (note the summation over i in (8.1)), while for a transmultiplexer, the constituent channels should satisfy biorthogonality constraints so that they can be reconstructed from the composite signal (note the biorthogonality conditions suggested by (8.2)).

Matrix characterization of PR

The second viewpoint is linear-algebraic in that it considers all signals as vectors and all filtering operations as matrix-vector multiplications [Vet87]. In Figure 8.1 and Figure 8.3 the signals $x(n)$, $d_i(n)$ and $y(n)$ can be naturally associated with infinite vectors $\mathbf{x}$, $\mathbf{d}_i$ and $\mathbf{y}$ respectively. For example, $\mathbf{x} = [\cdots, x(-1), x(0), x(1), \cdots]$. Then the analysis filtering operation can be expressed as

$$\mathbf{d}_i = \mathbf{H}_i\mathbf{x}, \qquad \text{for } i \in \{0, 1, 2, \ldots, L-1\}, \tag{8.3}$$

where, for each i, $\mathbf{H}_i$ is a matrix with entries appropriately drawn from filter h_i. $\mathbf{H}_i$ is a block Toeplitz matrix (since its obtained by retaining every M^{th} row of the Toeplitz matrix representing convolution by h_i) with every row containing h_i in an index-reversed order. Then the synthesis filtering operation can be expressed as

$$\mathbf{y} = \sum_i \mathbf{G}_i^T\mathbf{d}_i \tag{8.4}$$

where, for each i, $\mathbf{G}_i$ is a matrix with entries appropriately drawn from filter g_i. $\mathbf{G}_i$ is also a block Toeplitz matrix (since it is obtained by retaining every M^{th} row of the Toeplitz matrix whose transpose represents convolution by g_i) with every row containing g_i in its natural order. Define $\mathbf{d}$ to be the vector obtained by interlacing the entries of each of the vectors $\mathbf{d}_i$: $\mathbf{d} = [\cdots, d_0(0), d_1(0), \cdots, d_{M-1}(0), d_0(1), d_1(1), \cdots]$. Also define the matrices $\mathbf{H}$ and $\mathbf{G}$ (in terms of $\mathbf{H}_i$ and $\mathbf{G}_i$) so that

$$\mathbf{d} = \mathbf{Hx}, \quad \text{and} \quad \mathbf{y} = \mathbf{G}^T\mathbf{d}. \tag{8.5}$$

$\mathbf{H}$ is obtained by interlacing the rows of $\mathbf{H}_i$ and $\mathbf{G}$ is obtained by interlacing the rows of $\mathbf{G}_i$. For example, in the FIR case if the filters are all of length N,

$$\mathbf{d} = \mathbf{Hx} \stackrel{\text{def}}{=} \begin{bmatrix} & \vdots & \vdots & \vdots & \vdots & \vdots & \vdots & \\ & h_0(N-1) & \cdots & h_0(N-M-1) & \cdots & h_0(0) & 0 & \\ & \vdots & \cdots & \cdots & \cdots & \cdots & \cdots & \\ \ddots & h_{L-1}(N-1) & \cdots & h_{L-1}(N-M-1) & \cdots & h_{L-1}(0) & 0 & \ddots \\ & 0 & 0 & h_0(N-1) & \cdots & \cdots & \cdots & \\ & \vdots & \vdots & \vdots & \cdots & \cdots & \cdots & \\ & 0 & 0 & h_{L-1}(N-1) & \cdots & \cdots & \cdots & \\ & \vdots & \vdots & \vdots & \vdots & \vdots & \vdots & \end{bmatrix} \mathbf{x}. \tag{8.6}$$

From this development, we have the following result:

Theorem 39 *A filter bank is PR iff*

$$\mathbf{G}^T\mathbf{H} = \mathbf{I}. \tag{8.7}$$

A transmultiplexer is PR iff

$$\mathbf{H}\mathbf{G}^T = \mathbf{I}. \tag{8.8}$$

Moreover, when $L = M$, both conditions are equivalent.

One can also write the PR conditions for filter banks and transmultiplexers in the following form, which explicitly shows the formal relationship between the direct and matrix characterizations. For a PR filter bank we have

$$\sum_i \mathbf{G}_i^T\mathbf{H}_i = \mathbf{I}. \tag{8.9}$$

Correspondingly for a PR transmultiplexer we have

$$\mathbf{H}_i\mathbf{G}_j^T = \delta(i - j)\mathbf{I}. \tag{8.10}$$

Polyphase (Transform-Domain) Characterization of PR

We finally look at the analysis and synthesis filter banks from a polyphase representation viewpoint. Here subsequences of the input and output signals and the filters are represented in the z-transform domain. Indeed let the z-transforms of the signals and filters be expressed in terms of the z-transforms of their subsequences as follows:

$$X(z) = \sum_{k=0}^{M-1} z^k X_k(z^M) \tag{8.11}$$

$$Y(z) = \sum_{k=}^{M-1} z^k Y_k(z^M) \tag{8.12}$$

$$H_i(z) = \sum_{k=0}^{M-1} z^{-k} H_{i,k}(z^M) \tag{8.13}$$

$$G_i(z) = \sum_{k=0}^{M-1} z^k G_{i,k}(z^M) \tag{8.14}$$

Then, along each branch of the analysis bank we have

$$
\begin{aligned}
D_i(z) &= [\downarrow M]\{H_i(z)X(z)\} \\[2mm]
&= [\downarrow M]\left\{ \left(\sum_{k=0}^{M-1} z^{-k} H_{i,k}(z)\right)\left(\sum_{l=0}^{M-1} z^l X_l(z^M)\right)\right\} \\[2mm]
&= [\downarrow M]\left\{ \sum_{k,l=0}^{M-1} z^{l-k} H_{i,k}(z^M) X_l(z^M)\right\} \\[2mm]
&= \sum_{k,l=0}^{M-1} \delta(l - k) H_{i,k}(z) X_l(z) = \sum_k^{M-1} H_{i,k}(z) X_k(z).
\end{aligned}
\tag{8.15}
$$

Similarly, from the synthesis bank, we have

$$Y(z) = \sum_{i=0}^{L-1} D_i(z^M) G_i(z) \tag{8.16}$$

$$= \sum_{i=0}^{L-1} D_i(z^M) \left\{ \sum_{k=0}^{M-1} z^k G_{i,k}(z^M) \right\}$$

$$= \sum_{k=0}^{M-1} z^k \left\{ \sum_{i=0}^{L-1} G_{i,k}(z^M) D_i(z^M). \right\}$$

and therefore (from (8.12))

$$Y_k(z) = \sum_{i=0}^{L-1} G_{i,k}(z) D_i(z). \tag{8.17}$$

For $i \in \{0, 1, \ldots, L-1\}$ and $k \in \{0, 1, \ldots, M-1\}$, define the *polyphase* component matrices $(H_p(z))_{i,k} = H_{i,k}(z)$ and $(G_p(z))_{i,k} = G_{i,k}(z)$. Let $X_p(z)$ and $Y_p(z)$ denote the z-transforms of the polyphase signals $x_p(n)$ and $y_p(n)$, and let $D_p(z)$ be the vector whose components are $D_i(z)$. Equations (8.16) and (8.17) can be written compactly as

$$D_p(z) = H_p(z) X_p(z), \tag{8.18}$$

$$Y_p(z) = G_p^T(z) D_p(z), \tag{8.19}$$

and

$$Y_p(z) = G_p^T(z) H_p(z) X_p(z). \tag{8.20}$$

Thus, the analysis filter bank is represented by the multi-input (the polyphase components of $X(z)$), multi-output (the signals $D_i(z)$) linear-shift-invariant system $H_p(z)$ that takes in $X_p(z)$ and gives out $D_p(z)$. Similarly, the synthesis filter bank can be interpreted as a multi-input (the signals $D_i(z)$), multi-output (the polyphase components of $Y(z)$) system $G_p^T(z)$, which maps $D_p(z)$ to $Y_p(z)$. Clearly we have PR iff $Y_p(z) = X_p(z)$. This occurs precisely when $G_p^T(z) H_p(z) = I$.

For the transmultiplexer problem, let $Y_p(z)$ and $X_p(z)$ be vectorized versions of the input and output signals respectively and let $D_p(z)$ be the generalized polyphase representation of the signal $D(z)$. Now $D_p(z) = G_p^T(z) Y_p(z)$ and $X_p(z) = H_p(z) D_p(z)$. Hence $X_p(z) = H_p(z) G_p^T(z) Y_p(z)$, and for PR $H_p(z) G_p^T(z) = I$.

Theorem 40 *A filter bank has the PR property if and only if*

$$G_p^T(z) H_p(z) = I. \tag{8.21}$$

A transmultiplexer has the PR property if and only if

$$H_p(z) G_p^T(z) = I \tag{8.22}$$

where $H_p(z)$ and $G_p(z)$ are as defined above.

Remark: If $G_p^T(z)H_p(z) = I$, then $H_p(z)$ must have at least as many rows as columns (i.e., $L \geq M$ is necessary for a filter bank to be PR). If $H_p(z)G_p^T(z) = I$ then $H_p(z)$ must have at least as many columns as rows (i.e., $M \geq L$ is necessary for a tranmultiplexer to be PR). If $L = M$, $G_p^T(z)H_p(z) = I = H_p^T(z)G_p(z)$ and hence a filter bank is PR iff the corresponding transmultiplexer is PR. This equivalence is trivial with the polyphase representation, while it is not in the direct and matrix representations.

Notice that the PR property of a filter bank or transmultiplexer is unchanged if all the analysis and synthesis filters are shifted by the same amount in opposite directions. Also any one analysis/synthesis filter pair can be shifted by multiples of M in opposite directions without affecting the PR property. Using these two properties (and assuming all the filters are FIR), we can assume without loss of generality that the analysis filters are supported in $[0, N-1]$ (for some integer N). This also implies that $H_p(z)$ is a polynomial in z^{-1}, a fact that we will use in the parameterization of an important class of filter banks—*unitary filter banks*.

All through the discussion of the PR property of filter banks, we have deliberately not said anything about the length of the filters. The bilinear PR constraints are completely independent of filter lengths and hold for arbitrary sequences. However, if the sequences are infinite then one requires that the infinite summations over n in (8.1) and (8.2) converge. Clearly, assuming that these filter sequences are in $\ell^2(\mathbf{Z})$ is sufficient to ensure this since inner products are then well-defined.

8.2 Unitary Filter Banks

From (8.5) it follows that a filter bank can be ensured to be PR if the analysis filters are chosen such that $\mathbf{H}$ is left-unitary, i.e., $\mathbf{H}^T\mathbf{H} = \mathbf{I}$. In this case, the synthesis matrix $\mathbf{G} = \mathbf{H}$ (from (8.7)) and therefore $\mathbf{G}_i = \mathbf{H}_i$ for all i. Recall that the rows of $\mathbf{G}_i$ contain g_i in natural order while the rows of $\mathbf{H}_i$ contains h_i in index-reversed order. Therefore, for such a filter bank, since $\mathbf{G}_i = \mathbf{H}_i$, the synthesis filters are reflections of the analysis filters about the origin; i.e., $g_i(n) = h_i(-n)$. Filter banks where the analysis and synthesis filters satisfy this reflection relationship are called unitary (or orthogonal) filter banks for the simple reason that $\mathbf{H}$ is left-unitary. In a similar fashion, it is easy to see that if $\mathbf{H}$ is right-unitary (i.e., $\mathbf{H}\mathbf{H}^T = \mathbf{I}$), then the transmultiplexer associated with this set of analysis filters is PR with $g_i(n) = h_i(-n)$. This defines unitary transmultiplexers.

We now examine how the three ways of viewing PR filter banks and transmultiplexers simplify when we focus on unitary ones. Since $g_i(n) = h_i(-n)$, the direct characterization becomes the following:

Theorem 41 *A filter bank is unitary iff*

$$\sum_i \sum_n h_i(Mn + n_1)h_i(Mn + n_2) = \delta(n_1 - n_2). \tag{8.23}$$

A transmultiplexer is unitary iff

$$\sum_n h_i(n)h_j(Ml + n) = \delta(l)\delta(i - j). \tag{8.24}$$

If the number of channels is equal to the downsampling factor, then a filter bank is unitary iff the corresponding transmultiplexer is unitary.

The matrix characterization of unitary filter banks/transmultiplexers should be clear from the above discussion:

Theorem 42 *A filter bank is unitary iff* $\mathbf{H}^T\mathbf{H} = \mathbf{I}$, *and a transmultiplexer is unitary iff* $\mathbf{H}\mathbf{H}^T = \mathbf{I}$.

The z-transform domain characterization of unitary filter banks and transmultiplexers is given below:

Theorem 43 *A filter bank is unitary iff* $H_p^T(z^{-1})H_p(z) = I$, *and a transmultiplexer is unitary iff* $H_p(z)H_p^T(z^{-1}) = I$.

In this book (as in most of the work in the literature) one primarily considers the situation where the number of channels equals the downsampling factor. For such a unitary filter bank (transmultiplexer), (8.9) and (8.10) become:

$$\sum_i \mathbf{H}_i^T\mathbf{H}_i = \mathbf{I}, \tag{8.25}$$

and

$$\mathbf{H}_i\mathbf{H}_j^T = \delta(i - j)\mathbf{I}. \tag{8.26}$$

The matrices $\mathbf{H}_i$ are pairwise orthogonal and form a resolution of the identity matrix. In other words, for each i, $\mathbf{H}_i^T\mathbf{H}_i$ is an orthogonal projection matrix and the filter bank gives an orthogonal decomposition of a given signal. Recall that for a matrix P to be an orthogonal projection matrix, $P^2 = P$ and $P \geq 0$; in our case, indeed, we do have $\mathbf{H}_i^T\mathbf{H}_i \geq 0$ and $\mathbf{H}_i^T\mathbf{H}_i\mathbf{H}_i^T\mathbf{H}_i = \mathbf{H}_i^T\mathbf{H}_i$.

Unitarity is a very useful constraint since it leads to orthogonal decompositions. Besides, for a unitary filter bank, one does not have to design both the analysis and synthesis filters since $h_i(n) = g_i(-n)$. But perhaps the most important property of unitary filter banks and transmultiplexers is that they can be parameterized. As we have already seen, filter bank design is a nonlinear optimization (of some goodness criterion) problem subject to PR constraints. If the PR constraints are unitary, then a parameterization of unitary filters leads to an unconstrained optimization problem. Besides, for designing wavelets with high-order vanishing moments, nonlinear equations can be formulated and solved in this parameter space. A similar parameterization of nonunitary PR filter banks and transmultiplexers seems impossible and it is not too difficult to intuitively see why. Consider the following analogy: a PR filter bank is akin to a left-invertible matrix and a PR transmultiplexer to a right-invertible matrix. If $L = M$, the PR filter bank is akin to an invertible matrix. A unitary filter bank is akin to a left-unitary matrix, a unitary transmultiplexer to a right-unitary matrix, and when $L = M$, either of them to a unitary matrix. Left-unitary, right-unitary and in particular unitary matrices can be parameterized using Givens' rotations or Householder transformations [GL93]. However, left-invertible, right-invertible and, in particular, invertible matrices have no general parameterization. Also, unitariness allows explicit parameterization of filter banks and transmultiplexers which just PR alone precludes. The analogy is even more appropriate: There are two parameterizations of unitary filter banks and transmultiplexers that correspond to Givens' rotation and Householder transformations, respectively.

All our discussions on filter banks and transmultiplexers carry over naturally with very small notational changes to the multi-dimensional case where downsampling is by some integer matrix [GB94a]. However, the parameterization result we now proceed to develop is not known in the multi-dimensional case. In the two-dimensional case, however, an implicit, and perhaps not too practical (from a filter-design point of view), parameterization of unitary filter banks is described in [BC92].

Consider a unitary filter bank with finite-impulse response filters (i.e., for all i, h_i is a finite sequence). Recall that without loss of generality, the filters can be shifted so that $H_p(z)$ is a polynomial in z^{-1}. In this case $G_p(z) = H_p(z^{-1})$ is a polynomial in z. Let

$$H_p(z) = \sum_{k=0}^{K-1} h_p(k) z^{-k}. \tag{8.27}$$

That is, $H_p(z)$ is a matrix polynomial in z^{-1} with coefficients $h_p(k)$ and degree $K - 1$. Since $H_p^T(z^{-1})H_p(z) = I$, from (8.27) we must have $h_p^T(0)h_p(K-1) = 0$ as it is the coefficient of z^{K-1} in the product $H_p^T(z^{-1})H_p(z)$. Therefore $h_p(0)$ is singular. Let P_{K-1} be the *unique* projection matrix onto the range of $h_p(K-1)$ (say of dimension δ_{K-1}). Then $h_p(0)^T P_{K-1} = 0 = P_{K-1}h_p(0)$. Also $P_{K-1}h(K-1) = h(K-1)$ and hence $(I - P_{K-1})h(K-1) = 0$. Now $[I - P_{K-1} + zP_{K-1}] H_p(z)$ is a matrix polynomial of degree at most $K - 2$. If $h(0)$ and $h(K-1)$ are nonzero (an assumption one makes without loss of generality), the degree is *precisely* $K - 2$. Also it is unitary since $I - P_{K-1} + zP_{K-1}$ is unitary. Repeated application of this procedure $(K - 1)$ times gives a degree zero (constant) unitary matrix V_0. The discussion above shows that an arbitrary unitary polynomial matrix of degree $K - 1$ can be expressed *algorithmically uniquely* as described in the following theorem:

Theorem 44 *For a polynomial matrix $H_p(z)$, unitary on the unit circle (i.e., $H_p^T(z^{-1})H_p(z) = I$), and of polynomial degree $K - 1$, there exists a unique set of projection matrices P_k (each of rank some integer δ_k), such that*

$$H_p(z) = \left\{ \prod_{k=K-1}^{1} \left[I - P_k + z^{-1}P_k \right] \right\} V_0. \tag{8.28}$$

Remark: Since the projection P_k is of rank δ_k, it can be written as $v_1 v_1^T + \ldots + v_{\delta_k} v_{\delta_k}^T$, for a *nonunique* set of orthonormal vectors v_i. Using the fact that

$$\left[I - v_j v_j^T - v_{j-1}v_{j-1}^T + z^{-1}(v_j v_j^T + v_{j-1}v_{j-1}^T) \right] = \prod_{i=j}^{j-1} \left[I - v_i v_i^T + z^{-1}v_i v_i^T \right], \tag{8.29}$$

defining $\Delta = \sum_k \delta_k$ and collecting all the v_j's that define the P_k's into a single pool (and reindexing) we get the following factorization:

$$H_p(z) = \left\{ \prod_{k=\Delta}^{1} \left[I - v_k v_k^T + z^{-1}v_k v_k^T \right] \right\} V_0. \tag{8.30}$$

If $H_p(z)$ is the analysis bank of a filter bank, then notice that Δ (from (8.30)) is the number of storage elements required to implement the analysis bank. The minimum number of storage elements to implement any transfer function is called the McMillan degree and in this case Δ is indeed the McMillan degree [Vai92]. Recall that P_K is chosen to be the projection matrix onto the *range* of $h_p(K-1)$. Instead we could have chosen P_K to be the projection onto the nullspace of $h_p(0)$ (which contains the range of $h_p(K-1)$) or any space sandwiched between the two. Each choice leads to a different sequence of factors P_K and corresponding δ_k (except when the range and nullspaces in question coincide at some stage during the order reduction process). However, Δ, the McMillan degree is constant.

Equation (8.30) can be used as a starting point for filter bank design. It parameterizes all unitary filter banks with McMillan degree Δ. If $\Delta = K$, then all unitary filter banks with filters of length $N \leq MK$ are parameterized using a collection of $K-1$ unitary vectors, v_k, and a unitary matrix, V_0. Each unitary vector has $(M-1)$ free parameters, while the unitary matrix has $M(M-1)/2$ free parameters for a total of $(K-1)(M-1) + \binom{M}{2}$ free parameters for $H_p(z)$. The filter bank design problem is to choose these free parameters to optimize the "usefulness" criterion of the filter bank.

If $L > M$, and $H_p(z)$ is left-unitary, a similar analysis leads to exactly the same factorization as before except that V_0 is a left unitary matrix. In this case, the number of free parameters is given by $(K-1)(L-1) + \binom{L}{2} - \binom{M}{2}$. For a transmultiplexer with $L < M$, one can use the same factorization above for $H_p^T(z)$ (which is left unitary). Even for a filter bank or transmultiplexer with $L = M$, factorizations of left-/right-unitary $H_p(z)$ is useful for the following reason. Let us assume that a subset of the analysis filters has been predesigned (for example in wavelet theory one sometimes independently designs h_0 to be a K-regular scaling filter, as in Chapter 6). The submatrix of $H_p(z)$ corresponding to this subset of filters is right-unitary, hence its transpose can be parameterized as above with a collection of vectors v_i and a left-unitary V_0. Each choice for the remaining columns of V_0 gives a choice for the remaining filters in the filter bank. In fact, all possible *completions* of the original subset *with fixed McMillan degree* are given this way.

Orthogonal filter banks are sometimes referred to as lossless filter banks because the collective energy of the subband signals is the same as that of the original signal. If U is an orthogonal matrix, then the signals $x(n)$ and $Ux(n)$ have the same energy. If P is an orthogonal projection matrix, then

$$\|x\|^2 = \|Px\|^2 + \|(I-P)x\|^2 .$$

For any give $X(z)$, $X(z)$ and $z^{-1}X(z)$ have the same energy. Using the above facts, we find that for any projection matrix, P,

$$D_p(z) = \left[I - P + z^{-1}P\right] X_p(z) \stackrel{\text{def}}{=} T(z)X_p(z)$$

has the same energy as $X_p(z)$. This is equivalent to the fact that $T(z)$ is unitary on the unit circle (one can directly verify this). Therefore (from (8.28)) it follows that the subband signals have the same energy as the original signal.

In order to make the free parameters explicit for filter design, we now describe V_0 and $\{v_i\}$ using *angle* parameters. First consider v_i, with $\|v_i\| = 1$. Clearly, v_i has $(M - 1)$ degrees of freedom. One way to parameterize v_i using $(M - 1)$ angle parameters $\theta_{i,k}$, $k \in \{0, 1, \ldots, M - 2\}$ would be to define the components of v_i as follows:

$$
(v_i)_j =
\begin{cases}
\left\{ \displaystyle\prod_{l=0}^{j-1} \sin(\theta_{i,l}) \right\} \cos(\theta_{i,j}) & \text{for } j \in \{0, 1, \ldots, M - 2\} \\[3ex]
\left\{ \displaystyle\prod_{l=0}^{M-1} \sin(\theta_{i,l}) \right\} & \text{for } j = M - 1.
\end{cases}
$$

As for V_0, it being an $M \times M$ orthogonal matrix, it has $\binom{M}{2}$ degrees of freedom. There are two well known parameterizations of constant orthogonal matrices, one based on Givens' rotation (well known in QR factorization etc. [DS83]), and another based on Householder reflections. In the Householder parameterization

$$
V_0 = \prod_{i=0}^{M-1} \left[I - 2v_i v_i^T \right],
$$

where v_i are unit norm vectors with the first i components of v_i being zero. Each matrix factor $\left[I - 2v_i v_i^T \right]$ when multiplied by a vector q, reflects q about the plane perpendicular to v_i, hence the name Householder reflections. Since the first i components of v_i is zero, and $\|v_i\| = 1$, v_i has $M - i - 1$ degrees of freedom. Each being a unit vector, they can be parameterized as before using $M - i - 1$ angles. Therefore, the total degrees of freedom are

$$
\sum_{i=0}^{M-1} (M - 1 - i) = \sum_{i=0}^{M-1} i = \binom{M}{2}.
$$

In summary, any orthogonal matrix can be factored into a cascade of M reflections about the planes perpendicular to the vectors v_i.

Notice the similarity between Householder reflection factors for V_0 and the factors of $H_p(z)$ in (8.30). Based on this similarity, the factorization of unitary matrices and vectors in this section is called the Householder factorization. Analogous to the Givens' factorization for constant unitary matrices, also one can obtain a factorization of unitary matrices $H_p(z)$ and unitary vectors $V(z)$ [DVN88]. However, from the points of view of filter bank theory and wavelet theory, the Householder factorization is simpler to understand and implement except when $M = 2$.

Perhaps the simplest and most popular way to represent a 2×2 unitary matrix is by a rotation parameter (not by a Householder reflection parameter). Therefore, the simplest way to represent a unitary 2×2 matrix $H_p(z)$ is using a lattice parameterization using Given's rotations. Since two-channel unitary filter banks play an important role in the theory and design of unitary modulated filter banks (that we will shortly address), we present the lattice parameterization [VH88]. The lattice parameterization is also obtained by an order-reduction procedure we saw while deriving the Householder-type factorization in (8.28).

Theorem 45 *Every unitary 2×2 matrix $H_p(z)$ (in particular the polyphase matrix of a two channel FIR unitary filter bank) is of the form*

$$H_p(z) = \begin{bmatrix} 1 & 0 \\ 0 & \pm 1 \end{bmatrix} R(\theta_{K-1})ZR(\theta_{K-2})Z \ldots ZR(\theta_1)ZR(\theta_0), \tag{8.31}$$

where

$$R(\theta) = \begin{bmatrix} \cos\theta & \sin\theta \\ -\sin\theta & \cos\theta \end{bmatrix} \quad \text{and} \quad Z = \begin{bmatrix} 1 & 0 \\ 0 & z^{-1} \end{bmatrix} \tag{8.32}$$

Equation (8.31) is the *unitary lattice* parameterization of $H_p(z)$. The filters $H_0(z)$ and $H_1(z)$ are given by

$$\begin{bmatrix} H_0(z) \\ H_1(z) \end{bmatrix} = H_p(z^2) \begin{bmatrix} 1 \\ z^{-1} \end{bmatrix}.$$

By changing the sign of the filter $h_1(n)$, if necessary, one can always write $H_p(z)$ in the form

$$H_p(z) = R(\theta_{K-1})ZR(\theta_{K-2})Z \ldots ZR(\theta_0).$$

Now, if $H_{0,j}^R(z)$ is the reflection of $H_{0,j}(z)$ (i.e., $H_{0,j}(z) = z^{-K+1}H_{0,j}(z^{-1})$), then (from the algebraic form of $R(\theta)$)

$$\begin{bmatrix} H_{0,0}(z) & H_{0,1}(z) \\ H_{1,0}(z) & H_{1,1}(z) \end{bmatrix} = \begin{bmatrix} H_{0,0}(z) & H_{0,1}(z) \\ -H_{0,1}^R(z) & H_{0,0}^R(z) \end{bmatrix}. \tag{8.33}$$

With these parameterizations, filter banks can be designed as an unconstrained optimization problem. The parameterizations described are important for another reason. It turns out that the most efficient (from the number of arithmetic operations) implementation of unitary filter banks is using the Householder parameterization. With arbitrary filter banks, one can organize the computations so as capitalize on the rate-change operations of upsampling and downsampling. For example, one need not compute values that are thrown away by downsampling. The gain from using the parameterization of unitary filter banks is over and above this obvious gain (for example, see pages 330-331 and 386-387 in [Vai92]). Besides, with small modifications these parameterizations allow for unitariness to be preserved, even under filter coefficient quantization—with this having implications for fixed-point implementation of these filter banks in hardware digital signal processors [Vai92].

8.3   Unitary Filter Banks—Some Illustrative Examples

A few concrete examples of M-band unitary filter banks and their parameterizations should clarify our discussion.

First consider the two-band filter bank associated with Daubechies' four-coefficient scaling function and wavelet that we saw in Section 5.9. Recall that the lowpass filter (the scaling filter) is given by

n	0	1	2	3
$h_0(n)$	$\frac{1+\sqrt{3}}{4\sqrt{2}}$	$\frac{3+\sqrt{3}}{4\sqrt{2}}$	$\frac{3-\sqrt{3}}{4\sqrt{2}}$	$\frac{1-\sqrt{3}}{4\sqrt{2}}$.

The highpass filter (wavelet filter) is given by $h_1(n) = (-1)^n h_0(3-n)$, and both (8.1) and (8.2) are satisfied with $g_i(n) = h_i(-n)$. The matrix representation of the analysis bank of this filter bank is given by

$$
\mathbf{d} = \mathbf{Hx} =
\begin{bmatrix}
\vdots & \vdots & \vdots & \vdots & \vdots & \vdots \\
\frac{1-\sqrt{3}}{4\sqrt{2}} & \frac{3-\sqrt{3}}{4\sqrt{2}} & \frac{3+\sqrt{3}}{4\sqrt{2}} & \frac{1+\sqrt{3}}{4\sqrt{2}} & 0 & 0 \\
-\frac{1+\sqrt{3}}{4\sqrt{2}} & \frac{3+\sqrt{3}}{4\sqrt{2}} & -\frac{3-\sqrt{3}}{4\sqrt{2}} & \frac{1-\sqrt{3}}{4\sqrt{2}} & 0 & 0 \\
0 & 0 & \frac{1-\sqrt{3}}{4\sqrt{2}} & \frac{3-\sqrt{3}}{4\sqrt{2}} & \frac{3+\sqrt{3}}{4\sqrt{2}} & \frac{1+\sqrt{3}}{4\sqrt{2}} \\
0 & 0 & -\frac{1+\sqrt{3}}{4\sqrt{2}} & \frac{3+\sqrt{3}}{4\sqrt{2}} & -\frac{3-\sqrt{3}}{4\sqrt{2}} & \frac{1-\sqrt{3}}{4\sqrt{2}} \\
\vdots & \vdots & \vdots & \vdots & \vdots & \vdots
\end{bmatrix}
\mathbf{x}.
\qquad (8.34)
$$

One readily verifies that $\mathbf{H}^T\mathbf{H} = \mathbf{I}$ and $\mathbf{HH}^T = \mathbf{I}$. The polyphase representation of this filter bank is given by

$$
H_p(z) = \frac{1}{4\sqrt{2}}
\begin{bmatrix}
(1+\sqrt{3}) + z^{-1}(3-\sqrt{3}) & (3+\sqrt{3}) + z^{-1}(1-\sqrt{3}) \\
(3+\sqrt{3}) + z^{-1}(1-\sqrt{3}) & (-3+\sqrt{3}) - z^{-1}(1+\sqrt{3})
\end{bmatrix},
\qquad (8.35)
$$

and one can show that $H_p^T(z^{-1})H_p(z) = I$ and $H_p(z)H_p^T(z^{-1}) = I$. The Householder factorization of $H_p(z)$ is given by

$$
H_p(z) = \left[I - v_1 v_1^T + z^{-1} v_1 v_1^T \right] V_0,
\qquad (8.36)
$$

where

$$
v_1 = \begin{bmatrix} \sin(\pi/12) \\ \cos(\pi/12) \end{bmatrix}
\quad \text{and} \quad
V_0 = \begin{bmatrix} 1/\sqrt{2} & 1/\sqrt{2} \\ 1/\sqrt{2} & -1/\sqrt{2} \end{bmatrix}.
$$

Incidentally, all two-band unitary filter banks associated with wavelet tight frames have the same value of V_0. Therefore, all filter banks associated with two-band wavelet tight frames are completely specified by a set of orthogonal vectors v_i, $K-1$ of them if the h_0 is of length $2K$. Indeed, for the six-coefficient Daubechies wavelets (see Section 5.9), the parameterization of $H_p(z)$ is associated with the following two unitary vectors (since $K = 3$): $v_1^T = [-.3842 .9232]$ and $v_2^T = [-.1053 - .9944]$.

The Givens' rotation based factorization of $H_p(z)$ for the 4-coefficient Daubechies filters given by:

$$
H_p(z) =
\begin{bmatrix}
\cos\theta_0 & z^{-1}\sin\theta_0 \\
-\sin\theta_0 & z^{-1}\cos\theta_0
\end{bmatrix}
\begin{bmatrix}
\cos\theta_1 & \sin\theta_1 \\
-\sin\theta_1 & \cos\theta_1
\end{bmatrix},
\qquad (8.37)
$$

where $\theta_0 = \frac{\pi}{3}$ and $\theta_1 = -\frac{\pi}{12}$. The fact that the filter bank is associated with wavelets is precisely because $\theta_0 + \theta_1 = \frac{\pi}{4}$. More generally, for a filter bank with filters of length $2K$ to be associated with wavelets, $\sum_{k=0}^{K-1} \theta_k = \frac{\pi}{4}$. This is expected since for filters of length $2K$ to be associated with wavelets we have seen (from the Householder factorization) that there are $K-1$ parameters v_k.

Our second example belongs to a class of unitary filter banks called *modulated filter banks*, which is described in a following section. A Type 1 modulated filter bank with filters of length $N = 2M$ and associated with a wavelet orthonormal basis is defined by

$$
h_i(n) = \sqrt{\frac{1}{2M}} \left[\sin\left(\frac{\pi(i+1)(n+.5)}{M} - (2i+1)\frac{\pi}{4} \right) - \sin\left(\frac{\pi i(n+.5)}{M} - (2i+1)\frac{\pi}{4} \right) \right],
\qquad (8.38)
$$

where $i \in \{0, \ldots, M-1\}$ and $n \in \{0, \ldots, 2M-1\}$ [GB95b, GB95c]. Consider a three-band example with length six filters. In this case, $K = 2$, and therefore one has one projection P_1 and the matrix V_0. The projection is one-dimensional and given by the Householder parameter

$$v_1^T = \frac{1}{\sqrt{6}} \begin{bmatrix} 1 \\ -\sqrt{2} \\ 1 \end{bmatrix} \quad \text{and} \quad V_0 = \frac{1}{\sqrt{3}} \begin{bmatrix} 1 & 1 & 1 \\ -\frac{\sqrt{3}+1}{2} & 1 & \frac{\sqrt{3}-1}{2} \\ \frac{\sqrt{3}-1}{2} & 1 & -\frac{\sqrt{3}+1}{2} \end{bmatrix}. \tag{8.39}$$

The third example is another Type 1 modulated filter bank with $M = 4$ and $N = 8$. The filters are given in (8.38). $H_p(z)$ had the following factorization

$$H_p(z) = \left[I - P_1 + z^{-1} P_1 \right] V_0, \tag{8.40}$$

where P_1 is a two-dimensional projection $P_1 = v_1 v_1^T + v_2 v_2^T$ (notice the arbitrary choice of v_1 and v_2) given by

$$v_1 = \begin{bmatrix} 0.41636433418450 \\ -0.78450701561376 \\ 0.32495344564406 \\ 0.32495344564406 \end{bmatrix}, v_2 = \begin{bmatrix} 0.00000000000000 \\ -0.14088210492943 \\ 0.50902478635868 \\ -0.84914427477499 \end{bmatrix}$$

and

$$V_0 = \frac{1}{2} \begin{bmatrix} 1 & 1 & 1 & 1 \\ -\sqrt{2} & 0 & \sqrt{2} & 0 \\ 0 & \sqrt{2} & 0 & -\sqrt{2} \\ 1 & -1 & 1 & -1 \end{bmatrix}.$$

Notice that there are infinitely many choices of v_1 and v_2 that give rise to the same projection P_1.

8.4 M-band Wavelet Tight Frames

In Section 7.2, Theorem (7.7), while discussing the properties of M-band wavelet systems, we saw that the lowpass filter h_0 (h in the notation used there) must satisfy the linear constraint $\sum_n h_0(n) = \sqrt{M}$. Otherwise, a scaling function with nonzero integral could not exit. It turns out that this is precisely the *only* condition that an FIR unitary filter bank has to satisfy in order for it to generate an M-band wavelet system [Law90, GB92c]. Indeed, if this linear constraint is not satisfied the filter bank *does not* generate a wavelet system. This single linear constraint (for unitary filter banks) also implies that $\sum_h h_i(n) = 0$ for $i \in \{1, 2, \ldots, M-1\}$ (because of Eqn. 8.1). We now give the precise result connecting FIR unitary filter banks and wavelet tight frames.

Theorem 46 *Given an FIR unitary filter bank with $\sum_n h_0(n) = \sqrt{M}$, there exists an unique, compactly supported, scaling function $\psi_0(t) \in L^2(\mathbb{R})$ (with support in $[0, \frac{N-1}{M-1}]$, assuming h_0 is supported in $[0, N-1]$) determined by the scaling recursion:*

$$\psi_0(t) = \sqrt{M} \sum_k h_0(k) \psi_0(Mt - k). \tag{8.41}$$

Define wavelets, $\psi_i(t)$,

$$\psi_i(t) = \sqrt{M} \sum_k h_i(k)\psi_0(Mt - k) \qquad i \in \{1, 2, \ldots, M - 1\}, \tag{8.42}$$

and functions, $\psi_{i,j,k}(t)$,

$$\psi_{i,j,k}(t) = M^{j/2}\psi_i(M^j t - k). \tag{8.43}$$

Then $\{\psi_{i,j,k}\}$ forms a tight frame for $L^2(\mathbb{R})$. That is, for all $f \in L^2(\mathbb{R})$

$$f(t) = \sum_{i=1}^{M-1} \sum_{j,k=-\infty}^{\infty} \langle f, \psi_{i,j,k} \rangle \, \psi_{i,j,k}(t). \tag{8.44}$$

Also,

$$f(t) = \sum_k \langle f, \psi_{0,0,k} \rangle \, \psi_{0,0,k}(t) + \sum_{i=1}^{M-1} \sum_{j=1}^{\infty} \sum_{k=-\infty}^{\infty} \langle f, \psi_{i,j,k} \rangle \, \psi_{i,j,k}(t). \tag{8.45}$$

Remark: A similar result relates general FIR (not necessarily unitary) filter banks and M-band wavelet frames [CDF92, GB92c, Gop92].

Starting with (8.41), one can calculate the scaling function using either successive approximation or interpolation on the M-adic rationals—i.e., exactly as in the two-band case in Chapter 7.2. Equation (8.42) then gives the wavelets in terms of the scaling function. As in the two-band case, the functions $\psi_i(t)$, so constructed, invariably turn out highly irregular and sometimes fractal. The solution, once again, is to require that several moments of the scaling function (or equivalently the moments of the scaling filter h_0) are zero. This motivates the definition of K-regular M-band scaling filters: A unitary scaling filter h_0 is said to be K regular if its $\mathcal{Z}$-transform can be written in the form

$$H_0(z) = \left[\frac{1 + z^{-1} + \ldots + z^{-(M-1)}}{M}\right]^K Q(z), \tag{8.46}$$

for maximal possible K. By default, every unitary scaling filter h_0 is one-regular (because $\sum_n h_0(n) = \sqrt{M}$ - see Section 7.2, Theorem 7.7 for equivalent characterizations of K-regularity). Each of the K-identical factors in Eqn. 8.46 adds an extra linear constraint on h_0 (actually, it is one linear constraint on each of the M polyphase subsequences of h_0 - see Section 5.10).

There is no simple relationship between the *smoothness* of the scaling function and K-regularity. However, the smoothness of the *maximally* regular scaling filter, h_0, with fixed filter length N, tends to be an increasing function of N. Perhaps one can argue that K-regularity is an important concept independent of the smoothness of the associated wavelet system. K-regularity implies that the moments of the wavelets vanish up to order $K - 1$, and therefore, functions can be better approximated by using just the scaling function and its translates at a given scale. Formulae exist for M-band maximally regular K-regular scaling filters (i.e., only the sequence h_0) [SHGB93]. Using the Householder parameterization, one can then design the remaining filters in the filter bank.

The linear constraints on h_0 that constitute K-regularity become nonexplicit nonlinear constraints on the Householder parameterization of the associated filter bank. However, one-regularity can be explicitly incorporated and this gives a parameterization of all M-band compactly supported wavelet tight frames. To see, this consider the following two factorizations of $H_p(z)$ of a unitary filter bank.

$$H_p(z) = \prod_{k=K-1}^{1} \left[I - P_k + z^{-1} P_k \right] V_0, \qquad (8.47)$$

and

$$H_p^T(z) = \prod_{k=K-1}^{1} \left[I - Q_k + z^{-1} Q_k \right] W_0. \qquad (8.48)$$

Since $H_p(1) = V_0$ and $H_p^T(1) = W_0$, $V_0 = W_0^T$. The first column of W_0 is the unit vector $[H_{0,0}(1), H_{0,1}(1), \ldots, H_{0,M-1}(1)]^T$. Therefore,

$$\sum_{k=0}^{M-1} H_{0,k}(1)^2 = 1.$$

But since $\sum_n h_0(n) = H_0(1) = \sqrt{M}$,

$$\sum_{k=0}^{M-1} H_{0,k}(1) = \sqrt{M}.$$

Therefore, for all k, $H_{0,k}(1) = \frac{1}{\sqrt{M}}$. Hence, the first row of V_0 is $[1/\sqrt{M}, 1/\sqrt{M}, \ldots, 1/\sqrt{M}]$. In other words, a unitary filter bank gives rise to a WTF iff the first row of V_0 in the Householder parameterization is the vector with all entries $1/\sqrt{M}$.

Alternatively, consider the Given's factorization of $H_p(z)$ for a two-channel unitary filter bank.

$$H_p(z) = \left\{ \prod_{i=K-1}^{1} \begin{bmatrix} \cos\theta_i & z^{-1}\sin\theta_i \\ -\sin\theta_i & z^{-1}\cos\theta_i \end{bmatrix} \right\} \begin{bmatrix} \cos\theta_0 & \sin\theta_0 \\ -\sin\theta_0 & \cos\theta_0 \end{bmatrix}. \qquad (8.49)$$

Since for a WTF we require

$$\begin{bmatrix} \frac{1}{\sqrt{2}} & \frac{1}{\sqrt{2}} \\ -\frac{1}{\sqrt{2}} & \frac{1}{\sqrt{2}} \end{bmatrix} = \begin{bmatrix} H_{0,0}(1) & H_{1,0}(1) \\ H_{0,1}(1) & H_{1,1}(1) \end{bmatrix} = \begin{bmatrix} \cos(\Theta) & \sin(\Theta) \\ -\sin(\Theta) & \cos(\Theta) \end{bmatrix}, \qquad (8.50)$$

we have $\Theta = \sum_{k=0}^{K-1} \theta_k = \frac{\pi}{4}$. This is the condition for the lattice parameterization to be associated with wavelets.

8.5   Modulated Filter Banks

Filter bank design typically entails optimization of the filter coefficients to maximize some goodness measure subject to the perfect reconstruction constraint. Being a constrained (or unconstrained for parameterized unitary filter bank design) nonlinear programming problem, numerical optimization leads to local minima, with the problem exacerbated when there are a large number of filter coefficients. To alleviate this problem one can try to impose structural constraints on the filters. For example, if Fig. 8.2 is the desired ideal response, one can impose the constraint that all analysis (synthesis) filters are obtained by modulation of a single "prototype" analysis (synthesis) filter. This is the basic idea behind modulated filter banks [Mal92, KV92, Vai92, NK92, GB95b, GB93, Gop96b, LV95]. In what follows, we only consider the case where the number of filters is equal to the downsampling factor; i.e., $L = M$.

The frequency responses in Fig. 8.2 can be obtained by shifting an the response of an ideal lowpass filter (supported in $[-\frac{\pi}{2M}, \frac{\pi}{2M}]$) by $(i + \frac{1}{2})\frac{\pi}{M}$, $i \in \{0, \dots, M-1\}$. This can be achieved by modulating with a cosine (or sine) with appropriate frequency and arbitrary phase. However, some choices of phase may be incompatible with perfect reconstruction. A general choice of phase (and hence modulation, that covers all modulated filter banks of this type) is given by the following definition of the analysis and synthesis filters:

$$h_i(n) = h(n)\cos\left(\frac{\pi}{M}(i + \frac{1}{2})(n - \frac{\alpha}{2})\right), \quad i \in \mathcal{R}(M) \tag{8.51}$$

$$g_i(n) = g(n)\cos\left(\frac{\pi}{M}(i + \frac{1}{2})(n + \frac{\alpha}{2})\right), \quad i \in \mathcal{R}(M) \tag{8.52}$$

Here α is an integer parameter called the modulation phase. Now one can substitute these forms for the filters in (8.1) to explicit get PR constraints on the prototype filters h and g. This is a straightforward algebraic exercise, since the summation over i in (8.1) is a trigonometric sum that can be easily computed. It turns out that the PR conditions depend only on the parity of the modulation phase α. Hence without loss of generality, we choose $\alpha \in \{M-1, M-2\}$—other choices being incorporated as a preshift into the prototype filters h and g.

Thus there are two types of MFBs depending on the choice of modulation phase:

$$\alpha = \begin{cases} M-1 & \text{Type 1 Filter Bank} \\ M-2 & \text{Type 2 Filter Bank} \end{cases} \tag{8.53}$$

The PR constraints on h and g are quite messy to write down without more notational machinery. But the basic nature of the constraints can be easily understood pictorially. Let the M polyphase components of h and g respectively be partitioned into pairs as suggested in Fig. 8.4. Each polyphase pair from h and an associated polyphase pair g (i.e., those four sequences) satisfy the PR conditions for a two-channel filter bank. In other words, these subsequences could be used as analysis and synthesis filters respectively in a two-channel PR filter bank. As seen in Figure 8.4, some polyphase components are not paired. The constraints on these sequences that PR imposes will be explicitly described soon. Meanwhile, notice that the PR constraints on the coefficients are decoupled into roughly $M/2$ independent sets of constraints (since there are roughly $M/2$ PR pairs in Fig. 8.4). To quantify this, define J:

$$J = \begin{cases} \frac{M}{2} & \text{Type 1, } M \text{ even} \\ \frac{M-1}{2} & \text{Type 1, } M \text{ odd} \\ \frac{M-2}{2} & \text{Type 2, } M \text{ even} \\ \frac{M-1}{2} & \text{Type 2, } M \text{ odd.} \end{cases} \tag{8.54}$$

In other words, the MFB PR constraint decomposes into a set of J two-channel PR constraints and a few additional conditions on the unpaired polyphase components of h and g.

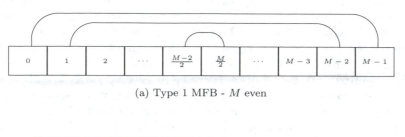

(a) Type 1 MFB - M even

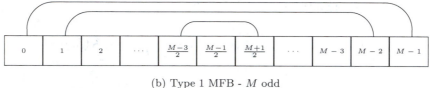

(b) Type 1 MFB - M odd

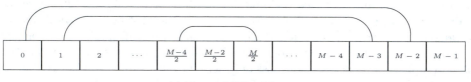

(c) Type 2 MFB - M even

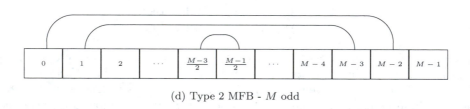

(d) Type 2 MFB - M odd

Figure 8.4. Two-Channel PR Pairs in a PR MFB

We first define M polyphase components of the analysis and synthesis prototype filters, viz., $P_l(z)$ and $Q_l(z)$ respectively. We split these sequences further into their even and odd components to give $P_{l,0}(z)$, $P_{l,1}(z)$, $Q_{l,0}(z)$ and $Q_{l,1}(z)$ respectively. More precisely, let

$$H(z) = \sum_{l=0}^{M-1} z^{-l} P_l(z^M) = \sum_{l=0}^{M-1} z^{-l} (P_{l,0}(z^{2M}) + z^{-M} P_{l,1}(z^{2M})), \qquad (8.55)$$

$$G(z) = \sum_{l=0}^{M-1} z^{l} Q_l(z^M) = \sum_{l=0}^{M-1} z^{l} (Q_{l,0}(z^{2M}) + z^{M} Q_{l,1}(z^{2M})), \qquad (8.56)$$

and let

$$\mathcal{P}(z) = \begin{bmatrix} P_{l,0}(z) & P_{l,1}(z) \\ P_{\alpha-l,0}(z) & -P_{\alpha-l,1}(z) \end{bmatrix} \tag{8.57}$$

with $\mathcal{Q}(z)$ defined similarly. Let $\mathcal{I}$ be the 2×2 identity matrix.

Theorem 47 (Modulated Filter Banks PR Theorem) *An modulated filter bank (Type 1 or Type 2) (as defined in (8.51) and (8.52)) is PR iff for $l \in \mathcal{R}(J)$*

$$\mathcal{P}_l(z)\mathcal{Q}_l^T(z) = \tag{8.58}$$

and furthermore if α is even $P_{\frac{\alpha}{2},0}(z)Q_{\frac{\alpha}{2},0}(z) = \frac{1}{M}$. In the Type 2 case, we further require $P_{M-1}(z)Q_{M-1}(z) = \frac{2}{M}$.

The result says that $P_l, P_{\alpha-l}, Q_l$ and $Q_{\alpha-l}$ form analysis and synthesis filters of a two-channel PR filter bank ((8.1) in Z-transform domain).

Modulated filter bank design involves choosing h and g to optimize some goodness criterion while subject to the constraints in the theorem above.

Unitary Modulated Filter Bank

In a unitary bank, the filters satisfy $g_i(n) = h_i(-n)$. From (8.13) and (8.14), it is clear that in a modulated filter bank if $g(n) = h(-n)$, then $g_i(n) = h_i(-n)$. Imposing this restriction (that the analysis and synthesis prototype filters are reflections of each other) gives PR conditions for unitary modulated filter banks. That $g(n) = h(-n)$ means that $P_l(z) = Q_l(z^{-1})$ and therefore $Q_l(z) = P_l(z^{-1})$. Indeed, for PR, we require

$$\mathcal{P}_l(z)\mathcal{P}_l^T(z^{-1}) = \frac{2}{M}\mathcal{I}. \tag{8.59}$$

This condition is equivalent to requiring that P_l and $P_{\alpha-l}$ are analysis filters of a two-channel unitary filter bank. Equivalently, for $l \in \mathcal{R}(M)$, $P_{l,0}$ and $P_{l,1}$ are power-complementary.

Corollary 6 (Unitary MFB PR Theorem) *A modulated filter bank (Type 1 or Type 2) is unitary iff for $l \in \mathcal{R}(J)$, $P_{l,0}(z)$ and $P_{l,1}(z)$ are power complementary.*

$$P_{l,0}(z)P_{l,0}(z^{-1}) + P_{l,1}(z)P_{l,1}(z^{-1}) = \frac{2}{M}, l \in \mathcal{R}(M) \tag{8.60}$$

Furthermore, when α is even $P_{\frac{\alpha}{2},0}(z)P_{\frac{\alpha}{2},0}(z^{-1}) = \frac{1}{M}$ (i.e., $P_{\frac{\alpha}{2},0}(z)$ has to be $\frac{1}{\sqrt{M}}z^k$ for some integer k). In the Type 2 case, we further require $P_{M-1}(z)P_{M-1}(z^{-1}) = \frac{2}{M}$ (i.e., $P_{M-1}(z)$ has to be $\frac{2}{\sqrt{M}}z^k$ for some integer k).

Unitary modulated filter bank design entails the choice of h, the analysis prototype filter. There are J associated two-channel unitary filter banks each of which can be parameterized using the lattice parameterization. Besides, depending on whether the filter is Type 2 and/or *alpha* is even one has to choose the locations of the delays.

For the prototype filter of a unitary MFB to be linear phase, it is necessary that

$$P_{\alpha-l}(z) = z^{-2k+1}P_l(z^{-1}), \tag{8.61}$$

for some integer k. In this case, the prototype filter (if FIR) is of length $2Mk$ and symmetric about $(Mk - \frac{1}{2})$ in the Type 1 case and of length $2Mk - 1$ and symmetric about $(Mk - 1)$ (for both Class A and Class B MFBs). In the FIR case, one can obtain linear-phase prototype filters by using the lattice parameterization [VH88] of two-channel unitary filter banks. Filter banks with FIR linear-phase prototype filters will be said to be *canonical*. In this case, $P_l(z)$ is typically a filter of length $2k$ for all l. For *canonical* modulated filter banks, one has to check power complementarity only for $l \in \mathcal{R}(J)$.

8.6   Modulated Wavelet Tight Frames

For all M, there exist M-band modulated WTFs. The simple linear constraint on h_0 becomes a set of J linear constraints, one each, on each of the J two-channel unitary lattices associated with the MFB.

Theorem 48 (Modulated Wavelet Tight Frames Theorem) *Every compactly supported modulated WTF is associated with an FIR unitary MFB and is parameterized by J unitary lattices such that the sum of the angles in the lattices satisfy (for $l \in \mathcal{R}(J)$) Eqn. 8.62.*

$$\sum_k \theta_{l,k} \overset{\text{def}}{=} \Theta_l = \frac{\pi}{4} + \frac{\pi}{2M}\left(\frac{\alpha}{2} - l\right). \tag{8.62}$$

If a canonical MFB has Jk parameters, the corresponding WTF has $J(k-1)$ parameters.

Notice that even though the PR conditions for MFBs depended on whether it is Type 1 or Type 2, the MWTF conditions are identical. Now consider a Type 1 or Type 2 MFB with one angle parameter per lattice; i.e., $N = 2M$ (Type 1) or $N = 2M - 1$ (Type 2). This angle parameter is specified by the MWTF theorem above if we want associated wavelets. This choice of angle parameters leads to a particularly simple form for the prototype filter.

In the Type 1 case [GB95b, GB93],

$$h(n) = \sqrt{\frac{2}{M}} \sin\left(\frac{\pi}{4M}(2n+1)\right). \tag{8.63}$$

and therefore

$$h_i(n) = \sqrt{\frac{1}{2M}}\left[\sin\left(\frac{\pi(i+1)(n+.5)}{M} - (2i+1)\frac{\pi}{4}\right) - \sin\left(\frac{\pi i(n+.5)}{M} - (2i+1)\frac{\pi}{4}\right)\right]. \tag{8.64}$$

In the Type 2 case [GB95b],

$$h(n) = \sqrt{\frac{2}{M}} \sin\left(\frac{\pi}{2M}(n+1)\right), \tag{8.65}$$

and hence

$$h_i(n) = \sqrt{\frac{1}{2M}}\left[\sin\left(\frac{\pi(i+1)(n+1)}{M} - (2i+1)\frac{\pi}{4}\right) - \sin\left(\frac{\pi i(n+1)}{M} - (2i+1)\frac{\pi}{4}\right)\right]. \tag{8.66}$$

8.7 Linear Phase Filter Banks

In some applications. it is desirable to have filter banks with linear-phase filters [SVN93]. The linear-phase constraint (like the modulation constraint studied earlier) reduces the number of free parameters in the design of a filter bank. Unitary linear phase filter banks have been studied recently [SVN93, GB94b]. In this section we develop algebraic characterizations of certain types of linear filter banks that can be used as a starting point for designing such filter banks.

In this section, we assume that the desired frequency responses are as in (8.2). For simplicity we also assume that the number of channels, M, is an even integer and that the filters are FIR. It should be possible to extend the results that follow to the case when M is an odd integer in a straightforward manner.

Consider an M-channel FIR filter bank with filters whose passbands approximate ideal filters. Several transformations relate the M ideal filter responses. We have already seen one example where all the ideal filters are obtained by modulation of a prototype filter. We now look at other types of transformations that relate the filters. Specifically, the ideal frequency response of h_{M-1-i} can be obtained by shifting the response of the h_i by π. This either corresponds to the restriction that

$$h_{M-1-i}(n) = (-1)^n h_i(n) \quad ; \quad H_{M-1-i}(z) = H_i(-z) \quad ; \quad H_{M-1-i}(\omega) = H_i(\omega + \pi), \qquad (8.67)$$

or to the restriction that

$$h_{M-1-i}(n) = (-1)^n h_i(N-1-n) \quad ; \quad H_{M-1-i}(z) = H_i^R(-z) \quad ; \quad H_{M-1-i}(\omega) = H_i^\star(\omega + \pi) \quad (8.68)$$

where N is the filter length and for polynomial $H(z)$, $H^R(z)$ denotes its reflection polynomial (i.e. the polynomial with coefficients in the reversed order). The former will be called *pairwise-shift* (or PS) symmetry (it is also known as pairwise-mirror image symmetry [NV88]) , while the latter will be called *pairwise-conjugated-shift* (or PCS) symmetry (also known as pairwise-symmetry [NV88]). Both these symmetries relate pairs of filters in the filter bank. Another type of symmetry occurs when the filters themselves are symmetric or linear-phase. The only type of linear-phase symmetry we will consider is of the form

$$h_i(n) = \pm h_i(N-1-n) \quad ; \quad H_i(z) = \pm H_i^R(z),$$

where the filters are all of fixed length N, and the symmetry is about $\frac{N-1}{2}$. For an M-channel linear-phase filter bank (with M an even integer), $M/2$ filters each are even-symmetric and odd-symmetric respectively [SVN93].

We now look at the structural restrictions on $H_p(z)$, the polyphase component matrix of the analysis bank that these three types of symmetries impose. Let J denote the *exchange matrix* with ones on the antidiagonal. Postmultiplying a matrix A by J is equivalent to reversing the order of the columns of A, and premultiplying is equivalent to reversing the order of the rows of A. Let V denote the *sign-alternating* matrix, the diagonal matrix of alternating ± 1's. Postmultiplying by

V, alternates the signs of the columns of A, while premultiplying alternates the signs of the rows of A. The polyphase components of $H(z)$ are related to the polyphase components of $H^R(z)$ by reflection and reversal of the ordering of the components. Indeed, if $H(z)$ is of length Mm, and $H(z) = \sum_{l=0}^{M-1} z^{-l} H_l(z^M)$, then,

$$
\begin{aligned}
H^R(z) &= z^{-Mm+1} \sum_{l=0}^{M-1} z^l H_l(z^{-M}) \\
&= \sum_{l=0}^{M-1} z^{-(M-1-l)} \left(z^{-mM+M} H_l(z^{-M}) \right) \\
&= \sum_{l=0}^{M-1} z^{-l} H_{M-1-l}^R(z^M).
\end{aligned}
$$

Therefore

$$
(H^R)_l(z) = (H_{M-1-l})^R(z) \tag{8.69}
$$

and for linear-phase $H(z)$, since $H^R(z) = \pm H(z)$,

$$
H_l(z) = \pm (H^R)_{M-1-l}(z). \tag{8.70}
$$

Lemma 1 *For even M, $H_p(z)$ is of the form*

PS Symmetry

$$
\begin{bmatrix} W_0(z) & W_1(z) \\ J W_0(z) V & (-1)^{M/2} J W_1(z) V \end{bmatrix} = \begin{bmatrix} I & 0 \\ 0 & J \end{bmatrix} \begin{bmatrix} W_0(z) & W_1(z) \\ W_0(z) V & (-1)^{M/2} W_1(z) V \end{bmatrix} \tag{8.71}
$$

PCS Symmetry

$$
\begin{bmatrix} W_0(z) & W_1(z) J \\ J W_1^R(z) V & (-1)^{M/2} J W_0^R(z) J V \end{bmatrix} = \begin{bmatrix} I & 0 \\ 0 & J \end{bmatrix} \begin{bmatrix} W_0(z) & W_1(z) J \\ W_1^R(z) V & (-1)^{M/2} W_0^R(z) J V \end{bmatrix} \tag{8.72}
$$

Linear Phase

$$
\begin{bmatrix} W_0(z) & D_0 W_0^R(z) J \\ W_1(z) & D_1 W_1^R(z) J \end{bmatrix} = \begin{bmatrix} W_0(z) & D_0 W_0^R(z) \\ W_1(z) & D_1 W_1^R(z) \end{bmatrix} \begin{bmatrix} I & 0 \\ 0 & J \end{bmatrix} \tag{8.73}
$$

or

$$
Q \begin{bmatrix} W_0(z) & W_0^R(z) J \\ W_1(z) & -W_1^R(z) J \end{bmatrix} = Q \begin{bmatrix} W_0(z) & W_0^R(z) \\ W_1(z) & -W_1^R(z) \end{bmatrix} \begin{bmatrix} I & 0 \\ 0 & J \end{bmatrix} \tag{8.74}
$$

Linear Phase and PCS

$$\begin{bmatrix} W_0(z) & DW_0^R(z)J \\ JDW_0(z)V & (-1)^{M/2}JW_0^R(z)JV \end{bmatrix} = \begin{bmatrix} I & 0 \\ 0 & J \end{bmatrix} \begin{bmatrix} W_0(z) & DW_0^R(z)J \\ DW_0(z)V & (-1)^{M/2}W_0^R(z)JV \end{bmatrix}$$

(8.75)

Linear Phase and PS

$$\begin{bmatrix} W_0(z) & DW_0^R(z)J \\ JW_0(z)V & (-1)^{M/2}JDW_0^R(z)JV \end{bmatrix} = \begin{bmatrix} I & 0 \\ 0 & JD \end{bmatrix} \begin{bmatrix} W_0(z) & DW_0^R(z)J \\ DW_0(z)V & (-1)^{M/2}W_0^R(z)JV \end{bmatrix}$$

(8.76)

Thus in order to *generate* $H_p(z)$ for all symmetries *other than PS*, we need a mechanism that generates a pair of matrices and their reflection (i.e., $W_0(z)$, $W_1(z)$ $W_0^R(z)$ and $W_1^R(z)$). In the scalar case, there are two well-known lattice structures, that generate such pairs. The first case is the orthogonal lattice [VH88], while the second is the linear-prediction lattice [RS83]. A Kth order orthogonal lattice is generated by the product

$$\left\{ \prod_{i=0}^{K} \begin{bmatrix} a_i & z^{-1}b_i \\ -b_i & z^{-1}a_i \end{bmatrix} \right\} \begin{bmatrix} a_0 & b_0 \\ -b_0 & a_0 \end{bmatrix} \overset{def}{=} \begin{bmatrix} Y_0(z) & Y_1(z) \\ -Y_1^R(z) & Y_0^R(z) \end{bmatrix} \overset{def}{=} X(z).$$

This lattice is always invertible (unless a_i and b_i are both zero!), and the inverse is anticausal since

$$\begin{bmatrix} a_i & z^{-1}b_i \\ -b_i & z^{-1}a_i \end{bmatrix}^{-1} = \frac{1}{a_1^2 + b_i^2} \begin{bmatrix} a_i & -b_i \\ zb_i & za_i \end{bmatrix}.$$

As we have seen, this lattice plays a fundamental role in the theory of two-channel FIR unitary modulated filter banks. The hyperbolic lattice of order K generates the product

$$\left\{ \prod_{i=0}^{K} \begin{bmatrix} a_i & z^{-1}b_i \\ b_i & z^{-1}a_i \end{bmatrix} \right\} \begin{bmatrix} a_0 & b_0 \\ b_0 & a_0 \end{bmatrix} \overset{def}{=} \begin{bmatrix} Y_0(z) & Y_1(z) \\ Y_1^R(z) & Y_0^R(z) \end{bmatrix} \overset{def}{=} X(z).$$

where $Y_0(z)$ and $Y_1(z)$ are of order K. This lattice is invertible only when $a_i^2 \neq b_i^2$ (or equivalently $(a_i + b_i)/2$ and $(a_i - b_i)/2$ are nonzero) in which case the inverse is noncausal since

$$\begin{bmatrix} a_i & z^{-1}b_i \\ b_i & z^{-1}a_i \end{bmatrix}^{-1} = \frac{1}{a_1^2 - b_i^2} \begin{bmatrix} a_i & -b_i \\ -zb_i & za_i \end{bmatrix}.$$

Since the matrix $\begin{bmatrix} a_i & b_i \\ b_i & a_i \end{bmatrix}$ can be orthogonal iff $\{a_i, b_i\} = \{\pm 1, 0\}$, or $\{a_i, b_i\} = \{0, \pm 1\}$, the (2×2) matrix generated by the lattice can never be unitary.

Formally, it is clear that if we replace the scalars a_i and b_i with square matrices of size $M/2 \times M/2$ then we would be able to generate matrix versions of these two lattices which can then be used to generate filter banks with the symmetries we have considered. We will shortly see that both the lattices can generate unitary matrices, and this will lead to a parameterization of FIR unitary $H_p(z)$ for PCS, linear-phase, and PCS plus linear-phase symmetries. We prefer to call the generalization of the orthogonal lattice, the *antisymmetric* lattice and to call the generalization of the hyperbolic lattice, the *symmetric* lattice, which should be obvious from the form of the product. The reason for this is that the antisymmetric lattice *may not* generate a unitary matrix transfer function (in the scalar case, the 2×2 transfer function generated is always unitary). The antisymmetric lattice is defined by the product

$$X(z) \overset{\text{def}}{=} \left\{ \prod_{i=1}^{K} \begin{bmatrix} A_i & z^{-1}B_i \\ -B_i & z^{-1}A_i \end{bmatrix} \right\} \begin{bmatrix} A_0 & B_0 \\ -B_0 & A_0 \end{bmatrix} \tag{8.77}$$

where A_i and B_i are constant square matrices of size $M/2 \times M/2$. It is readily verified that $X(z)$ is of the form

$$X(z) = \begin{bmatrix} Y_0(z) & Y_1(z) \\ -Y_1^R(z) & Y_0^R(z) \end{bmatrix} \tag{8.78}$$

Given $X(z)$, its invertibility is equivalent to the invertibility of the constant matrices,

$$\begin{bmatrix} A_i & B_i \\ -B_i & A_i \end{bmatrix} \quad \text{since} \quad \begin{bmatrix} A_i & z^{-1}B_i \\ -B_i & z^{-1}A_i \end{bmatrix} = \begin{bmatrix} A_i & B_i \\ -B_i & A_i \end{bmatrix} \begin{bmatrix} I & 0 \\ 0 & z^{-1}I \end{bmatrix}, \tag{8.79}$$

which, in turn is related to the invertibility of the *complex* matrices $C_i = (A_i + \imath B_i)$ and $D_i = (A_i - \imath B_i)$, since,

$$\frac{1}{2} \begin{bmatrix} I & I \\ \imath I & -\imath I \end{bmatrix} \begin{bmatrix} C_i & 0 \\ 0 & D_i \end{bmatrix} \begin{bmatrix} I & -\imath I \\ I & \imath I \end{bmatrix} = \begin{bmatrix} A_i & B_i \\ -B_i & A_i \end{bmatrix}.$$

Moreover, the orthogonality of the matrix is equivalent to the unitariness of the complex matrix C_i (since D_i is just its Hermitian conjugate). Since an arbitrary complex matrix of size $M/2 \times M/2$ is determined by precisely $2 \binom{M/2}{2}$ parameters, each of the matrices $\begin{bmatrix} A_i & B_i \\ -B_i & A_i \end{bmatrix}$ has that many degrees of freedom. Clearly when these matrices are orthogonal $X(z)$ is unitary (on the unit circle) and $X^T(z^{-1})X(z) = I$. For unitary $X(z)$ the converse is also true as will be shortly proved.

The symmetric lattice is defined by the product

$$X(z) \overset{\text{def}}{=} \left\{ \prod_{i=1}^{K} \begin{bmatrix} A_i & z^{-1}B_i \\ B_i & z^{-1}A_i \end{bmatrix} \right\} \begin{bmatrix} A_0 & B_0 \\ B_0 & A_0 \end{bmatrix} \tag{8.80}$$

Once again A_i and B_i are constant square matrices, and it is readily verified that $X(z)$ written as a product above is of the form

$$X(z) = \begin{bmatrix} Y_0(z) & Y_1(z) \\ Y_1^R(z) & Y_0^R(z) \end{bmatrix} \tag{8.81}$$

The invertibility of $X(z)$ is equivalent to the invertibility of

$$\begin{bmatrix} A_i & B_i \\ B_i & A_i \end{bmatrix} \quad \text{since} \quad \begin{bmatrix} A_i & z^{-1}B_i \\ B_i & z^{-1}A_i \end{bmatrix} = \begin{bmatrix} A_i & B_i \\ B_i & A_i \end{bmatrix} \begin{bmatrix} I & 0 \\ 0 & z^{-1}I \end{bmatrix}, \tag{8.82}$$

which in turn is equivalent to the invertibility of $C_i = (A_i + B_i)$ and $D_i = (A_i - B_i)$ since

$$\frac{1}{2} \begin{bmatrix} I & I \\ I & -I \end{bmatrix} \begin{bmatrix} C_i & 0 \\ 0 & D_i \end{bmatrix} \begin{bmatrix} I & I \\ I & -I \end{bmatrix} = \begin{bmatrix} A_i & B_i \\ B_i & A_i \end{bmatrix}.$$

The orthogonality of the constant matrix is equivalent to the orthogonality of the *real* matrices C_i and D_i, and since each real orthogonal matrix of size $M/2 \times M/2$ is determined by $\binom{M/2}{2}$ parameters, the constant orthogonal matrices have $2\binom{M/2}{2}$ degrees of freedom. Clearly when the matrices are orthogonal $X^T(z^{-1})X(z) = I$. For the hyperbolic lattice too, the converse is true.

We now give a theorem that leads to a parameterization of unitary filter banks with the symmetries we have considered (for a proof, see [GB94b]).

Theorem 49 *Let $X(z)$ be a unitary $M \times M$ polynomial matrix of degree K. Depending on whether $X(z)$ is of the form in (8.78), or (8.81), it is generated by an order K antisymmetric or symmetric lattice.*

Characterization of Unitary $H_p(z)$ — PS Symmetry

The form of $H_p(z)$ for PS symmetry in (8.71) can be simplified by a permutation. Let P be the permutation matrix that exchanges the first column with the last column, the third column with the last but third, etc. That is,

$$
P = \begin{bmatrix}
0 & 0 & 0 & \cdots & 0 & 0 & 1 \\
0 & 1 & 0 & \cdots & 0 & 0 & 0 \\
0 & 0 & 0 & \cdots & 1 & 0 & 0 \\
\vdots & \vdots & \vdots & \cdots & \vdots & \vdots & \vdots \\
0 & 0 & 1 & \cdots & 0 & 0 & 0 \\
0 & 0 & 0 & \cdots & 0 & 1 & 0 \\
1 & 0 & 0 & \cdots & 0 & 0 & 0
\end{bmatrix}.
$$

Then the matrix $\begin{bmatrix} W_0(z) & W_1(z) \\ W_0(z)V & (-1)^{M/2}W_1(z)V \end{bmatrix}$ in (8.71) can be rewritten as $\dfrac{1}{\sqrt{2}}\begin{bmatrix} W_0'(z) & W_1'(z) \\ -W_0'(z) & W_1'(z) \end{bmatrix}P$, and therefore

$$
\begin{aligned}
H_p(z) &= \begin{bmatrix} I & 0 \\ 0 & J \end{bmatrix}\begin{bmatrix} W_0(z) & W_1(z) \\ W_0(z)V & (-1)^{M/2}W_1(z)V \end{bmatrix} \\[2mm]
&= \frac{1}{\sqrt{2}}\begin{bmatrix} I & 0 \\ 0 & J \end{bmatrix}\begin{bmatrix} W_0'(z) & W_1'(z) \\ -W_0'(z) & W_1'(z) \end{bmatrix}P \\[2mm]
&= \frac{1}{\sqrt{2}}\begin{bmatrix} I & 0 \\ 0 & J \end{bmatrix}\begin{bmatrix} I & I \\ -I & I \end{bmatrix}\begin{bmatrix} W_0'(z) & 0 \\ 0 & W_1'(z) \end{bmatrix}P \\[2mm]
&= \frac{1}{\sqrt{2}}\begin{bmatrix} I & I \\ -J & J \end{bmatrix}\begin{bmatrix} W_0'(z) & 0 \\ 0 & W_1'(z) \end{bmatrix}P.
\end{aligned}
$$

For PS symmetry, one has the following parameterization of unitary filter banks.

Theorem 50 (Unitary PS Symmetry) $H_p(z)$ *of order* K *forms a unitary PR filter bank with PS symmetry iff there exist unitary, order* K, $M/2 \times M/2$ *matrices* $W'_0(z)$ *and* $W'_1(z)$, *such that*

$$H_p(z) = \frac{1}{\sqrt{2}} \begin{bmatrix} I & I \\ -J & J \end{bmatrix} \begin{bmatrix} W'_0(z) & 0 \\ 0 & W'_1(z) \end{bmatrix} P. \qquad (8.83)$$

A unitary H_p, *with PS symmetry is determined by precisely* $2(M/2-1)(L_0+L_1)+2\binom{M/2}{2}$ *parameters where* $L_0 \geq K$ *and* $L_1 \geq K$ *are the McMillan degrees of* $W'_0(z)$ *and* $W'_1(z)$ *respectively.*

Characterization of Unitary $H_p(z)$ — PCS Symmetry

In this case

$$\begin{aligned} H_p(z) &= \begin{bmatrix} I & 0 \\ 0 & J \end{bmatrix} \begin{bmatrix} W_0(z) & W_1(z)J \\ W_1^R(z)V & (-1)^{M/2}W_0^R(z)JV \end{bmatrix} \\ &\stackrel{\text{def}}{=} \begin{bmatrix} I & 0 \\ 0 & J \end{bmatrix} \begin{bmatrix} W'_0 & W'_1 J \\ -(W'_1)^R & (W'_0)^R J \end{bmatrix} P \\ &= \begin{bmatrix} I & 0 \\ 0 & J \end{bmatrix} \begin{bmatrix} W'_0 & W'_1 \\ -(W'_1)^R & (W'_0)^R \end{bmatrix} \begin{bmatrix} I & 0 \\ 0 & J \end{bmatrix} P. \end{aligned}$$

Hence from Lemma 1 $H_p(z)$ of unitary filter banks with PCS symmetry can be parameterized as follows:

Theorem 51 $H_p(z)$ *forms an order* K, *unitary filter bank with PCS symmetry iff*

$$H_p(z) = \begin{bmatrix} I & 0 \\ 0 & J \end{bmatrix} \left\{ \prod_{i=1}^{K} \begin{bmatrix} A_i & z^{-1}B_i \\ -B_i & z^{-1}A_i \end{bmatrix} \right\} \begin{bmatrix} A_0 & B_0 \\ -B_0 & A_0 \end{bmatrix} \begin{bmatrix} I & 0 \\ 0 & J \end{bmatrix} P \qquad (8.84)$$

where $\begin{bmatrix} A_i & B_i \\ -B_i & A_i \end{bmatrix}$ *are constant orthogonal matrices.* $H_p(z)$ *is characterized by* $2K\binom{M/2}{2}$ *parameters.*

Characterization of Unitary $H_p(z)$ — Linear-Phase Symmetry

For the linear-phase case,

$$\begin{aligned} H_p(z) &= Q \begin{bmatrix} W_0(z) & W_0^R(z) \\ W_1(z) & -W_1^R(z) \end{bmatrix} \begin{bmatrix} I & 0 \\ 0 & J \end{bmatrix} \\ &= \frac{1}{2}Q \begin{bmatrix} I & I \\ I & -I \end{bmatrix} \begin{bmatrix} W_0(z)+W_1(z) & W_0^R(z)-W_1^R(z) \\ W_0(z)-W_1(z) & W_0^R(z)+W_1^R(z) \end{bmatrix} \begin{bmatrix} I & 0 \\ 0 & J \end{bmatrix} \\ &\stackrel{\text{def}}{=} \frac{1}{\sqrt{2}}Q \begin{bmatrix} I & I \\ I & -I \end{bmatrix} \begin{bmatrix} W'_0(z) & W'_1(z) \\ (W'_1)^R(z) & (W'_0)^R(z) \end{bmatrix} \begin{bmatrix} I & 0 \\ 0 & J \end{bmatrix}. \end{aligned}$$

Therefore, we have the following Theorem:

Theorem 52 $H_p(z)$ *of order K, forms a unitary filter bank with linear-phase filters iff*

$$H_p(z) = \frac{1}{\sqrt{2}} Q \begin{bmatrix} I & I \\ I & -I \end{bmatrix} \left\{ \prod_{i=1}^{K} \begin{bmatrix} A_i & z^{-1}B_i \\ B_i & z^{-1}A_i \end{bmatrix} \right\} \begin{bmatrix} A_0 & B_0 \\ B_0 & A_0 \end{bmatrix} \begin{bmatrix} I & 0 \\ 0 & J \end{bmatrix}, \qquad (8.85)$$

where $\begin{bmatrix} A_i & B_i \\ B_i & A_i \end{bmatrix}$ *are constant orthogonal matrices. $H_p(z)$ is characterized by $2K \binom{M/2}{2}$ parameters.*

Characterization of Unitary $H_p(z)$ — Linear Phase and PCS Symmetry

In this case, $H_p(z)$ is given by

$$
\begin{aligned}
H_p(z) &= \begin{bmatrix} I & 0 \\ 0 & J \end{bmatrix} \begin{bmatrix} W_0(z) & DW_0^R(z)J \\ DW_0(z)V & (-1)^{M/2}W_0^R(z)JV \end{bmatrix} \\
&\stackrel{\text{def}}{=} \frac{1}{\sqrt{2}} \begin{bmatrix} I & 0 \\ 0 & J \end{bmatrix} \begin{bmatrix} W_0'(z) & D(W_0')^R(z)J \\ -DW_0'(z) & (W_0')^R(z)J \end{bmatrix} P \\
&= \frac{1}{\sqrt{2}} \begin{bmatrix} I & 0 \\ 0 & J \end{bmatrix} \begin{bmatrix} I & D \\ -D & I \end{bmatrix} \begin{bmatrix} W_0'(z) & 0 \\ 0 & (W_0')^R(z) \end{bmatrix} \begin{bmatrix} I & 0 \\ 0 & J \end{bmatrix} P.
\end{aligned}
$$

Therefore we have proved the following Theorem:

Theorem 53 $H_p(z)$ *of order K forms a unitary filter bank with linear-phase and PCS filters iff there exists a unitary, order K, $M/2 \times M/2$ matrix $W_0'(z)$ such that*

$$H_p(z) = \frac{1}{\sqrt{2}} \begin{bmatrix} I & D \\ -JD & J \end{bmatrix} \begin{bmatrix} W_0'(z) & 0 \\ 0 & (W_0')^R(z) \end{bmatrix} \begin{bmatrix} I & 0 \\ 0 & J \end{bmatrix} P. \qquad (8.86)$$

In this case $H_p(z)$ is determined by precisely $(M/2 - 1)L + \binom{M/2}{2}$ parameters where $L \geq K$ is the McMillan degree of $W_0'(z)$.

Characterization of Unitary $H_p(z)$ — Linear Phase and PS Symmetry

From the previous result we have the following result:

Theorem 54 $H_p(z)$ *of order K forms a unitary filter bank with linear-phase and PS filters iff there exists a unitary, order K, $M/2 \times M/2$ matrix $W_0'(z)$ such that*

$$H_p(z) = \frac{1}{\sqrt{2}} \begin{bmatrix} I & D \\ -J & JD \end{bmatrix} \begin{bmatrix} W_0'(z) & 0 \\ 0 & (W_0')^R(z) \end{bmatrix} \begin{bmatrix} I & 0 \\ 0 & J \end{bmatrix} P. \qquad (8.87)$$

H_p *is determined by precisely $(M/2 - 1)L + \binom{M/2}{2}$ parameters where $L \geq K$ is the McMillan degree of $W_0'(z)$.*

Notice that Theorems 50 through Theorem 54 give a completeness characterization for unitary filter banks with the symmetries in question (and the appropriate length restrictions on the filters). However, if one requires only the matrices $W_0'(z)$ and $W_1'(z)$ in the above theorems to be invertible on the unit circle (and not unitary), then the above results gives a method to generate nonunitary PR filter banks with the symmetries considered. Notice however, that in the nonunitary case this is *not* a complete parameterization of all such filter banks.

8.8   Linear-Phase Wavelet Tight Frames

A necessary and sufficient condition for a unitary (FIR) filter bank to give rise to a compactly supported wavelet tight frame (WTF) is that the lowpass filter h_0 in the filter bank satisfies the linear constraint [GB92c]

$$\sum_n h_0(n) = \sqrt{M}. \tag{8.88}$$

We now examine and characterize how $H_p(z)$ for unitary filter banks with symmetries can be constrained to give rise to wavelet tight frames (WTFs). First consider the case of PS symmetry in which case $H_p(z)$ is parameterized in (8.83). We have a WTF iff

$$\text{first row of } H_p(z)|_{z=1} = \left[\begin{array}{ccc} 1/\sqrt{M} & \cdots & 1/\sqrt{M} \end{array}\right]. \tag{8.89}$$

In (8.83), since P permutes the columns, the first row is unaffected. Hence (8.89) is equivalent to the first rows of both $W_0'(z)$ and $W_1'(z)$ when $z = 1$ is given by

$$\left[\begin{array}{ccc} \sqrt{2/M} & \cdots & \sqrt{2/M} \end{array}\right].$$

This is precisely the condition to be satisfied by a WTF of multiplicity $M/2$. Therefore both $W_0'(z)$ and $W_1'(z)$ give rise to multiplicity $M/2$ compactly supported WTFs. If the McMillan degree of $W_0'(z)$ and $W_1'(z)$ are L_0 and L_1 respectively, then they are parameterized respectively by $\binom{M/2-1}{2} + (M/2-1)L_0$ and $\binom{M/2-1}{2} + (M/2-1)L_1$ parameters. In summary, a WTF with PS symmetry can be *explicitly* parameterized by $2\binom{M/2-1}{2} + (M/2-1)(L_0+L_1)$ parameters. Both L_0 and L_1 are greater than or equal to K.

PS symmetry does not reflect itself as any simple property of the scaling function $\psi_0(t)$ and wavelets $\psi_i(t)$, $i \in \{1, \ldots, M-1\}$ of the WTF. However, from design and implementation points of view, PS symmetry is useful (because of the reduction in the number of parameters).

Next consider PCS symmetry. From (8.84) one sees that (8.89) is equivalent to the first rows of the matrices A and B defined by

$$\left[\begin{array}{cc} A & B \\ -B & A \end{array}\right] = \left\{ \prod_{i=K}^{0} \left[\begin{array}{cc} A_i & B_i \\ -B_i & A_i \end{array}\right] \right\}$$

are of the form $\left[\begin{array}{ccc} 1/\sqrt{M} & \cdots & 1/\sqrt{M} \end{array}\right]$. Here we only have an implicit parameterization of WTFs, unlike the case of PS symmetry. As in the case of PS symmetry, there is no simple symmetry relationships between the wavelets.

Now consider the case of linear phase. In this case, it can be seen [SVN93] that the wavelets are also linear phase. If we define

$$\left[\begin{array}{cc} A & B \\ B & A \end{array}\right] = \left\{ \prod_{i=K}^{0} \left[\begin{array}{cc} A_i & B_i \\ B_i & A_i \end{array}\right] \right\},$$

then it can be verified that one of the rows of the matrix $A + B$ has to be of the form $\left[\begin{array}{ccc} \sqrt{2/M} & \cdots & \sqrt{2/M} \end{array}\right]$. This is an implicit parameterization of the WTF.

Finally consider the case of linear phase with PCS symmetry. In this case, also the wavelets are linear-phase. From (8.86) it can be verified that we have a WTF iff the first row of $W_0'(z)$ for $z = 1$, evaluates to the vector $\left[\ \sqrt{2/M}\quad \ldots \quad \sqrt{2/M}\ \right]$. Equivalently, $W_0'(z)$ gives rise to a multiplicity $M/2$ WTF. In this case, the WTF is parameterized by precisely $\binom{M/2 - 1}{2} + (M/2 - 1)L$ parameters where $L \geq K$ is the McMillan degree of $W_0'(z)$.

8.9   Linear-Phase Modulated Filter Banks

The modulated filter banks we described

1. have filters with nonoverlapping ideal frequency responses as shown in Figure 8.2.

2. are associated with DCT III/IV (or equivalently DST III/IV) in their implementation

3. and do not allow for linear phase filters (even though the prototypes could be linear phase).

In trying to overcome 3, Lin and Vaidyanathan introduced a new class of linear-phase modulated filter banks by giving up 1 and 2 [LV95]. We now introduce a generalization of their results from a viewpoint that unifies the theory of modulated filter banks as seen earlier with the new class of modulated filter banks we introduce here. For a more detailed exposition of this viewpoint see [Gop96b].

The new class of modulated filter banks have $2M$ analysis filters, but M bands—each band being shared by two overlapping filters. The M bands are the M-point Discrete Fourier Transform bands as shown in Figure 8.5.

$$k_n = \begin{cases} \frac{1}{\sqrt{2}} & n \in \{0, M\} \\ 1 & \text{otherwise.} \end{cases} \qquad (8.90)$$

Two broad classes of MFBs (that together are associated with all four DCT/DSTs [RY90]) can be defined.

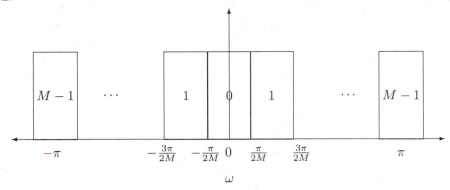

Figure 8.5. Ideal Frequency Responses in an M-band DFT-type Filter Bank

DCT/DST I/II based $2M$ Channel Filter Bank

$$h_i(n) = k_i h(n) \cos\left(\frac{\pi}{M}i(n - \frac{\alpha}{2})\right), \quad i \in S_1 \tag{8.91a}$$

$$h_{M+i}(n) = k_i h(n - M) \sin\left(\frac{\pi}{M}i(n - \frac{\alpha}{2})\right), \quad i \in S_2 \tag{8.91b}$$

$$g_i(n) = k_i g(n) \cos\left(\frac{\pi}{M}i(n + \frac{\alpha}{2})\right), \quad i \in S_1 \tag{8.91c}$$

$$g_{M+i}(n) = -k_i g(n + M) \sin\left(\frac{\pi}{M}i(n + \frac{\alpha}{2})\right), \quad i \in S_2 \tag{8.91d}$$

The sets S_1 and S_2 are defined depending on the parity of α as shown in Table 8.1. When α is even (i.e., Type 1 with odd M or Type 2 with even M), the MFB is associated with DCT I and DST I. When α is odd (i.e., Type 1 with even M or Type 2 with odd M), the MFB is associated with DCT II and DST II. The linear-phase MFBs introduced in [LV95] correspond to the special case where $h(n) = g(n)$ and α is even. The other cases above and their corresponding PR results are new.

	S_1	S_2
α even, DCT/DST I	$\mathcal{R}(M) \cup \{M\}$	$\mathcal{R}(M) \backslash \{0\}$
α odd, DCT/DST II	$\mathcal{R}(M)$	$\mathcal{R}(M) \backslash \{0\} \cup \{M\}$

Table 8.1. Class A MFB: The Filter Index Sets S_1 and S_2

The PR constraints on the prototype filters h and g (for both versions of the filter banks above) *are exactly the same as that for the modulated filter bank studied earlier* [Gop96b]. When the prototype filters are linear phase, these filter banks are also linear phase. An interesting consequence is that if one designs an M-channel Class B modulated filter bank, the prototype filter can also be used for a Class A $2M$ channel filter bank.

8.10 Linear Phase Modulated Wavelet Tight Frames

Under what conditions do linear phase modulated filter banks give rise to wavelet tight frames (WTFs)? To answer this question, it is convenient to use a slightly different lattice parameterization than that used for Class B modulated filter banks. A seemingly surprising result is that some Class A unitary MFBs *cannot* be associated with WTFs. More precisely, a Class A MFB is associated with a WTF only if it is Type 1.

$$\begin{bmatrix} P_{l,0}(z) & P_{l,1}^R(z) \\ -P_{l,1}(z) & P_{l,0}^R(z) \end{bmatrix} = \sqrt{\frac{2}{M}} \left\{ \prod_{k=k_l-1}^{1} T'(\theta_{l,k}) \right\} T'(\theta_{l,0}) \tag{8.92}$$

where

$$T'(\theta) = \begin{bmatrix} \cos\theta_{l,k} & z^{-1}\sin\theta_{l,k} \\ -\sin\theta_{l,k} & z^{-1}\cos\theta_{l,k} \end{bmatrix}.$$

With this parameterization we define Θ_l as follows:

$$\begin{bmatrix} P_{l,0}(1) & P_{l,1}^R(1) \\ -P_{l,1}(1) & P_{l,0}^R(1) \end{bmatrix} = \begin{bmatrix} \cos(\Theta_l) & \sin(\Theta_l) \\ -\sin(\Theta_l) & \cos(\Theta_l) \end{bmatrix}, \tag{8.93}$$

where in the FIR case $\Theta_l = \sum_{k=0}^{k_l-1} \theta_{l,k}$ as before. Type 1 Class A MFBs give rise to a WTF iff $\Theta_l = \frac{\pi}{4}$ for all $l \in \mathcal{R}(J)$.

Theorem 55 (Modulated Wavelet Tight Frames Theorem) *A class A MFB of Type 1 gives rise to a WTF iff* $\Theta_l = \frac{\pi}{4}$. *A class B MFB (Type 1 or Type 2) gives rise to a WTF iff* $\Theta_l = \frac{\pi}{4} + \frac{\pi}{2M}(\frac{\alpha}{2} - l)$.

8.11 Time-Varying Filter Bank Trees

Filter banks can be applied in cascade to give a rich variety of decompositions. By appropriately designing the filters one can obtain an arbitrary resolution in frequency. This makes them particularly useful in the analysis of stationary signals with phenomena residing in various frequency bands or scales. However, for the analysis of nonstationary or piecewise stationary signals such filter bank trees do not suffice. With this in mind we turn to filter banks for finite-length signals.

If we had filter bank theory for finite-length signals, then, piecewise stationary signals can be handled by considering each homogeneous segment separately. Several approaches to filter banks for finite-length signals exist and we follow the one in [GB95a]. If we consider the filter bank tree as a *machine* that takes an input sample(s) and produces an output sample(s) every instant then one can consider changing machine every instant (i.e., changing the filter coefficients every instant). Alternatively, we could use a fixed machine for several instants and then *switch* to another filter bank tree. The former approach is investigated in [PV96]. We follow the latter approach, which, besides leveraging upon powerful methods to design the constituent filter bank trees switched between, also leads to a theory of wavelet bases on an interval [HKRV93, GB95a].

Let $H_p(z)$, the polyphase component matrix of the analysis filter bank of a unitary filter bank be of the form (see (8.27))

$$H_p(z) = \sum_{k=0}^{K-1} h_p(k) z^{-k}. \tag{8.94}$$

It is convenient to introduce the sequence $\check{x} = [\cdots, x(0), x(-1), \cdots, x(-M+1), x(M), x(M-1), \cdots]$ obtained from $\mathbf{x}$ by a permutation. Then,

$$\mathbf{d} = \check{\mathbf{H}}\check{\mathbf{x}} \stackrel{def}{=} \begin{bmatrix} \ddots & \vdots & \vdots & \vdots & \vdots & \vdots & \\ \cdots & h_p(K-1) & h_p(K-2) & \cdots & h_p(0) & 0 & \ddots \\ & 0 & h_p(K-1) & \cdots & h_p(1) & h_p(0) & \\ & \vdots & \vdots & \vdots & \vdots & \vdots & \end{bmatrix} \check{\mathbf{x}}, \tag{8.95}$$

and $\mathbf{H}$ is unitary iff $\check{\mathbf{H}}$ is unitary. (8.28) induces a factorization of $\check{\mathbf{H}}$ (and hence $\mathbf{H}$). If $\mathbf{V}_0 = diag(V_0)$ and

$$
\mathbf{V}_i = \begin{bmatrix} & \vdots & \vdots & \vdots & \\ \ddots & P_i & I - P_i & 0 & \ddots \\ & 0 & P_i & I - P_i & \\ & \vdots & \vdots & \vdots & \end{bmatrix}, \quad \text{for } i \in \{1, \ldots, K-1\}, \tag{8.96}
$$

$$
\check{\mathbf{H}} = \prod_{i=K-1}^{0} \mathbf{V}_i. \tag{8.97}
$$

The factors $\mathbf{V}_i$, with appropriate modifications, will be used as fundamental building blocks for filter banks for finite-length signals.

Now consider a finite input signal $\mathbf{x} = [x(0), x(1), \cdots, x(L-1)]$, where L is a multiple of M and let $\check{\mathbf{x}} = [x(M-1), \ldots, x(0), x(M), \cdots, x(L-1), \cdots, x(L-M)]$. Then, the *finite* vector $\mathbf{d}$ (the output signal) is given by

$$
\mathbf{d} = \check{\mathbf{H}}\check{\mathbf{x}} \stackrel{\text{def}}{=} \begin{bmatrix} h_p(K-1) & \ldots & h_p(0) & 0 & \ldots & \ldots & \ldots & 0 \\ 0 & h_p(K-1) & \ldots & h_p(0) & \vdots & \vdots & \vdots & \vdots \\ \vdots & \vdots & \ddots & \ddots & \ddots & \ddots & \vdots & \vdots \\ \vdots & \ldots & \ldots & 0 & h_p(K-1) & \ldots & h_p(0) & 0 \\ 0 & \ldots & \ldots & \ldots & 0 & h_p(K-1) & \ldots & h_p(0) \end{bmatrix} \check{\mathbf{x}}. \tag{8.98}
$$

$\check{\mathbf{H}}$ is an $(L - N + M) \times L$ matrix, where $N = MK$ is the length of the filters. Now since the rows of $\check{\mathbf{H}}$ are mutually orthonormal (i.e., has rank L), one has to append $N - M = M(K-1)$ rows from the orthogonal complement of $\check{\mathbf{H}}$ to make the map from $\mathbf{x}$ to an augmented $\mathbf{d}$ unitary. To get a complete description of these rows, we turn to the factorization of $H_p(z)$. Define the $L \times L$ matrix $\mathbf{V}_0 = diag(V_0)$ and for $i \in \{1, \ldots, K-1\}$ the $(L - Mi) \times (L - Mi + M)$ matrices

$$
\mathbf{V}_i = \begin{bmatrix} P_i & I - P_i & 0 & \ldots & \ldots & 0 \\ 0 & P_i & I - P_i & \ldots & \ldots & \vdots \\ \vdots & \ddots & \ddots & \ddots & 0 & \vdots \\ \vdots & \ddots & & 0 & P_i & I - P_i & 0 \\ 0 & \ldots & & 0 & P_i & I - P_i \end{bmatrix}. \tag{8.99}
$$

then $\check{\mathbf{H}}$ is readily verified by induction to be $\prod_{i=K-1}^{0} \mathbf{V}_i$. Since each of the factors (except $\mathbf{V}_0$) has M more columns than rows, they can be made unitary by appending appropriate rows. Indeed, $\begin{bmatrix} B_i \\ \mathbf{V}_i \\ C_i \end{bmatrix}$ is unitary where, $B_j = \begin{bmatrix} \Upsilon_j^T(I - P_j) & 0 & \ldots & 0 \end{bmatrix}$, and $C_j = \begin{bmatrix} 0 & \ldots & 0 & \Xi_j^T P_j. \end{bmatrix}$.

Here Ξ is the $\delta_i \times M$ left unitary matrix that spans the range of the P_i; i.e., $P_i = \Xi\Xi^T$, and Υ is the $(M - \delta_i) \times M$ left unitary matrix that spans the range of the $I - P_i$; i.e., $I - P_i = \Upsilon\Upsilon^T$. Clearly $[\Upsilon_i\Xi_i]$ is unitary. Moreover, if we define $\mathbf{T}_0 = \mathbf{V}_0$ and for $i \in \{1, \dots, K-1\}$,

$$\mathbf{T}_i = \begin{bmatrix} I_{(M-\delta_i)(i-1)} & 0 & 0 \\ 0 & B_i & 0 \\ 0 & \mathbf{V}_i & 0 \\ 0 & C_i & 0 \\ 0 & 0 & I_{\delta_i(i-1)} \end{bmatrix}$$

then each of the factors $\mathbf{T}_i$ is a square unitary matrix of size $L - N + M$ and

$$\prod_{i=K-1}^{0} \mathbf{T}_i = \begin{bmatrix} B_1\mathbf{V}_0 \\ \vdots \\ \check{\mathbf{H}} \\ \vdots \\ C_1\mathbf{V}_0 \end{bmatrix}, \tag{8.100}$$

is the unitary matrix that acts on the data. The corresponding unitary matrix that acts on $\mathbf{x}$ (rather than $\check{\mathbf{x}}$) is of the form $\begin{bmatrix} \mathbf{U} \\ \mathbf{H} \\ \mathbf{W} \end{bmatrix}$, where $\mathbf{U}$ has $MK - M - \Delta$ rows of entry filters in $(K-1)$ sets given by (8.101), while $\mathbf{W}$ has Δ rows of exit filters in $(K-1)$ given by (8.102):

$$\Upsilon_j(I - P_j)\begin{bmatrix} h_p^j(j-1)J & h_p^j(j-2)J & \dots & h_p^j(0)J, \end{bmatrix} \tag{8.101}$$

$$\Xi_j P_j \begin{bmatrix} h_p^j(j-1)J & h_p^j(j-2)J & \dots & h_p^j(0)J \end{bmatrix}, \tag{8.102}$$

where J is the exchange matrix (i.e., permutation matrix of ones along the anti-diagonal) and

$$H_p^j(z) = \prod_{i=j-1}^{1} [I - P_i + z^{-1}P_i]V_0 \overset{\text{def}}{=} \sum_{i=0}^{j-1} h_p^j(i)z^{-i}. \tag{8.103}$$

The rows of $\mathbf{U}$ and $\mathbf{W}$ form the entry and exit filters respectively. Clearly they are nonunique. The input/output behavior is captured in

$$\begin{bmatrix} \mathbf{u} \\ \mathbf{d} \\ \mathbf{w} \end{bmatrix} = \begin{bmatrix} \mathbf{U} \\ \mathbf{H} \\ \mathbf{W} \end{bmatrix} \mathbf{x}. \tag{8.104}$$

For example, in the four-coefficient Daubechies' filters in [Dau88a] case, there is one entry filter and exit filter.

$$\begin{bmatrix} 0.8660 & 0.5000 & 0 & 0 \\ -0.1294 & 0.2241 & 0.8365 & 0.4830 \\ -0.4830 & 0.8365 & -0.2241 & -0.1294 \\ 0 & 0 & -0.5000 & 0.8660 \end{bmatrix}. \tag{8.105}$$

If the input signal is right-sided (i.e., supported in $\{0, 1, \dots\}$), then the corresponding filter bank would only have entry filters. If the filter bank is for left-sided signals one would only have exit filters. Based on the above, we can consider switching between filter banks (that operate on infinite extent input signals). Consider switching from a one-channel to an M channel filter bank. Until instant $n = -1$, the input is the same as the output. At $n = 0$, one switches into an M-channel filter bank as quickly as possible. The transition is accomplished by the entry filters (hence the name entry) of the M-channel filter bank. The input/output of this time-varying filter bank is

$$\mathbf{d} = \left[\begin{array}{c|c} \mathbf{I} & 0 \\ \hline 0 & \mathbf{U} \\ 0 & \mathbf{H} \end{array} \right] \mathbf{x}. \qquad (8.106)$$

Next consider switching from an M-channel filter bank to a one-channel filter bank. Until $n = -1$, the M-channel filter bank is operational. From $n = 0$ onwards the inputs leaks to the output. In this case, there are exit filters corresponding to flushing the states in the first filter bank implementation at $n = 0$.

$$\mathbf{d} = \left[\begin{array}{c|c} \mathbf{H} & 0 \\ \mathbf{W} & 0 \\ \hline 0 & \mathbf{I} \end{array} \right] \mathbf{x}. \qquad (8.107)$$

Finally, switching from an M_1-band filter bank to an M_2-band filter bank can be accomplished as follows:

$$\mathbf{d} = \left[\begin{array}{c|c} \mathbf{H_1} & 0 \\ \mathbf{W_1} & 0 \\ \hline 0 & \mathbf{U_2} \\ 0 & \mathbf{H_2} \end{array} \right] \mathbf{x}. \qquad (8.108)$$

The transition region is given by the exit filters of the first filter bank and the entry filters of the second. Clearly the transition filters are abrupt (they do not overlap). One can obtain overlapping transition filters as follows: replace them by any orthogonal basis for the row space of the matrix $\left[\begin{array}{c|c} \mathbf{W_1} & 0 \\ 0 & \mathbf{U_2} \end{array} \right]$. For example, consider switching between two-channel filter banks with length-4 and length-6 Daubechies' filters. In this case, there is one exit filter ($\mathbf{W_1}$) and two entry filters ($\mathbf{U_2}$).

Growing a Filter Bank Tree

Consider growing a filter bank tree at $n = 0$ by replacing a certain output channel in the tree (point of tree growth) by an M channel filter bank. This is equivalent to switching from a one-channel to an M-channel filter bank at the point of tree growth. The transition filters associated with this change are related to the entry filters of the M-channel filter bank. In fact, every transition filter is the net effect of an entry filter at the point of tree growth seen from the perspective of the input rather than the output point at which the tree is grown. Let the mapping from the input to the output "growth" channel be as shown in Figure 8.6. The transition filters are given by the system in Figure 8.7, which is driven by the entry filters of the newly added filter bank. Every transition filter is obtained by running the corresponding time-reversed entry filter through the synthesis bank of the corresponding branch of the extant tree.

Pruning a Filter Bank Tree

In the more general case of tree pruning, if the map from the input to the point of pruning is given as in Figure 8.6, then the transition filters are given by Figure 8.8.

Figure 8.6. A Branch of an Existing Tree

Wavelet Bases for the Interval

By taking the effective input/output map of an arbitrary unitary time-varying filter bank tree, one readily obtains time-varying discrete-time wavelet packet bases. Clearly we have such bases for one-sided and finite signals also. These bases are orthonormal because they are built from unitary building blocks. We now describe the construction of continuous-time time-varying wavelet bases. What follows is the most economical (in terms of number of entry/exit functions) continuous-time time-varying wavelet bases.

Figure 8.7. Transition Filter For Tree Growth

Wavelet Bases for $L^2([0, \infty))$

Recall that an M channel unitary filter bank (with synthesis filters $\{h_i\}$) such that $\sum_n h_0(n) = \sqrt{M}$ gives rise to an M-band wavelet tight frame for $L^2(\mathbb{R})$. If

$$W_{i,j} = Span\{\psi_{i,j,k}\} \overset{\text{def}}{=} \left\{M^{j/2}\psi_i(M^j t - k)\right\} \quad \text{for } k \in \mathbf{Z}, \tag{8.109}$$

then $W_{0,j}$ forms a multiresolution analysis of $L^2(\mathbb{R})$ with

$$W_{0,j} = W_{0,j-1} \oplus W_{1,j-1} \ldots \oplus W_{M-1,j-1} \quad \forall j \in \mathbf{Z}. \tag{8.110}$$

In [Dau92], Daubechies outlines an approach due to Meyer to construct a wavelet basis for $L^2([0, \infty))$. One projects $W_{0,j}$ onto $W_{0,j}^{half}$ which is the space spanned by the restrictions of $\psi_{0,j,k}(t)$ to $t > 0$. We give a different construction based on the following idea. For $k \in \mathbb{N}$, support of $\psi_{i,j,k}(t)$ is in $[0, \infty)$. With this restriction (in (8.109)) define the spaces $W_{i,j}^+$. As $j \to \infty$ (since $W_{0,j} \to L^2(\mathbb{R})$) $W_{0,j}^+ \to L^2([0, \infty))$. Hence it suffices to have a multiresolution

Figure 8.8. Transition Filter For Pruning

analysis for $W_{0,j}^+$ to get a wavelet basis for $L^2([0,\infty))$. (8.110) does not hold with $W_{i,j}$ replaced by $W_{i,j}^+$ because $W_{0,j}^+$ is *bigger* than the direct sum of the constituents at the next coarser scale. Let $\mathcal{U}_{j-1}$ be this *difference* space:

$$W_{0,j}^+ = W_{0,j-1}^+ \oplus W_{1,j-1}^+ \cdots \oplus W_{M-1,j-1}^+ \oplus \mathcal{U}_{j-1} \qquad (8.111)$$

If we can find an orthonormal basis for $\mathcal{U}_j$, then we have a multiresolution analysis for $L^2([0,\infty))$.

We proceed as follows. Construct entry filters (for the analysis filters) of the filter bank with synthesis filters $\{h_i\}$. Time-reverse them to obtain entry filters (for the synthesis filters). If Δ is the McMillan degree of the synthesis bank, there are Δ entry filters. Let $u_i(n)$ denote the ith synthesis entry filters. Define the *entry* functions

$$\mu_l(t) = \sqrt{M} \sum_{k=0}^{L_l-1} u_l(k)\psi_0(Mt - k), \qquad l \in \{0,\dots,\Delta-1\}. \qquad (8.112)$$

$\mu_i(t)$ is *compactly supported* in $[0, \frac{L_l-1}{M} + \frac{N-1}{M-1}]$. Let $\mathcal{U}_j = Span\{\mu_{l,j}\} \stackrel{\text{def}}{=} Span\{M^{j/2}\mu_i(M^j t)\}$. By considering one stage of the analysis and synthesis stages of this PR filter bank on right sided signals), it readily follows that (8.111) holds. Therefore

$$\{\psi_{i,j,k} \mid i \in \{1,\dots,M-1\}, j \in \mathbf{Z}, k \in \mathbb{N}\} \cup \{\mu_{i,j} \mid i \in \{0,\dots,\Delta-1\}, j \in \mathbf{Z}\}$$

forms a wavelet tight frame for $L^2([0,\infty))$. If one started with an ON basis for $L^2(\mathbb{R})$, the newly constructed basis is an ON basis for $L^2([0,\infty))$. Indeed if $\{\psi_0(t-k)\}$ is an orthonormal system

$$\int_{t \geq 0} \mu_l(t)\psi_i(t - n)\,dt = \sum_k u_l(k)h_i(Ml + k) = 0, \qquad (8.113)$$

and

$$\int_{t \geq 0} \mu_l(t)\mu_m(t)\,dt = \sum_k u_l(k)u_m(k) = 0. \qquad (8.114)$$

The dimension of $\mathcal{U}_j$ is precisely the McMillan degree of the polyphase component matrix of the scaling and wavelet filters *considered as the filters of the synthesis bank*. There are precisely as many entry functions as there are entry filters, and supports of these functions are explicitly given in terms of the lengths of the corresponding entry filters. Figure 8.9 shows the scaling function, wavelet, their integer translates and the single entry function corresponding to Daubechies four coefficient wavelets. In this case, $u_0 = \{-\sqrt{3}/2, 1/2\}$.

Wavelet Bases for $L^2((-\infty, 0])$

One could start with a wavelet basis for $L^2([0,\infty))$ and reflect all the functions about $t = 0$. This is equivalent to swapping the analysis and synthesis filters of the filter bank. We give an independent development. We start with a WTF for $L^2(\mathbb{R})$ with functions

$$\psi_i(t) = \sqrt{M} \sum_{k=0}^{N-1} h_i(k)\psi_0(Mt + k), \qquad (8.115)$$

supported in $[-\frac{N-1}{M-1}, 0]$. Scaling and wavelet filters constitute the analysis bank in this case. Let Δ be the McMillan degree of the analysis bank and let $\{w_i\}$ be the (analysis) exit filters. Define the *exit* functions

$$\nu_l(t) = \sqrt{M} \sum_{k=0}^{L_l - 1} w_l(k)\psi_0(Mt + k), \qquad l \in \{0, \ldots, \Delta - 1\}. \tag{8.116}$$

$\mathcal{W}_j = Span\left\{M^{j/2}\nu_i(M^j t)\right\}$, and $W_{i,j}^- = Span\{\psi_{i,j,k}\} \overset{\text{def}}{=} \left\{M^{j/2}\psi_i(M^j t + k)\right\}$ for $k \in \mathbb{N}$. Then as $j \to \infty$, $W_{0,j}^- \to L^2((-\infty, 0])$ and

$$\{\psi_{i,j,k} \mid i \in \{1, \ldots, M-1\}, j \in \mathbf{Z}, k \in \mathbb{N}\} \cup \{\nu_{i,j} \mid i \in \{0, \ldots, \Delta - 1\}, j \in \mathbf{Z}\}$$

forms a WTF for $L^2((-\infty, 0])$. Orthonormality of this basis is equivalent to the orthonormality of its parent basis on the line. An example with one exit function (corresponding to $M = 3$, $N = 6$) Type 1 modulated WTF obtained earlier is given in Figure 8.10.

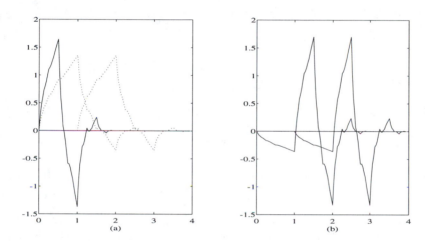

Figure 8.9. (a) Entry function $\mu_0(t)$, $\psi_0(t)$ and $\psi_0(t-1)$ (b) Wavelet $\psi_1(t)$ and $\psi_1(t-1)$

Segmented Time-Varying Wavelet Packet Bases

Using the ideas above, one can construct wavelet bases for the interval and consequently segmented wavelet bases for $L^2(\mathbb{R})$. One can write $\mathbb{R}$ as a disjoint union of intervals and use a different wavelet basis in each interval. Each interval is will be spanned by a combination of scaling functions, wavelets, and corresponding entry and exit functions. For instance Figure 8.10 and Figure 8.9 together correspond to a wavelet basis for $L^2(\mathbb{R})$, where a 3-band wavelet basis with length-6 filters is used for $t < 0$ and a 2-band wavelet basis with length-4 filters is used for $t > 0$. Certainly a degree of overlap between the exit functions on the left of a transition and entry functions on the right of the transition can be obtained by merely changing coordinates in the finite dimensional space corresponding to these functions. Extension of these ideas to obtain segmented *wavelet packet* bases is also immediate.

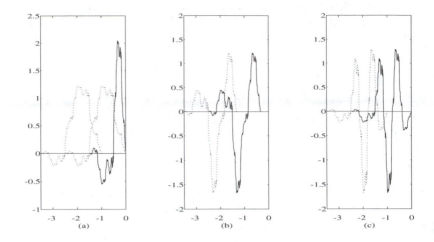

Figure 8.10. (a) Exit function $\nu_0(t)$, $\psi_0(t)$ and $\psi_0(t+1)$ (b) Wavelet $\psi_1(t)$ and $\psi_1(t+1)$ (c) Wavelet $\psi_2(t)$ and $\psi_2(t+1)$

8.12   Filter Banks and Wavelets—Summary

Filter banks are structures that allow a signal to be decomposed into subsignals—typically at a lower sampling rate. If the original signal can be reconstituted from the subsignals, the filter bank is said to be a perfect reconstruction (PR) filter bank. For PR, the analysis and synthesis filters have to satisfy a set of bilinear constraints. These constraints can be viewed from three perspectives, viz., the direct, matrix, and polyphase formulations. In PR filter bank design one chooses filters that maximize a "goodness" criterion and satisfy the PR constraints.

Unitary filter banks are an important class of PR filter banks—they give orthogonal decompositions of signals. For unitary filter banks, the PR constraints are quadratic in the analysis filters since the synthesis filters are index-reverses of the analysis filters. All FIR unitary filter banks can be explicitly parameterized. This leads to easy design (unconstrained optimization) and efficient implementation. Sometimes one can impose structural constraints compatible with the goodness criterion. For example, modulated filter banks require that the analysis and synthesis filters are modulates of single analysis and synthesis prototype filter respectively. Unitary modulated filter banks exist and can be explicitly parameterized. This allows one to design and implement filter banks with hundreds of channels easily and efficiently. Other structural constraints on the filters (e.g., linear phase filters) can be imposed and lead to parameterizations of associated unitary filter banks. Cascades of filter banks (used in a tree structure) can be used to recursively decompose signals.

Every unitary FIR filter bank with an additional linear constraint on the lowpass filter is associated with a wavelet tight frame. The lowpass filter is associated with the scaling function, and the remaining filters are each associated with wavelets. The coefficients of the wavelet expansion of a signal can be computed using a tree-structure where the filter bank is applied recursively along the lowpass filter channel. The parameterization of unitary filter banks, with a minor modification, gives a parameterization of all compactly supported wavelet tight frames.

In general, wavelets associated with a unitary filter bank are irregular (i.e., not smooth). By imposing further linear constraints (regularity constraints) on the lowpass filter, one obtains smooth wavelet bases. Structured filter banks give rise to associated structured wavelet bases; modulated filter banks are associated with modulated wavelet bases and linear phase filter banks are associated with linear-phase wavelet bases. Filter banks cascade—where all the channels are recursively decomposed, they are associated with wavelet packet bases.

From a time-frequency analysis point of view filter banks trees can be used to give arbitrary resolutions of the frequency. In order to obtain arbitrary temporal resolutions one has to use local bases or switch between filter bank trees at points in time. Techniques for time-varying filter banks can be used to generate segmented wavelet bases (i.e., a different wavelet bases for disjoint segments of the time axis). Finally, just as unitary filter banks are associated with wavelet tight frames, general PR filter banks, with a few additional constraints, are associated with wavelet frames (or biorthogonal bases).

Chapter 9

Calculation of the Discrete Wavelet Transform

Although when using the wavelet expansion as a tool in an abstract mathematical analysis, the infinite sum and the continuous description of $t \in \mathbf{R}$ are appropriate, as a practical signal processing or numerical analysis tool, the function or signal $f(t)$ in (9.1) is available only in terms of its samples, perhaps with additional information such as its being band-limited. In this chapter, we examine the practical problem of numerically calculating the discrete wavelet transform.

9.1   Finite Wavelet Expansions and Transforms

The wavelet expansion of a signal $f(t)$ as first formulated in (1.9) is repeated here by

$$f(t) = \sum_{k=-\infty}^{\infty} \sum_{j=-\infty}^{\infty} \langle f, \psi_{j,k} \rangle \psi_{j,k}(t) \tag{9.1}$$

where the $\{\psi_{j,k}(t)\}$ form a basis or tight frame for the signal space of interest (e.g., L_2). At first glance, this infinite series expansion seems to have the same practical problems in calculation that an infinite Fourier series or the Shannon sampling formula has. In a practical situation, this wavelet expansion, where the coefficients are called the discrete wavelet transform (DWT), is often more easily calculated. Both the time summation over the index k and the scale summation over the index j can be made finite with little or no error.

The Shannon sampling expansion [Pap77, Mar91] of a signal with infinite support in terms of $\mathrm{sinc}(t) = \frac{\sin(t)}{t}$ expansion functions

$$f(t) = \sum_{n=-\infty}^{\infty} f(Tn) \, \mathrm{sinc}(\frac{\pi}{T}t - \pi n) \tag{9.2}$$

requires an infinite sum to evaluate $f(t)$ at one point because the sinc basis functions have infinite support. This is not necessarily true for a wavelet expansion where it is possible for the wavelet basis functions to have finite support and, therefore, only require a finite summation over k in (9.1) to evaluate $f(t)$ at any point.

The lower limit on scale j in (9.1) can be made finite by adding the scaling function to the basis set as was done in (2.28). By using the scaling function, the expansion in (9.1) becomes

$$f(t) = \sum_{k=-\infty}^{\infty} \langle f, \varphi_{J_0,k} \rangle \varphi_{J_0,k}(t) + \sum_{k=-\infty}^{\infty} \sum_{j=J_0}^{\infty} \langle f, \psi_{j,k} \rangle \psi_{j,k}(t). \tag{9.3}$$

where $j = J_0$ is the coarsest scale that is separately represented. The level of resolution or coarseness to start the expansion with is arbitrary, as was shown in Chapter 2 in (2.19), (2.20), and (2.21). The space spanned by the scaling function contains all the spaces spanned by the lower resolution wavelets from $j = -\infty$ up to the arbitrary starting point $j = J_0$. This means $\mathcal{V}_{J_0} = \mathcal{W}_{-\infty} \oplus \cdots \oplus \mathcal{W}_{J_0-1}$. In a practical case, this would be the scale where separating detail becomes important. For a signal with finite support (or one with very concentrated energy), the scaling function might be chosen so that the support of the scaling function and the size of the features of interest in the signal being analyzed were approximately the same.

This choice is similar to the choice of period for the basis sinusoids in a Fourier series expansion. If the period of the basis functions is chosen much larger than the signal, much of the transform is used to describe the zero extensions of the signal or the edge effects.

The choice of a finite upper limit for the scale j in (9.1) is more complicated and usually involves some approximation. Indeed, for samples of $f(t)$ to be an accurate description of the signal, the signal should be essentially bandlimited and the samples taken at least at the Nyquist rate (two times the highest frequency in the signal's Fourier transform).

The question of how one can calculate the Fourier series coefficients of a continuous signal from the discrete Fourier transform of samples of the signal is similar to asking how one calculates the discrete wavelet transform from samples of the signal. And the answer is similar. The samples must be "dense" enough. For the Fourier series, if a frequency can be found above which there is very little energy in the signal (above which the Fourier coefficients are very small), that determines the Nyquist frequency and the necessary sampling rate. For the wavelet expansion, a scale must be found above which there is negligible detail or energy. If this scale is $j = J_1$, the signal can be written

$$f(t) \approx \sum_{k=-\infty}^{\infty} \langle f, \varphi_{J_1,k} \rangle \varphi_{J_1,k}(t) \tag{9.4}$$

or, in terms of wavelets, (9.3) becomes

$$f(t) \approx \sum_{k=-\infty}^{\infty} \langle f, \varphi_{J_0,k} \rangle \varphi_{J_0,k}(t) + \sum_{k=-\infty}^{\infty} \sum_{j=J_0}^{J_1-1} \langle f, \psi_{j,k} \rangle \psi_{j,k}(t). \tag{9.5}$$

This assumes that approximately $f \in \mathcal{V}_{J_1}$ or equivalently, $\|f - P_{J_1} f\| \approx 0$, where P_{J_1} denotes the orthogonal projection of f onto $\mathcal{V}_{J_1}$.

Given $f(t) \in \mathcal{V}_{J_1}$ so that the expansion in (9.5) is exact, one computes the DWT coefficients in two steps.

1. Projection onto finest scale: Compute $\langle f, \varphi_{J_1,k} \rangle$

2. Analysis: Compute $\langle f, \psi_{j,k} \rangle$, $j \in \{J_0, \ldots, J_1 - 1\}$ and $\langle f, \varphi_{J_0,k} \rangle$.

For J_1 large enough, $\varphi_{J_1,k}(t)$ can be approximated by a Dirac impulse at its center of mass since $\int \varphi(t)\,dt = 1$. For large j this gives

$$2^j \int f(t)\,\varphi(2^j t)\,dt \approx \int f(t)\,\delta(t - 2^{-j}m_0)\,dt = f(t - 2^{-j}m_0) \tag{9.6}$$

where $m_0 = \int t\,\varphi(t)\,dt$ is the first moment of $\varphi(t)$. Therefore the scaling function coefficients at the $j = J_1$ scale are

$$c_{J_1}(k) = \langle f, \varphi_{J_1,k}\rangle = 2^{J_1/2}\int f(t)\,\varphi(2^{J_1}t - k)\,dt = 2^{J_1/2}\int f(t + 2^{-J_1}k)\,\varphi(2^{J_1}t)\,dt \tag{9.7}$$

which are approximately

$$c_{J_1}(k) = \langle f, \varphi_{J_1,k}\rangle \approx 2^{-J_1/2} f(2^{-J_1}(m_0 + k)). \tag{9.8}$$

For all 2-regular wavelets (i.e., wavelets with two vanishing moments, regular wavelets other than the Haar wavelets—even in the M-band case where one replaces 2 by M in the above equations, $m_0 = 0$), one can show that the samples of the functions themselves form a third-order approximation to the scaling function coefficients of the signal [GB92b]. That is, if $f(t)$ is a quadratic polynomial, then

$$c_{J_1}(k) = \langle f, \varphi_{J_1,k}\rangle = 2^{-J_1/2} f(2^{-J_1}(m_0 + k)) \approx 2^{-J_1/2} f(2^{-J_1}k). \tag{9.9}$$

Thus, in practice, the finest scale J_1 is determined by the sampling rate. By rescaling the function and amplifying it appropriately, one can assume the samples of $f(t)$ are equal to the scaling function coefficients. These approximations are made better by setting some of the scaling function moments to zero as in the coiflets. These are discussed in Section 6.8.

Finally there is one other aspect to consider. If the signal has finite support and L samples are given, then we have L nonzero coefficients $\langle f, \varphi_{J_1,k}\rangle$. However, the DWT will typically have more than L coefficients since the wavelet and scaling functions are obtained by convolution and downsampling. In other words, the DWT of a L-point signal will have more than L points. Considered as a finite discrete transform of one vector into another, this situation is undesirable. The reason this "expansion" in dimension occurs is that one is using a basis for L^2 to represent a signal that is of finite duration, say in $L^2[0, P]$.

When calculating the DWT of a long signal, J_0 is usually chosen to give the wavelet description of the slowly changing or longer duration features of the signal. When the signal has finite support or is periodic, J_0 is generally chosen so there is a single scaling coefficient for the entire signal or for one period of the signal. To reconcile the difference in length of the samples of a finite support signal and the number of DWT coefficients, zeros can be appended to the samples of $f(t)$ or the signal can be made periodic as is done for the DFT.

9.2   Periodic or Cyclic Discrete Wavelet Transform

If $f(t)$ has finite support, create a periodic version of it by

$$\widetilde{f}(t) = \sum_n f(t + Pn) \tag{9.10}$$

where the period P is an integer. In this case, $\langle f, \varphi_{j,k} \rangle$ and $\langle f, \psi_{j,k} \rangle$ are periodic sequences in k with period $P\,2^j$ (if $j \geq 0$ and 1 if $j < 0$) and

$$d(j,k) = 2^{j/2} \int \widetilde{f}(t) \psi(2^j t - k)\, dt \tag{9.11}$$

$$d(j,k) = 2^{j/2} \int \widetilde{f}(t + 2^{-j}k) \psi(2^j t)\, dt = 2^{j/2} \int \widetilde{f}(t + 2^{-j}\ell) \psi(2^j t)\, dt \tag{9.12}$$

where $\ell = <k>_{P\,2^j}$ (k modulo $P\,2^j$) and $l \in \{0, 1, \ldots, P\,2^j - 1\}$. An obvious choice for J_0 is 1. Notice that in this case given $L = 2^{J_1}$ samples of the signal, $\langle f, \varphi_{J_1,k} \rangle$, the wavelet transform has exactly $1 + 1 + 2 + 2^2 + \cdots + 2^{J_1 - 1} = 2^{J_1} = L$ terms. Indeed, this gives a linear, invertible discrete transform which can be considered apart from any underlying continuous process similar the discrete Fourier transform existing apart from the Fourier transform or series.

There are at least three ways to calculate this cyclic DWT and they are based on the equations (9.25), (9.26), and (9.27) later in this chapter. The first method simply convolves the scaling coefficients at one scale with the time-reversed coefficients $h(-n)$ to give an $L + N - 1$ length sequence. This is aliased or wrapped as indicated in (9.27) and programmed in `dwt5.m` in Appendix 3. The second method creates a periodic $\widetilde{c}_j(k)$ by concatenating an appropriate number of $c_j(k)$ sections together then convolving $h(n)$ with it. That is illustrated in (9.25) and in `dwt.m` in Appendix 3. The third approach constructs a periodic $\widetilde{h}(n)$ and convolves it with $c_j(k)$ to implement (9.26). The MATLAB programs should be studied to understand how these ideas are actually implemented.

Because the DWT is not shift-invariant, different implementations of the DWT may appear to give different results because of shifts of the signal and/or basis functions. It is interesting to take a test signal and compare the DWT of it with different circular shifts of the signal.

Making $f(t)$ periodic can introduce discontinuities at 0 and P. To avoid this, there are several alternative constructions of orthonormal bases for $L^2[0, P]$ [CDV93, HKRV92, HKRV93, GB95a]. All of these constructions use (directly or indirectly) the concept of time-varying filter banks. The basic idea in all these constructions is to retain basis functions with support in $[0, P]$, remove ones with support outside $[0, P]$ and replace the basis functions that overlap across the endpoints with special *entry/exit* functions that ensure *completeness*. These boundary functions are chosen so that the constructed basis is orthonormal. This is discussed in Section 8.11. Another way to deal with edges or boundaries uses "lifting" as mentioned in Section 3.4.

9.3   Filter Bank Structures for Calculation of the DWT and Complexity

Given that the wavelet analysis of a signal has been posed in terms of the finite expansion of (9.5), the discrete wavelet transform (expansion coefficients) can be calculated using Mallat's algorithm implemented by a filter bank as described in Chapter 3 and expanded upon in Chapter 8. Using the direct calculations described by the one-sided tree structure of filters and down-samplers in Figure 3.4 allows a simple determination of the computational complexity.

If we assume the length of the sequence of the signal is L and the length of the sequence of scaling filter coefficients $h(n)$ is N, then the number of multiplications necessary to calculate each scaling function and wavelet expansion coefficient at the next scale, $c(J_1 - 1, k)$ and $d(J_1 - 1, k)$, from the samples of the signal, $f(Tk) \approx c(J_1, k)$, is LN. Because of the downsampling, only half are needed to calculate the coefficients at the next lower scale, $c(J_2 - 1, k)$ and $d(J_2 - 1, k)$, and repeats until there is only one coefficient at a scale of $j = J_0$. The total number of multiplications is, therefore,

$$\text{Mult} = LN + LN/2 + LN/4 + \cdots + N \tag{9.13}$$

$$= LN(1 + 1/2 + 1/4 + \cdots + 1/L) = 2NL - N \tag{9.14}$$

which is linear in L and in N. The number of required additions is essentially the same.

If the length of the signal is very long, essentially infinity, the coarsest scale J_0 must be determined from the goals of the particular signal processing problem being addressed. For this case, the number of multiplications required per DWT coefficient or per input signal sample is

$$\text{Mult/sample} = N(2 - 2^{-J_0}) \tag{9.15}$$

Because of the relationship of the scaling function filter $h(n)$ and the wavelet filter $h_1(n)$ at each scale (they are quadrature mirror filters), operations can be shared between them through the use of a lattice filter structure, which will almost halve the computational complexity. That is developed in Chapter 8 and [Vai92].

9.4   The Periodic Case

In many practical applications, the signal is finite in length (finite support) and can be processed as single "block," much as the Fourier Series or discrete Fourier transform (DFT) does. If the signal to be analyzed is finite in length such that

$$f(t) = \begin{cases} 0 & t < 0 \\ 0 & t > P \\ f(t) & 0 < t < P \end{cases} \tag{9.16}$$

we can construct a periodic signal $\widetilde{f}(t)$ by

$$\widetilde{f}(t) = \sum_n f(t + Pn) \tag{9.17}$$

and then consider its wavelet expansion or DWT. This creation of a meaningful periodic function can still be done, even if $f(t)$ does not have finite support, if its energy is concentrated and some overlap is allowed in (9.17).

Periodic Property 1: *If $\widetilde{f}(t)$ is periodic with integer period P such that $\widetilde{f}(t) = \widetilde{f}(t + Pn)$, then the scaling function and wavelet expansion coefficients (DWT terms) at scale J are periodic with period $2^J P$.*

$$\text{If} \quad \widetilde{f}(t) = \widetilde{f}(t + P) \quad \text{then} \quad \widetilde{d}_j(k) = \widetilde{d}_j(k + 2^j P) \tag{9.18}$$

This is easily seen from

$$\widetilde{d}_j(k) = \int_{-\infty}^{\infty} \widetilde{f}(t)\,\psi(2^j t - k)\,dt = \int \widetilde{f}(t + Pn)\,\psi(2^j t - k)\,dt \tag{9.19}$$

which, with a change of variables, becomes

$$= \int \widetilde{f}(x)\,\psi(2^j(x - Pn) - k)\,dx = \int \widetilde{f}(x)\,\psi(2^j x - (2^j Pn + k))\,dx = \widetilde{d}_j(k + 2^j Pn) \tag{9.20}$$

and the same is true for $\widetilde{c}_j(k)$.

Periodic Property 2: *The scaling function and wavelet expansion coefficients (DWT terms) can be calculated from the inner product of $\widetilde{f}(t)$ with $\varphi(t)$ and $\psi(t)$ or, equivalently, from the inner product of $f(t)$ with the periodized $\widetilde{\varphi}(t)$ and $\widetilde{\psi}(t)$.*

$$\widetilde{c}_j(k) = \left\langle \widetilde{f}(t), \varphi(t) \right\rangle = \langle f(t), \widetilde{\varphi}(t) \rangle \tag{9.21}$$

and

$$\widetilde{d}_j(k) = \left\langle \widetilde{f}(t), \psi(t) \right\rangle = \left\langle f(t), \widetilde{\psi}(t) \right\rangle \tag{9.22}$$

where $\widetilde{\varphi}(t) = \sum_n \varphi(t + Pn)$ and $\widetilde{\psi}(t) = \sum_n \psi(t + Pn)$.

This is seen from

$$\widetilde{d}_j(k) = \int_{-\infty}^{\infty} \widetilde{f}(t)\,\psi(2^j t - k)\,dt = \sum_n \int_0^P f(t)\,\psi(2^j(t + Pn) - k)\,dt = \int_0^P f(t) \sum_n \psi(2^j(t + Pn) - k)\,dt \tag{9.23}$$

$$\widetilde{d}_j(k) = \int_0^P f(t)\,\widetilde{\psi}(2^j t - k)\,dt \tag{9.24}$$

where $\widetilde{\psi}(2^j t - k) = \sum_n \psi(2^j(t + Pn) - k)$ is the periodized scaled wavelet.

Periodic Property 3: *If $\widetilde{f}(t)$ is periodic with period P, then Mallat's algorithm for calculating the DWT coefficients in (3.9) becomes*

$$\widetilde{c}_j(k) = \sum_m h(m - 2k)\,\widetilde{c}_{j+1}(m) \tag{9.25}$$

or

$$\widetilde{c}_j(k) = \sum_m \widetilde{h}(m - 2k)\,c_{j+1}(m) \tag{9.26}$$

or

$$\widetilde{c}_j(k) = \sum_n c_j(k + Pn) \tag{9.27}$$

where for (9.27)

$$c_j(k) = \sum_m h(m - 2k)\,c_{j+1}(m) \tag{9.28}$$

The corresponding relationships for the wavelet coefficients are

$$\tilde{d}_j(k) = \sum_m h_1(m - 2k)\,\tilde{c}_{j+1}(m) = \sum_m \tilde{h}_1(m - 2k)\,c_{j+1}(m) \tag{9.29}$$

or

$$\tilde{d}_j(k) = \sum_n d_j(k + 2^j Pn) \tag{9.30}$$

where

$$d_j(k) = \sum_m h_1(m - 2k)\,c_{j+1}(m) \tag{9.31}$$

These are very important properties of the DWT of a periodic signal, especially one artificially constructed from a nonperiodic signal in order to use a block algorithm. They explain not only the aliasing effects of having a periodic signal but how to calculate the DWT of a periodic signal.

9.5   Structure of the Periodic Discrete Wavelet Transform

If $f(t)$ is essentially infinite in length, then the DWT can be calculated as an ongoing or continuous process in time. In other words, as samples of $f(t)$ come in at a high enough rate to be considered equal to $c_{J1}(k)$, scaling function and wavelet coefficients at lower resolutions continuously come out of the filter bank. This is best seen from the simple two-stage analysis filter bank in Figure 3.4. If samples come in at what is called scale $J_1 = 5$, wavelet coefficients at scale $j = 4$ come out the lower bank at half the input rate. Wavelet coefficients at $j = 3$ come out the next lower bank at one quarter the input rate and scaling function coefficients at $j = 3$ come out the upper bank also at one quarter the input rate. It is easy to imagine more stages giving lower resolution wavelet coefficients at a lower and lower rate depending on the number of stages. The last one will always be the scaling function coefficients at the lowest rate.

For a continuous process, the number of stages and, therefore, the level of resolution at the coarsest scale is arbitrary. It is chosen to be the nature of the slowest features of the signals being processed. It is important to remember that the lower resolution scales correspond to a slower sampling rate and a larger translation step in the expansion terms at that scale. This is why the wavelet analysis system gives good time localization (but poor frequency localization) at high resolution scales and good frequency localization (but poor time localization) at low or coarse scales.

For finite length signals or block wavelet processing, the input samples can be considered as a finite dimensional input vector, the DWT as a square matrix, and the wavelet expansion coefficients as an output vector. The conventional organization of the output of the DWT places the output of the first wavelet filter bank in the lower half of the output vector. The output of the next wavelet filter bank is put just above that block. If the length of the signal is two to a power, the wavelet decomposition can be carried until there is just one wavelet coefficient and one scaling function coefficient. That scale corresponds to the translation step size being the length of the signal. Remember that the decomposition does not have to carried to that level. It can be stopped at any scale and is still considered a DWT, and it can be inverted using the appropriate synthesis filter bank (or a matrix inverse).

9.6 More General Structures

The one-sided tree structure of Mallet's algorithm generates the basic DWT. From the filter bank in Figure 3.4, one can imagine putting a pair of filters and downsamplers at the output of the lower wavelet bank just as is done on the output of the upper scaling function bank. This can be continued to any level to create a balanced tree filter bank. The resulting outputs are "wavelet packets" and are an alternative to the regular wavelet decomposition. Indeed, this "growing" of the filter bank tree is usually done adaptively using some criterion at each node to decide whether to add another branch or not.

Still another generalization of the basic wavelet system can be created by using a scale factor other than two. The multiplicity-M scaling equation is

$$\varphi(t) = \sum_k h(k)\, \varphi(Mt - k)$$

and the resulting filter bank tree structure has one scaling function branch and $M - 1$ wavelet branches at each stage with each followed by a downsampler by M. The resulting structure is called an M-band filter bank, and it too is an alternative to the regular wavelet decomposition. This is developed in Section 7.2.

In many applications, it is the continuous wavelet transform (CWT) that is wanted. This can be calculated by using numerical integration to evaluate the inner products in (1.11) and (7.105) but that is very slow. An alternative is to use the DWT to approximate samples of the CWT much as the DFT can be used to approximate the Fourier series or integral [nRGB91, RD92, VLU97].

As you can see from this discussion, the ideas behind wavelet analysis and synthesis are basically the same as those behind filter bank theory. Indeed, filter banks can be used calculate discrete wavelet transforms using Mallat's algorithm, and certain modifications and generalizations can be more easily seen or interpreted in terms of filter banks than in terms of the wavelet expansion. The topic of filter banks in developed in Chapter 3 and in more detail in Chapter 8.

Chapter 10

Wavelet-Based Signal Processing and Applications

This chapter gives a brief discussion of several areas of application. It is intended to show what areas and what tools are being developed and to give some references to books, articles, and conference papers where the topics can be further pursued. In other words, it is a sort of annotated bibliography that does not pretend to be complete. Indeed, it is impossible to be complete or up-to-date in such a rapidly developing new area and in an introductory book.

In this chapter, we briefly consider the application of wavelet systems from two perspectives. First, we look at wavelets as a tool for denoising and compressing a wide variety of signals. Second, we very briefly list several problems where the application of these tools shows promise or has already achieved significant success. References will be given to guide the reader to the details of these applications, which are beyond the scope of this book.

10.1 Wavelet-Based Signal Processing

To accomplish frequency domain signal processing, one can take the Fourier transform (or Fourier series or DFT) of a signal, multiply some of the Fourier coefficients by zero (or some other constant), then take the inverse Fourier transform. It is possible to completely remove certain components of a signal while leaving others completely unchanged. The same can be done by using wavelet transforms to achieve wavelet-based, wavelet domain signal processing, or filtering. Indeed, it is sometimes possible to remove or separate parts of a signal that overlap in both time and frequency using wavelets, something impossible to do with conventional Fourier-based techniques.

The classical paradigm for transform-based signal processing is illustrated in Figure 10.1 where the center "box" could be either a linear or nonlinear operation. The "dynamics" of the processing are all contained in the transform and inverse transform operation, which are linear. The transform-domain processing operation has no dynamics; it is an algebraic operation. By dynamics, we mean that a process depends on the present and past, and by algebraic, we mean it depends only on the present. An FIR (finite impulse response) filter such as is part of a filter bank is dynamic. Each output depends on the current and a finite number of past inputs (see (3.11)). The process of operating point-wise on the DWT of a signal is static or algebraic. It does not depend on the past (or future) values, only the present. This structure, which separates the linear,

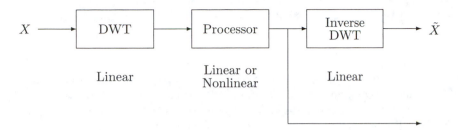

Figure 10.1. Transform-Based Signal Processor

dynamic parts from the nonlinear static parts of the processing, allows practical and theoretical results that are impossible or very difficult using a completely general nonlinear dynamic system.

Linear wavelet-based signal processing consists of the processor block in Figure 10.1 multiplying the DWT of the signal by some set of constants (perhaps by zero). If undesired signals or noise can be separated from the desired signal in the wavelet transform domain, they can be removed by multiplying their coefficients by zero. This allows a more powerful and flexible processing or filtering than can be achieved using Fourier transforms. The result of this total process is a linear, time-varying processing that is far more versatile than linear, time-invariant processing. The next section gives an example of using the concentrating properties of the DWT to allow a faster calculation of the FFT.

10.2 Approximate FFT using the Discrete Wavelet Transform

In this section, we give an example of wavelet domain signal processing. Rather than computing the DFT from the time domain signal using the FFT algorithm, we will first transform the signal into the wavelet domain, then calculate the FFT, and finally go back to the signal domain which is now the Fourier domain.

Most methods of approximately calculating the discrete Fourier transform (DFT) involve calculating only a few output points (pruning), using a small number of bits to represent the various calculations, or approximating the kernel, perhaps by using cordic methods. Here we use the characteristics of the signal being transformed to reduce the amount of arithmetic. Since the wavelet transform concentrates the energy of many classes of signals onto a small number of wavelet coefficients, this can be used to improve the efficiency of the DFT [GB96a, GB97c, Guo97, GB97a] and convolution [GB96b].

Introduction

The DFT is probably the most important computational tool in signal processing. Because of the characteristics of the basis functions, the DFT has enormous capacity for the improvement of its arithmetic efficiency [BP85]. The classical Cooley-Tukey fast Fourier transform (FFT) algorithm has the complexity of $O(N \log_2 N)$. Thus the Fourier transform and its fast algorithm, the FFT, are widely used in many areas, including signal processing and numerical analysis. Any scheme to speed up the FFT would be very desirable.

Although the FFT has been studied extensively, there are still some desired properties that are not provided by the classical FFT. Here are some of the disadvantages of the FFT algorithm:

1. Pruning is not easy.

   When the number of input points or output points are small compared to the length of the DWT, a special technique called *pruning* [SB93] is often used. However, this often requires that the nonzero input data are grouped together. Classical FFT pruning algorithms do not work well when the few nonzero inputs are randomly located. In other words, a sparse signal may not necessarily give rise to faster algorithm.

2. No speed versus accuracy tradeoff.

   It is common to have a situation where some error would be allowed if there could be a significant increase in speed. However, this is not easy with the classical FFT algorithm. One of the main reasons is that the twiddle factors in the butterfly operations are unit magnitude complex numbers. So all parts of the FFT structure are of equal importance. It is hard to decide which part of the FFT structure to omit when error is allowed and the speed is crucial. In other words, the FFT is a single speed and single accuracy algorithm.

3. No built-in noise reduction capacity.

   Many real world signals are noisy. What people are really interested in are the DFT of the signals without the noise. The classical FFT algorithm does not have built-in noise reduction capacity. Even if other denoising algorithms are used, the FFT requires the same computational complexity on the denoised signal. Due to the above mentioned shortcomings, the fact that the signal has been denoised cannot be easily used to speed up the FFT.

Review of the Discrete Fourier Transform and FFT

The discrete Fourier transform (DFT) is defined for a length-N complex data sequence by

$$X(k) = \sum_{n=0}^{N-1} x(n) \, e^{-j2\pi \, nk/N}, \qquad k = 0, \ldots, N-1 \tag{10.1}$$

where we use $j = \sqrt{-1}$. There are several ways to derive the different fast Fourier transform (FFT) algorithms. It can be done by using index mapping [BP85], by matrix factorization, or by polynomial factorization. In this chapter, we only discuss the matrix factorization approach, and only discuss the so-called *radix-2 decimation in time* (DIT) variant of the FFT.

Instead of repeating the derivation of the FFT algorithm, we show the block diagram and matrix factorization, in an effort to highlight the basic idea and gain some insight. The block diagram of the last stage of a length-8 radix-2 DIT FFT is shown in Figure 10.2. First, the input data are separated into even and odd groups. Then, each group goes through a length-4 DFT block. Finally, *butterfly operations* are used to combine the shorter DFTs into longer DFTs.

The details of the *butterfly operations* are shown in Figure 10.3, where $W_N^i = e^{-j2\pi i/N}$ is called the *twiddle factor*. All the twiddle factors are of magnitude one on the unit circle. This is the main reason that there is no complexity versus accuracy tradeoff for the classical FFT. Suppose some of the twiddle factors had very small magnitude, then the corresponding branches

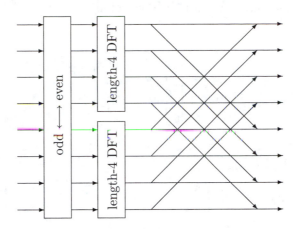

Figure 10.2. Last Stage of a Length-8 Radix-2 DIT FFT

of the butterfly operations could be dropped (pruned) to reduce complexity while minimizing the error to be introduced. Of course the error also depends on the value of the data to be multiplied with the twiddle factors. When the value of the data is unknown, the best way is to cutoff the branches with small twiddle factors.

The computational complexity of the FFT algorithm can be easily established. If we let $C_{FFT}(N)$ be the complexity for a length-N FFT, we can show

$$C_{FFT}(N) = O(N) + 2C_{FFT}(N/2), \qquad (10.2)$$

where $O(N)$ denotes linear complexity. The solution to Equation 10.2 is well known:

$$C_{FFT}(N) = O(N \log_2 N). \qquad (10.3)$$

This is a classical case where the *divide and conquer* approach results in very effective solution.

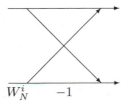

Figure 10.3. Butterfly Operations in a Radix-2 DIT FFT

The matrix point of view gives us additional insight. Let $\mathbf{F}_N$ be the $N{\times}N$ DFT matrix; i.e., $\mathbf{F}_N(m,n) = e^{-j2\pi mn/N}$, where $m, n \in \{0, 1, \ldots, N{-}1\}$. Let $\mathbf{S}_N$ be the $N{\times}N$ even-odd separation matrix; e.g.,

$$\mathbf{S}_4 = \begin{bmatrix} 1 & 0 & 0 & 0 \\ 0 & 0 & 1 & 0 \\ 0 & 1 & 0 & 0 \\ 0 & 0 & 0 & 1 \end{bmatrix}. \tag{10.4}$$

Clearly $\mathbf{S}'_N \mathbf{S}_N = \mathbf{I}_N$, where $\mathbf{I}_N$ is the $N{\times}N$ identity matrix. Then the DIT FFT is based on the following matrix factorization,

$$\mathbf{F}_N = \mathbf{F}_N \mathbf{S}'_N \mathbf{S}_N = \begin{bmatrix} \mathbf{I}_{N/2} & \mathbf{T}_{N/2} \\ \mathbf{I}_{N/2} & -\mathbf{T}_{N/2} \end{bmatrix} \begin{bmatrix} \mathbf{F}_{N/2} & 0 \\ 0 & \mathbf{F}_{N/2} \end{bmatrix} \mathbf{S}_N, \tag{10.5}$$

where $\mathbf{T}_{N/2}$ is a diagonal matrix with W_N^i, $i \in \{0, 1, \ldots, N/2{-}1\}$ on the diagonal. We can visualize the above factorization as

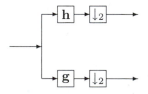

$$\tag{10.6}$$

where we image the real part of DFT matrices, and the magnitude of the matrices for butterfly operations and even-odd separations. N is taken to be 128 here.

Review of the Discrete Wavelet Transform

In this section, we briefly review the fundamentals of the discrete wavelet transform and introduce the necessary notation for future sections. The details of the DWT have been covered in other chapters.

At the heart of the discrete wavelet transform are a pair of filters $\mathbf{h}$ and $\mathbf{g}$ — lowpass and highpass respectively. They have to satisfy a set of constraints (5.1) [VK95, SN96, Vai92]. The block diagram of the DWT is shown in Figure 10.4. The input data are first filtered by $\mathbf{h}$ and $\mathbf{g}$ then downsampled. The same building block is further iterated on the lowpass outputs.

Figure 10.4. Building Block for the Discrete Wavelet Transform

The computational complexity of the DWT algorithm can also be easily established. Let $C_{DWT}(N)$ be the complexity for a length-N DWT. Since after each scale, we only further operate on half of the output data, we can show

$$C_{DWT}(N) = O(N) + C_{DWT}(N/2), \tag{10.7}$$

which gives rise to the solution

$$C_{DWT}(N) = O(N). \tag{10.8}$$

The operation in Figure 10.4 can also be expressed in matrix form $\mathbf{W}_N$; e.g., for Haar wavelet,

$$\mathbf{W}_4^{Haar} = \sqrt{2}/2 \begin{bmatrix} 1 & 1 & 0 & 0 \\ 0 & 0 & 1 & 1 \\ 1 & -1 & 0 & 0 \\ 0 & 0 & 1 & -1 \end{bmatrix}. \tag{10.9}$$

The orthogonality conditions on $\mathbf{h}$ and $\mathbf{g}$ ensure $\mathbf{W}'_N \mathbf{W}_N = \mathbf{I}_N$. The matrix for multiscale DWT is formed by $\mathbf{W}_N$ for different N; e.g., for three scale DWT,

$$\begin{bmatrix} \begin{bmatrix} \mathbf{W}_{N/4} & \\ & \mathbf{I}_{N/4} \end{bmatrix} & \\ & \mathbf{I}_{N/2} \end{bmatrix} \begin{bmatrix} \mathbf{W}_{N/2} & \\ & \mathbf{I}_{N/2} \end{bmatrix} \mathbf{W}_N. \tag{10.10}$$

We could further iterate the building block on some of the highpass outputs. This generalization is called the wavelet packets [CW92].

The Algorithm Development

The key to the fast Fourier transform is the factorization of $\mathbf{F}_N$ into several sparse matrices, and one of the sparse matrices represents two DFTs of half the length. In a manner similar to the DIT FFT, the following matrix factorization can be made:

$$\mathbf{F}_N = \mathbf{F}_N \mathbf{W}_N^T \mathbf{W}_N = \begin{bmatrix} \mathbf{A}_{N/2} & \mathbf{B}_{N/2} \\ \mathbf{C}_{N/2} & \mathbf{D}_{N/2} \end{bmatrix} \begin{bmatrix} \mathbf{F}_{N/2} & 0 \\ 0 & \mathbf{F}_{N/2} \end{bmatrix} \mathbf{W}_N, \tag{10.11}$$

where $\mathbf{A}_{N/2}, \mathbf{B}_{N/2}, \mathbf{C}_{N/2}$, and $\mathbf{D}_{N/2}$ are all diagonal matrices. The values on the diagonal of $\mathbf{A}_{N/2}$ and $\mathbf{C}_{N/2}$ are the length-N DFT (*i.e.*, *frequency response*) of $\mathbf{h}$, and the values on the diagonal of $\mathbf{B}_{N/2}$ and $\mathbf{D}_{N/2}$ are the length-N DFT of $\mathbf{g}$. We can visualize the above factorization as

$$\tag{10.12}$$

where we image the real part of DFT matrices, and the magnitude of the matrices for butterfly operations and the one-scale DWT using length-16 Daubechies' wavelets [Dau88a, Dau92]. Clearly we can see that the new twiddle factors have non-unit magnitudes.

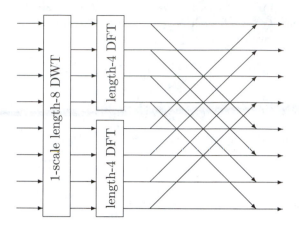

Figure 10.5. Last stage of a length-8 DWT based FFT.

The above factorization suggests a DWT-based FFT algorithm. The block diagram of the last stage of a length-8 algorithm is shown in Figure 10.5. This scheme is iteratively applied to shorter length DFTs to get the full DWT based FFT algorithm. The final system is equivalent to a full binary tree wavelet packet transform [CW90] followed by classical FFT butterfly operations, where the new twiddle factors are the frequency response of the wavelet filters.

The detail of the butterfly operation is shown in Figure 10.6, where $i \in \{0, 1, \ldots, N/2-1\}$. Now the twiddle factors are length-N DFT of $\mathbf{h}$ and $\mathbf{g}$. For well defined wavelet filters, they have well known properties; e.g., for Daubechies' family of wavelets, their frequency responses are monotone, and nearly half of which have magnitude close to zero. This fact can be exploited to achieve speed vs. accuracy tradeoff. The classical radix-2 DIT FFT is a special case of the above algorithm when $\mathbf{h} = [1, 0]$ and $\mathbf{g} = [0, 1]$. Although they do not satisfy some of the conditions required for wavelets, they do constitute a legitimate (and trivial) orthogonal filter bank and are often called the *lazy* wavelets in the context of lifting.

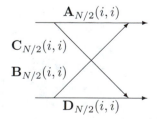

Figure 10.6. Butterfly Operations in a Radix-2 DIT FFT

Computational Complexity

For the DWT-based FFT algorithm, the computational complexity is on the same order of the FFT — $O(N \log_2 N)$, since the recursive relation in (10.2) is again satisfied. However, the constant appearing before $N \log_2 N$ depends on the wavelet filters used.

Fast Approximate Fourier Transform

The basic idea of the fast approximate Fourier transform (FAFT) is *pruning*; i.e., cutting off part of the diagram. Traditionally, when only part of the inputs are nonzero, or only part of the outputs are required, the part of the FFT diagram where either the inputs are zero or the outputs are undesired is pruned [SB93], so that the computational complexity is reduced. However, the classical pruning algorithm is quite restrictive, since for a majority of the applications, both the inputs and the outputs are of full length.

The structure of the DWT-based FFT algorithm can be exploited to generalize the classical pruning idea for arbitrary signals. From the input data side, the signals are made sparse by the wavelet transform [Mey93, Mey87, Mey90, Dau92]; thus approximation can be made to speed up the algorithm by *dropping* the insignificant data. In other words, although the input signal are normally not sparse, DWT creates the sparse inputs for the butterfly stages of the FFT. So any scheme to prune the butterfly stages for the classical FFT can be used here. Of course, the price we have to pay here is the computational complexity of the DWT operations. In actual implementation, the wavelets in use have to be carefully chosen to balance the benefit of the pruning and the price of the transform. Clearly, the optimal choice depends on the class of the data we would encounter.

From the transform side, since the twiddle factors of the new algorithm have decreasing magnitudes, approximation can be made to speed up the algorithm by *pruning* the sections of the algorithm which correspond to the insignificant twiddle factors. The frequency response of the Daubechies' wavelets are shown in Figure 10.7. We can see that they are monotone decreasing. As the length increases, more and more points are close to zero. It should be noted that those filters are not designed for frequency responses. They are designed for flatness at 0 and π. Various methods can be used to design wavelets or orthogonal filter banks [Ode96, Sel96, Vai92] to achieve better frequency responses. Again, there is a tradeoff between the good frequency response of the longer filters and the higher complexity required by the longer filters.

Computational Complexity

The wavelet coefficients are mostly sparse, so the input of the shorter DFTs are sparse. If the implementation scales well with respect to the percentage of the significant input (e.g., it uses half of the time if only half of the inputs are significant), then we can further lower the complexity. Assume for N inputs, αN of them are significant ($\alpha \leq 1$), we have

$$C_{FAFT}(N) = O(N) + 2\alpha C_{FAFT}(N/2). \tag{10.13}$$

For example, if $\alpha = \frac{1}{2}$, Equation (10.13) simplifies to

$$C_{FAFT}(N) = O(N) + C_{FAFT}(N/2), \tag{10.14}$$

which leads to

$$C_{FAFT}(N) = O(N). \tag{10.15}$$

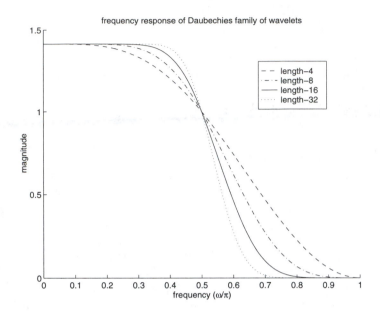

Figure 10.7. The Frequency Responses of Daubechies' Family of Wavelets

So under the above conditions, we have a linear complexity approximate FFT. Of course, the complexity depends on the input data, the wavelets we use, the threshold value used to drop insignificant data, and the threshold value used to prune the butterfly operations. It remains to find a good tradeoff. Also the implementation would be more complicated than the classical FFT.

Noise Reduction Capacity

It has been shown that the thresholding of wavelet coefficients has near optimal noise reduction property for many classes of signals [Don95]. The thresholding scheme used in the approximation in the proposed FAFT algorithm is exactly the hard thresholding scheme used to denoise the data. Soft thresholding can also be easily embedded in the FAFT. Thus the proposed algorithm also reduces the noise while doing approximation. If we need to compute the DFT of noisy signals, the proposed algorithm not only can reduce the numerical complexity but also can produce cleaner results.

Summary

In the past, the FFT has been used to calculate the DWT [VK95, SN96, Vai92], which leads to an efficient algorithm when filters are infinite impulse response (IIR). In this chapter, we did just the opposite – using DWT to calculate FFT. We have shown that when no intermediate coefficients are dropped and no approximations are made, the proposed algorithm computes the exact result, and its computational complexity is on the same order of the FFT; i.e., $O(N \log_2 N)$. The advantage of our algorithm is two fold. From the input data side, the signals are made sparse by the wavelet transform, thus approximation can be made to speed up the algorithm by *dropping* the insignificant data. From the transform side, since the twiddle factors of the new algorithm have decreasing magnitudes, approximation can be made to speed up the algorithm by *pruning* the section of the algorithm which corresponds to the insignificant twiddle factors. Since wavelets are an unconditional basis for many classes of signals [SN96, Mey90, Dau92], the algorithm is very efficient and has built-in denoising capacity. An alternative approach has been developed by Shentov, Mitra, Heute, and Hossen [SMHH95, HHSM95] using subband filter banks.

10.3   Nonlinear Filtering or Denoising with the DWT

Wavelets became known to most engineers and scientists with the publication of Daubechies' important paper [Dau88a] in 1988. Indeed, the work of Daubechies [Dau92], Mallat [Mal89b, Mal89c, MZ93], Meyer [Mey92b, Mey93], and others produced beautiful and interesting structures, but many engineers and applied scientist felt they had a "solution looking for a problem." With the recent work of Donoho and Johnstone together with ideas from Coifman, Beylkin and others, the field is moving into a second phase with a better understanding of why wavelets work. This new understanding combined with *nonlinear processing* not only solves currently important problems, but gives the potential of formulating and solving completely new problems. We now have a coherence of approach and a theoretical basis for the success of our methods that should be extraordinarily productive over the next several years. Some of the Donoho and Johnstone references are [DJ94b, Don93b, Don95, Don93a, Donar, DJKP95b, DJ95, DJ94a, CD94, Don93c, Don94, CD95a] and related ones are [Sai94b, Mou93, TM94b, TM94a, BDGM94]. Ideas from Coifman are in [CW90, CW92, CM94, CMW92, CMQW92, CD95b, BCR91].

These methods are based on taking the discrete wavelet transform (DWT) of a signal, passing this transform through a threshold, which removes the coefficients below a certain value, then taking the inverse DWT, as illustrated in Figure 10.1. They are able to remove noise and achieve high compression ratios because of the "concentrating" ability of the wavelet transform. If a signal has its energy concentrated in a small number of wavelet dimensions, its coefficients will be relatively large compared to any other signal or noise that has its energy spread over a large number of coefficients. This means that thresholding or shrinking the wavelet transform will remove the low amplitude noise or undesired signal in the wavelet domain, and an inverse wavelet transform will then retrieve the desired signal with little loss of detail. In traditional Fourier-based signal processing, we arrange our signals such that the signals and any noise overlap as little as possible in the frequency domain and linear time-invariant filtering will approximately separate them. Where their Fourier spectra overlap, they cannot be separated. Using linear wavelet or other time-frequency or time-scale methods, one can try to choose basis systems such that in that coordinate system, the signals overlap as little as possible, and separation is possible.

The new nonlinear method is entirely different. The spectra can overlap as much as they want. The idea is to have the amplitude, rather than the location of the spectra be as different as possible. This allows clipping, thresholding, and shrinking of the amplitude of the transform to separate signals or remove noise. It is the localizing or concentrating properties of the wavelet transform that makes it particularly effective when used with these nonlinear methods. Usually the same properties that make a system good for denoising or separation by nonlinear methods, makes it good for compression, which is also a nonlinear process.

Denoising by Thresholding

We develop the basic ideas of thresholding the wavelet transform using Donoho's formulations [Don95, DJ94b, LGO*95]. Assume a finite length signal with additive noise of the form

$$y_i = x_i + \epsilon n_i, \quad i = 1, \ldots, N \tag{10.16}$$

as a finite length signal of observations of the signal x_i that is corrupted by i.i.d. zero mean, white Gaussian noise n_i with standard deviation ϵ, i.e., $n_i \overset{iid}{\sim} \mathcal{N}(0, 1)$. The goal is to recover the signal x from the noisy observations y. Here and in the following, v denotes a vector with the ordered elements v_i if the index i is omitted. Let W be a left invertible wavelet transformation matrix of the discrete wavelet transform (DWT). Then Eq. (10.16) can be written in the transformation domain

$$Y = X + N, \quad \text{or,} \quad Y_i = X_i + N_i, \tag{10.17}$$

where capital letters denote variables in the transform domain, i.e., $Y = Wy$. Then the inverse transform matrix W^{-1} exists, and we have

$$W^{-1}W = I. \tag{10.18}$$

The following presentation follows Donoho's approach [DJ94b, Don93b, Don95, Don93a, LGO*95] that assumes an orthogonal wavelet transform with a square W; i.e., $W^{-1} = W^T$. We will use the same assumption throughout this section.

Let $\hat{X}$ denote an estimate of X, based on the observations Y. We consider diagonal linear projections

$$\Delta = \text{diag}(\delta_1, \ldots, \delta_N), \quad \delta_i \in \{0, 1\}, \quad i = 1, \ldots, N, \tag{10.19}$$

which give rise to the estimate

$$\hat{x} = W^{-1}\hat{X} = W^{-1}\Delta Y = W^{-1}\Delta Wy. \tag{10.20}$$

The estimate $\hat{X}$ is obtained by simply keeping or zeroing the individual wavelet coefficients. Since we are interested in the l_2 error we define the risk measure

$$\mathcal{R}(\hat{X}, X) = E\left[\|\hat{x} - x\|_2^2\right] = E\left[\|W^{-1}(\hat{X} - X)\|_2^2\right] = E\left[\|\hat{X} - X\|_2^2\right]. \tag{10.21}$$

Notice that the last equality in Eq. (10.21) is a consequence of the orthogonality of W. The optimal coefficients in the diagonal projection scheme are $\delta_i = 1_{X_i > \epsilon}$;[1] i.e., only those values of Y where the corresponding elements of X are larger than ϵ are kept, all others are set to zero. This leads to the ideal risk

$$\mathcal{R}_{id}(\hat{X}, X) = \sum_{n=1}^{N} \min(X^2, \epsilon^2). \tag{10.22}$$

The ideal risk cannot be attained in practice, since it requires knowledge of X, the wavelet transform of the unknown vector x. However, it does give us a lower limit for the l_2 error.

---

[1]It is interesting to note that allowing arbitrary $\delta_i \in \mathbb{R}$ improves the ideal risk by at most a factor of 2[DJ94a]

Donoho proposes the following scheme for denoising:

1. compute the DWT $Y = Wy$

2. perform thresholding in the wavelet domain, according to so-called *hard thresholding*

$$\hat{X} = T_h(Y,t) = \begin{cases} Y, & |Y| \geq t \\ 0, & |Y| < t \end{cases} \tag{10.23}$$

or according to so-called *soft thresholding*

$$\hat{X} = T_S(Y,t) = \begin{cases} \text{sgn}(Y)(|Y| - t), & |Y| \geq t \\ 0, & |Y| < t \end{cases} \tag{10.24}$$

3. compute the inverse DWT $\hat{x} = W^{-1}\hat{X}$

This simple scheme has several interesting properties. It's risk is within a logarithmic factor ($\log N$) of the ideal risk for both thresholding schemes and properly chosen thresholds $t(N, \epsilon)$. If one employs soft thresholding, then the estimate is with high probability at least as smooth as the original function. The proof of this proposition relies on the fact that wavelets are unconditional bases for a variety of smoothness classes and that soft thresholding guarantees (with high probability) that the shrinkage condition $|\hat{X}_i| < |X_i|$ holds. The shrinkage condition guarantees that $\hat{x}$ is in the same smoothness class as is x. Moreover, the soft threshold estimate is the optimal estimate that satisfies the shrinkage condition. The smoothness property guarantees an estimate free from spurious oscillations which may result from hard thresholding or Fourier methods. Also, it can be shown that it is not possible to come closer to the ideal risk than within a factor $\log N$. Not only does Donoho's method have nice theoretical properties, but it also works very well in practice.

Some comments have to be made at this point. Similar to traditional approaches (e.g., low pass filtering), there is a trade-off between suppression of noise and oversmoothing of image details, although to a smaller extent. Also, hard thresholding yields better results in terms of the l_2 error. That is not surprising since the observation value y_i itself is clearly a better estimate for the real value x_i than a shrunk value in a zero mean noise scenario. However, the estimated function obtained from hard thresholding typically exhibits undesired, spurious oscillations and does not have the desired smoothness properties.

Shift-Invariant or Nondecimated Discrete Wavelet Transform

As is well known, the discrete wavelet transform is not shift invariant; i.e., there is no "simple" relationship between the wavelet coefficients of the original and the shifted signal[2]. In this section we will develop a shift-invariant DWT using ideas of a nondecimated filter bank or a redundant DWT [LGO*95, LGO*96, LGOB95]. Because this system is redundant, it is not a basis but will be a frame or tight frame (see Section 7.6). Let $X = Wx$ be the (orthogonal) DWT of x and S_R be a matrix performing a circular right shift by R with $R \in \mathbf{Z}$. Then

$$X_s = Wx_s = WS_Rx = WS_RW^{-1}X, \tag{10.25}$$

which establishes the connection between the wavelet transforms of two shifted versions of a signal, x and x_s, by the orthogonal matrix WS_RW^{-1}. As an illustrative example, consider Fig. 10.8.

---

[2]Since we deal with finite length signals, we really mean circular shift.

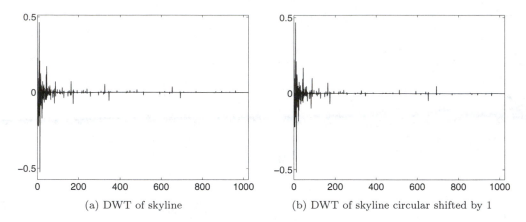

(a) DWT of skyline   (b) DWT of skyline circular shifted by 1

Figure 10.8. Shift Variance of the Wavelet Transform

The first and most obvious way of computing a shift invariant discrete wavelet transform (SIDWT) is simply computing the wavelet transform of *all* shifts. Usually the two band wavelet transform is computed as follows: 1) filter the input signal by a low-pass and a high-pass filter, respectively, 2) downsample each filter output, and 3) iterate the low-pass output. Because of the downsampling, the number of output values at each stage of the filter bank (corresponding to coarser and coarser scales of the DWT) is equal to the number of the input values. Precisely N values have to be stored. The computational complexity is $O(N)$. Directly computing the wavelet transform of all shifts therefore requires the storage of N^2 elements and has computational complexity $O(N^2)$.

Beylkin [Bey92], Shensa [She92], and the Rice group[3] independently realized that 1) there are only $N \log N$ different coefficient values among those corresponding to all shifts of the input signal and 2) those can be computed with computational complexity $N \log N$. This can be easily seen by considering one stage of the filter bank. Let

$$y = [y_0\, y_1\, y_2\, \cdots\, y_N]^T = hx \qquad (10.26)$$

where y is the output of either the high-pass or the low-pass filter in the analysis filter bank, x the input and the matrix h describes the filtering operation. Downsampling of y by a factor of two means keeping the even indexed elements and discarding the odd ones. Consider the case of an input signal shifted by one. Then the output signal is shifted by one as well, and sampling with the same operator as before corresponds to keeping the odd-indexed coefficients as opposed to the even ones. Thus, the set of data points to be further processed is completely different. However, for a shift of the input signal by two, the downsampled output signal differs from the output of the nonshifted input only by a shift of one. This is easily generalized for any odd and even shift and we see that the set of wavelet coefficients of the first stage of the filter bank for arbitrary shifts consists of only $2N$ different values. Considering the fact that only the low-pass component (N values) is iterated, one recognizes that after L stages exactly LN values result. Using the same arguments as in the shift variant case, one can prove that the computational complexity is $O(N \log N)$. The derivation for the synthesis is analogous.

---

[3]Those are the ones we are aware of.

Mallat proposes a scheme for computing an approximation of the continuous wavelet transform [Mal91] that turns out to be equivalent to the method described above. This has been realized and proved by Shensa[She92]. Moreover, Shensa shows that Mallat's algorithm exhibits the same structure as the so-called algorithm à trous. Interestingly, Mallat's intention in [Mal91] was not in particular to overcome the shift variance of the DWT but to get an approximation of the continuous wavelet transform.

In the following, we shall refer to the algorithm for computing the SIDWT as the Beylkin algorithm[4] since this is the one we have implemented. Alternative algorithms for computing a shift-invariant wavelet transform[LP94b] are based on the scheme presented in [Bey92]. They explicitly or implicitly try to find an optimal, signal-dependent shift of the input signal. Thus, the transform becomes shift-invariant and orthogonal but signal dependent and, therefore, nonlinear. We mention that the generalization of the Beylkin algorithm to the multidimensional case, to an M-band multiresolution analysis, and to wavelet packets is straightforward.

Combining the Shensa-Beylkin-Mallat-à trous Algorithms and Wavelet Denoising

It was Coifman who suggested that the application of Donoho's method to several shifts of the observation combined with averaging yields a considerable improvement.[5] This statement first lead us to the following algorithm: 1) apply Donoho's method not only to "some" but to *all* circular shifts of the input signal 2) average the adjusted output signals. As has been shown in the previous section, the computation of all possible shifts can be effectively done using Beylkin's algorithm. Thus, instead of using the algorithm just described, one simply applies thresholding to the SIDWT of the observation and computes the inverse transform.

Before going into details, we want to briefly discuss the differences between using the traditional orthogonal and the shift-invariant wavelet transform. Obviously, by using more than N wavelet coefficients, we introduce redundancy. Several authors stated that redundant wavelet transforms, or frames, add to the numerical robustness [Dau92] in case of adding white noise in the transform domain; e.g., by quantization. This is, however, different from the scenario we are interested in, since 1) we have correlated noise due to the redundancy, and 2) we try to remove noise in the transform domain rather than considering the effect of adding some noise [LGO*95, LGO*96].

Performance Analysis

The analysis of the ideal risk for the SIDWT is similar to that by Guo[Guo95]. Define the sets A and B according to

$$A = \{i|\ |X_i| \geq \epsilon\} \tag{10.27}$$

$$B = \{i|\ |X_i| < \epsilon\} \tag{10.28}$$

and an ideal diagonal projection estimator, or oracle,

$$\widetilde{X} = \begin{cases} Y_i = X_i + N_i & i \in A \\ 0 & i \in B. \end{cases} \tag{10.29}$$

---

[4]However, it should be noted that Mallat published his algorithm earlier.
[5]A similar remark can be found in [Sai94a], p. 53.

The pointwise estimation error is then

$$\widetilde{X}_i - X_i = \begin{cases} N_i & i \in A \\ -X_i & i \in B. \end{cases} \tag{10.30}$$

In the following, a vector or matrix indexed by A (or B) indicates that only those rows are kept that have indices out of A (or B). All others are set to zero. With these definitions and (10.21), the ideal risk for the SIDWT can be derived

$$
\begin{aligned}
\mathcal{R}_{id}(\widetilde{X}, X) &= E\left[\|W^{-1}(\widetilde{X} - X)\|_2^2\right] & (10.31) \\
&= E\left[\|W^{-1}(N_A - X_B)\|_2^2\right] & (10.32) \\
&= E\left[(N_A - X_B)^T \underbrace{W^{-1^T}W^{-1}}_{C_{W^{-1}}}(N_A - X_B)\right] & (10.33) \\
&= E\left[N_A^T W^{-1^T} W^{-1} N_A\right] - 2X_B^T C_{W^{-1}} E\left[N_A\right] + X_B^T C_{W^{-1}} X_B & (10.34) \\
&= \mathrm{tr}\left[E\left[W^{-1} N_A N_A^T W^{-1^T}\right]\right] + X_B^T C_{W^{-1}} X_B & (10.35) \\
&= \mathrm{tr}\left[W^{-1} E\left[W_A \epsilon n \epsilon n^T W_A^T\right] W^{-1^T}\right] + X_B^T C_{W^{-1}} X_B & (10.36) \\
&= \epsilon^2 \mathrm{tr}\left[W^{-1} W_A W_A^T W^{-1^T}\right] + X_B^T C_{W^{-1}} X_B. & (10.37)
\end{aligned}
$$

where $\mathrm{tr}(X)$ denotes the trace of X. For the derivation we have used, the fact that $N_A = \epsilon W_A n$ and consequently the N_{Ai} have zero mean. Notice that for orthogonal W the Eq. (10.37) immediately specializes to Eq. (10.22). Eq. (10.37) depends on the particular signal X_B, the transform, W^{-1}, and the noise level ϵ.

It can be shown that when using the SIDWT introduced above and the thresholding scheme proposed by Donoho (including his choice of the threshold) then there exists the same upper bound for the actual risk as for case of the orthogonal DWT. That is the ideal risk times a logarithmic (in N) factor. We give only an outline of the proof. Johnstone and Silverman state[JS94b] that for colored noise an oracle chooses $\delta_i = 1_{X_i \geq \epsilon_i}$, where ϵ_i is the standard deviation of the ith component. Since Donoho's method applies uniform thresholding to all components, one has to show that the diagonal elements of $C_{W^{-1}}$ (the variances of the components of N) are identical. This can be shown by considering the reconstruction scheme of the SIDWT. With these statements, the rest of the proof can be carried out in the same way as the one given by Donoho and Johnstone [DJ94b].

Examples of Denoising

The two examples illustrated in Figure 10.9 show how wavelet based denoising works. The first shows a chirp or doppler signal which has a changing frequency and amplitude. Noise is added to this chirp in (b) and the result of basic Donoho denoising is shown in (c) and of redundant DWT denoising in (d). First, notice how well the noise is removed and at almost no sacrifice in the signal. This would be impossible with traditional linear filters.

The second example is the Houston skyline where the improvement of the redundant DWT is more obvious.

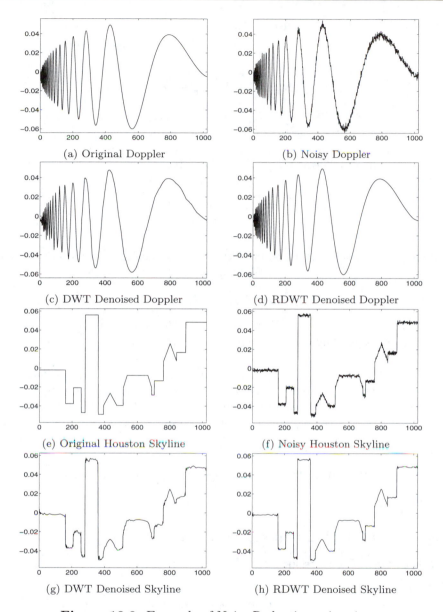

Figure 10.9. Example of Noise Reduction using $\psi_{D8'}$

10.4   Statistical Estimation

This problem is very similar to the signal recovery problem; a signal has to be estimated from additive white Gaussian noise. By linearity, additive noise is additive in the transform domain where the problem becomes: estimate θ from $y = \theta + \epsilon z$, where z is a noise vector (with each component being a zero mean variance one Gaussian random variable) and $\epsilon > 0$ is a scalar noise level. The performance measured by the mean squared error (by Parseval) is given by

$$R_\epsilon(\hat{\theta}, \theta) = E \left\| \hat{\theta}(y) - \theta \right\|_2^2.$$

It depends on the signal (θ), the estimator $\hat{\theta}$, the noise level ϵ, and the basis.

For a fixed ϵ, the optimal minmax procedure is the one that minimizes the error for the worst possible signal from the coefficient body Θ.

$$R_\epsilon^*(\Theta) = \inf_{\hat{\theta}} \sup_{\theta \in \Theta} R_\epsilon(\hat{\theta}, \theta).$$

Consider the particular nonlinear procedure $\hat{\theta}$ that corresponds to *soft-thresholding* of every noisy coefficient y_i:

$$T_\epsilon(x_i) = \text{sgn}(y_i)(|y_i| - \epsilon)_+.$$

Let $r_\epsilon(\theta)$ be the corresponding error for signal θ and let $r_\epsilon^\star(\Theta)$ be the worst-case error for the coefficient body Θ.

If the coefficient body is solid, orthosymmetric in a particular basis, then asymptotically ($\epsilon \to 0$) the error decays at least as fast in this basis as in any other basis. That is $r_\epsilon(\Theta)$ approaches zero at least as fast as $r_\epsilon(U\Theta)$ for any orthogonal matrix U. Therefore, unconditional bases are nearly optimal asymptotically. Moreover, for small ϵ we can relate this procedure to any other procedure as follows [Don93b]:

$$R^*(\epsilon, \Theta) \leq r^*(\epsilon, \Theta) \leq O(\log(1/\epsilon)) \cdot R^*(\epsilon, \Theta), \qquad \epsilon \to 0.$$

10.5   Signal and Image Compression

Fundamentals of Data Compression

From basic information theory, we know the minimum average number of bits needed to represent realizations of a independent and identically distributed discrete random variable X is its *entropy* $H(X)$ [CT91]. If the distribution $p(X)$ is known, we can design Huffman codes or use the arithmetic coding method to achieve this minimum [BCW90]. Otherwise we need to use adaptive method [WNC87].

Continuous random variables require an infinite number of bits to represent, so quantization is always necessary for practical finite representation. However, quantization introduces error. Thus the goal is to achieve the best rate-distortion tradeoff [JN84, CT91, BW94]. Text compression [BCW90], waveform coding [JN84] and subband coding [VK95] have been studied extensively over the years. Here we concentrate on wavelet compression, or more general, transform coding. Also we concentrate on low bitrate.

Figure 10.10. Prototype Transform Coder

Prototype Transform Coder

The simple three-step structure of a prototype transform coder is shown in Figure 10.10. The first step is the transform of the signal. For a length-N discrete signal $f(n)$, we expand it using a set of orthonormal basis functions as

$$f(n) = \sum_{1}^{N} c_i \psi_i(n), \tag{10.38}$$

where

$$c_i = \langle f(n), \psi_i(n) \rangle. \tag{10.39}$$

We then use the uniform scalar quantizer Q as in Figure 10.11, which is widely used for wavelet based image compression [Sha93, SP96a],

$$\hat{c}_i = Q(c_i). \tag{10.40}$$

Denote the quantization step size as T. Notice in the figure that the quantizer has a dead zone, so if $|c_i| < T$, then $Q(c_i) = 0$. We define an index set for those insignificant coefficients

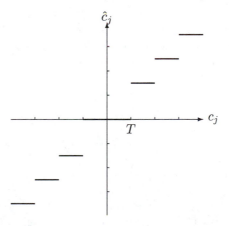

Figure 10.11. Uniform Scalar Quantizer

$\mathcal{I} = \{i : |c_i| < T\}$. Let M be the number of coefficients with magnitudes greater than T (significant coefficients). Thus the size of $\mathcal{I}$ is $N - M$. The squared error caused by the quantization is

$$\sum_{i=1}^{N}(c_i - \hat{c}_i)^2 = \sum_{i \in \mathcal{I}} c_i^2 + \sum_{i \notin \mathcal{I}}(c_i - \hat{c}_i)^2. \tag{10.41}$$

Since the transform is orthonormal, it is the same as the reconstruction error. Assume T is small enough, so that the significant coefficients are uniformly distributed within each quantization bins. Then the second term in the error expression is

$$\sum_{i \notin \mathcal{I}}(c_i - \hat{c}_i)^2 = M\frac{T^2}{12}. \tag{10.42}$$

For the first term, we need the following standard approximation theorem [DL93] that relates it to the l_p norm of the coefficients,

$$\|f\|_p = \left(\sum_{i=1}^{N} |c_i|^p\right)^{1/p}. \tag{10.43}$$

Theorem 56 *Let* $\lambda = \frac{1}{p} > \frac{1}{2}$ *then*

$$\sum_{i \in \mathcal{I}} c_i^2 \leq \frac{\|f\|_p^2}{2\lambda - 1} M^{1-2\lambda} \tag{10.44}$$

This theorem can be generalized to infinite dimensional space if $\|f\|_p^2 < +\infty$. It has been shown that for functions in a Besov space, $\|f\|_p^2 < +\infty$ does not depend on the particular choice of the wavelet as long as each wavelet in the basis has $q > \lambda - \frac{1}{2}$ vanishing moments and is q times continuously differentiable [Mey92b]. The Besov space includes piece-wise regular functions that may include discontinuities. This theorem indicates that the first term of the error expression decreases very fast when the number of significant coefficient increases.

The bit rate of the prototype compression algorithm can also be separated in two parts. For the first part, we need to indicate whether the coefficient is significant, also known as the significant map. For example, we could use 1 for significant, and 0 for insignificant. We need a total of N these indicators. For the second part, we need to represent the values of the significant coefficients. We only need M values. Because the distribution of the values and the indicators are not known in general, adaptive entropy coding is often used [WNC87].

Energy concentration is one of the most important properties for low bitrate transform coding. Suppose for the sample quantization step size T, we have a second set of basis that generate less significant coefficients. The distribution of the significant map indicators is more skewed, thus require less bits to code. Also, we need to code less number of significant values, thus it may require less bits. In the mean time, a smaller M reduces the second error term as in Equation 10.42. Overall, it is very likely that the new basis improves the rate-distortion performance. Wavelets have better energy concentration property than the Fourier transform for signals with discontinuities. This is one of the main reasons that wavelet based compression methods usually out perform DCT based JPEG, especially at low bitrate.

Improved Wavelet Based Compression Algorithms

The above prototype algorithm works well [MF97, Guo97], but can be further improved for its various building blocks [GB97b]. As we can see from Figure 10.12, the significant map still has considerable structure, which could be exploited. Modifications and improvements use the following ideas:

- Insignificant coefficients are often clustered together. Especially, they often cluster around the same location across several scales. Since the distance between nearby coefficients doubles for every scale, the insignificant coefficients often form a tree shape, as we can see from Figure 2.5. These so called *zero-trees* can be exploited [Sha93, SP96a] to achieve excellent results.

- The choice of basis is very important. Methods have been developed to adaptively choose the basis for the signal [RV93, XHRO95]. Although they could be computationally very intensive, substantial improvement can be realized.

- Special run-length codes could be used to code significant map and values [TW96, TVC96].

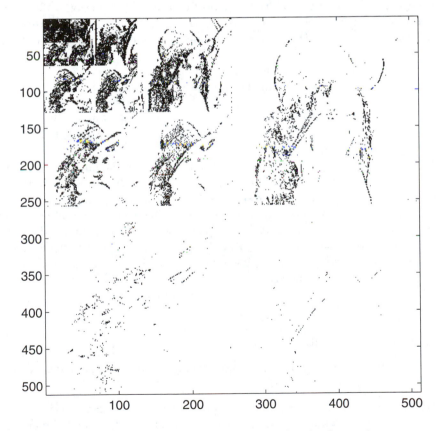

Figure 10.12. The Significant Map for the Lenna image.

- Advanced quantization methods could be used to replace the simple scalar quantizer [JCF95].

- Method based on statistical analysis like classification, modeling, estimation, and prediction also produces impressive result [LRO97].

- Instead of using one fixed quantization step size, we can successively refine the quantization by using smaller and smaller step sizes. These embedded schemes allow both the encoder and the decoder to stop at any bit rate [Sha93, SP96a].

- The wavelet transform could be replaced by an integer-to-integer wavelet transform, no quantization is necessary, and the compression is lossless [SP96a].

Other references are:[Don95, DJ94b, Don93b, Sai94b, GOL*94b, Sha93, SC95, SC96, Sha93, VBL95, SP96a, SP96b, BW94, Guo97].

10.6   Why are Wavelets so Useful?

The basic wavelet in wavelet analysis can be chosen so that it is *smooth*, where smoothness is measured in a variety of ways [Mey87]. To represent $f(t)$ with K derivatives, one can choose a wavelet $\psi(t)$ that is K (or more) times continuously differentiable; the penalty for imposing greater smoothness in this sense is that the supports of the basis functions, the filter lengths and hence the computational complexity all increase. Besides, smooth wavelet bases are also the "best bases" for representing signals with arbitrarily many singularities [Don93b], a remarkable property.

The usefulness of wavelets in representing functions in these and several other classes stems from the fact that for most of these spaces the wavelet basis is an *unconditional basis*, which is a near-optimal property.

To complete this discussion, we have to motivate the property of an unconditional basis being asymptotically optimal for a particular problem, say data compression [Don93b]. Figure 10.13 suggests why a basis in which the coefficients are solid and orthosymmetric may be desired. The signal class is defined to be the interior of the rectangle bounded by the lines $x = \pm a$ and $y = \pm b$. The signal corresponding to point A is the worst-case signal for the two bases shown in the figure; the residual error (with $n = 1$) is given by $a\sin(\theta) + b\cos(\theta)$ for $\theta \in \{0, \alpha\}$ and is minimized by $\theta = 0$, showing that the orthosymmetric basis is preferred. This result is really a consequence of the fact that $a \neq b$ (which is typically the case why one uses transform coding—if $a = b$, it turns out that the "diagonal" basis with $\theta = \frac{\pi}{4}$ is optimal for $n = 1$). The closer the coefficient body is to a solid, orthosymmetric body with varying side lengths, the less the individual coefficients are correlated with each other and the greater the compression in this basis.

In summary, the wavelet bases have a number of useful properties:

1. They can represent smooth functions.

2. They can represent singularities

3. The basis functions are local. This makes most coefficient-based algorithms naturally adaptive to inhomogeneities in the function.

4. They have the unconditional basis (or near optimal in a minimax sense) property for a variety of function classes implying that if one knows very little about a signal, the wavelet basis is usually a reasonable choice.

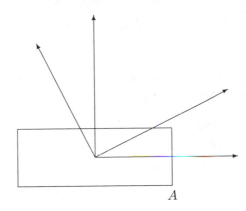

A

Figure 10.13. Optimal Basis for Data Compression

5. They are computationally inexpensive—perhaps one of the few really useful linear transform with a complexity that is $O(N)$—as compared to a Fourier transform, which is $N \log(N)$ or an arbitrary linear transform which is $O(N^2)$.

6. Nonlinear soft-thresholding is near optimal for statistical estimation.

7. Nonlinear soft-thresholding is near optimal for signal recovery.

8. Coefficient vector truncation is near optimal for data compression.

10.7 Applications

Listed below are several application areas in which wavelet methods have had some success.

Numerical Solutions to Partial Differential Equations

The use of wavelets as basis functions for the discretization of PDEs has had excellent success. They seem to give a generalization of finite element methods with some characteristics of multigrid methods. It seems to be the localizing ability of wavelet expansions that give rise to sparse operators and good numerical stability of the methods [GLRT90, RBSB94, RW97, BCR91, Bey94, HPW94, BCR92, LPT92, BK97, BKV93].

Seismic and Geophysical Signal Processing

One of the exciting applications areas of wavelet-based signal processing is in seismic and geophysical signal processing. Applications of denoising, compression, and detection are all important here, especially with higher-dimensional signals and images. Some of the references can be found in [RD86, OY87, Sca94, FK94, Sai94b, LM89, GGT89, GB96c] [GGM84b, GGM84a, DDO97, Bey97].

Medical and Biomedical Signal and Image Processing

Another exciting application of wavelet-based signal processing is in medical and biomedical signal and image processing. Again, applications of denoising, compression, and detection are all important here, especially with higher dimensional signals and images. Some of the references can be found in [AU96, Tur86, IRP*96].

Application in Communications

Some applications of wavelet methods to communications problems are in [TM94b, LKC*94, Lin95, WWJ95, SL96].

Fractals

Wavelet-based signal processing has been combined with fractals and to systems that are chaotic [FHV93, Mas94, Wor96, DL92, AAEG89, Aus89, WO92a, WO92b]. The multiresolution formulation of the wavelet and the self-similar characteristic of certain fractals make the wavelet a natural tool for this analysis. An application to noise removal from music is in [BCG94].

Other applications are to the automatic target recognition (ATR) problem, and many other questions.

10.8   Wavelet Software

There are several software packages available to study, experiment with, and apply wavelet signal analysis. There are several MATLAB programs at the end of this book. MathWorks, Inc. has a Wavelet Toolbox [MMOP96]; Donoho's group at Stanford has WaveTool; the Yale group has XWPL and WPLab [Wic95]; Taswell at Stanford has WavBox [Tas96], a group in Spain has Uvi-Wave; MathSoft, Inc. has S+WAVELETS; Aware, Inc. has WaveTool; and the DSP group at Rice has a MATLAB wavelet toolbox available over the internet at http://www-dsp.rice.edu. There is a good description and list of several wavelet software packages in [BDG96]. There are several MATLAB programs in Appendix C of this book. They were used to create the various examples and figures in this book and should be studied when studying the theory of a particular topic.

Chapter 11

Summary Overview

This chapter presents a brief summary of the results of this book to show how the basic ideas developed around the multiplier-2 (2-band) wavelet analysis and the DWT generalize into a very rich set of systems and choices that can become bewildering.

11.1 Properties of the Basic Multiresolution Scaling Function

The first summary is given in four tables of the basic relationships and equations, primarily developed in Chapter 5, for the scaling function $\varphi(t)$, scaling coefficients $h(n)$, and their Fourier transforms $\Phi(\omega)$ and $H(\omega)$ for the multiplier $M = 2$ or two-band multiresolution system. The various assumptions and conditions are omitted in order to see the "big picture" and to see the effects of increasing constraints.

Case	Condition	$\varphi(t)$	$\Phi(\omega)$	Signal Space
1	Multiresolution	$\varphi(t) = \sum h(n)\sqrt{2}\varphi(2t-n)$	$\Phi(\omega) = \prod \frac{1}{\sqrt{2}}H(\frac{\omega}{2^k})$	distribution
2	Partition of 1	$\sum \varphi(t-n) = 1$	$\Phi(2\pi k) = \delta(k)$	distribution
3	Orthogonal	$\int \varphi(t)\,\varphi(t-k)\,dt = \delta(k)$	$\sum |\Phi(\omega+2\pi k)|^2 = 1$	L^2
5	SF Smoothness	$\frac{d^{(\ell)}\varphi}{dt} < \infty$		poly $\in \mathcal{V}_j$
6	SF Moments	$\int t^k \varphi(t)\,dt = 0$		Coiflets

Table 11.1. Properties of $M = 2$ Scaling Functions (SF) and their Fourier Transforms

Case	Condition	$h(n)$	$H(\omega)$	Eigenval.$\{\mathbf{T}\}$
1	Existence	$\sum h(n) = \sqrt{2}$	$H(0) = \sqrt{2}$	
2	Fundamental	$\sum h(2n) = \sum h(2n+1)$	$H(\pi) = 0$	EV $= 1$
3	QMF	$\sum h(n)\,h(n-2k) = \delta(k)$	$|H(\omega)|^2 + |H(\omega+\pi)|^2 = 2$	EV ≤ 1
4	Orthogonal L^2 Basis	$\sum h(n)\,h(n-2k) = \delta(k)$	$|H(\omega)|^2 + |H(\omega+\pi)|^2 = 2$ and $H(\omega) \neq 0, |\omega| \leq \pi/3$	one EV $= 1$ others < 1
6	Coiflets	$\sum n^k h(n) = 0$		

Table 11.2. Properties of $M = 2$ Scaling Coefficients and their Fourier Transforms

219

Case	Condition	$\psi(t)$	$\Psi(\omega)$	Signal Space
1	MRA	$\psi(t) = \sum h_1(n)\sqrt{2}\varphi(2t-n)$	$\Psi(\omega) = \prod \frac{1}{\sqrt{2}}H_1(\frac{\omega}{2^k})$	distribution
3	Orthogonal	$\int \varphi(t)\,\psi(t-k)\,dt = 0$		L^2
3	Orthogonal	$\int \psi(t)\,\psi(t-k)\,dt = \delta(k)$		L^2
5	W Moments	$\int t^k\,\psi(t)\,dt = 0$		poly $not \in \mathcal{W}_j$

Table 11.3. Properties of $M = 2$ Wavelets (W) and their Fourier Transforms

Case	Condition	$h_1(n)$	$H_1(\omega)$	Eigenval.$\{\mathbf{T}\}$
2	Fundamental	$\sum h_1(n) = 0$	$H_1(0) = 0$	
3	Orthogonal	$h_1(n) = (-1)^n h(1-n)$	$\|H_1(\omega)\| = \|H(\omega+\pi)\|$	
3	Orthogonal	$\sum h_1(n)h_1(2m-n) = \delta(m)$	$\|H_1(\omega)\|^2 + \|H(\omega)\|^2 = 2$	
5	Smoothness	$\sum n^k h_1(n) = 0$	$H(\omega) = (\omega-\pi)^k \tilde{H}(\omega)$	$1, \frac{1}{2}, \frac{1}{4}, \cdots$

Table 11.4. Properties of $M = 2$ Wavelet Coefficients and their Fourier Transforms

The different "cases" represent somewhat similar conditions for the stated relationships. For example, in Case 1, Table 1, the multiresolution conditions are stated in the time and frequency domains while in Table 2 the corresponding necessary conditions on $h(n)$ are given for a scaling function in L^1. However, the conditions are not sufficient unless general distributions are allowed. In Case 1, Table 3, the definition of a wavelet is given to span the appropriate multiresolution signal space but nothing seems appropriate for Case 1 in Table 4. Clearly the organization of these tables are somewhat subjective.

If we "tighten" the restrictions by adding one more linear condition, we get Case 2 which has consequences in Tables 1, 2, and 4 but does not guarantee anything better that a distribution. Case 3 involves orthogonality, both across scales and translations, so there are two rows for Case 3 in the tables involving wavelets. Case 4 adds to the orthogonality a condition on the frequency response $H(\omega)$ or on the eigenvalues of the transition matrix to guarantee an L^2 basis rather than a tight frame guaranteed for Case 3. Cases 5 and 6 concern zero moments and scaling function smoothness and symmetry.

In some cases, columns 3 and 4 are equivalent and others, they are not. In some categories, a higher numbered case assumes a lower numbered case and in others, they do not. These tables try to give a structure without the details. It is useful to refer to them while reading the earlier chapters and to refer to the earlier chapters to see the assumptions and conditions behind these tables.

11.2   Types of Wavelet Systems

Here we try to present a structured list of the various classes of wavelet systems in terms of modification and generalizations of the basic $M = 2$ system. There are some classes not included here because the whole subject is still an active research area, producing new results daily. However, this list plus the table of contents, index, and references will help guide the reader through the maze. The relevant section or chapter is given in parenthesis for each topic.

- Signal Expansions   (1)

    - General Expansion Systems   (7.6)
    - Multiresolution Systems   (2)

- Multiresolution Wavelet Systems   (2)

    - $M = 2$ or two-band wavelet systems   (2-6)
    - $M > 2$ or M-band wavelet systems   (7.2)
    - wavelet packet systems   (7.3)
    - multiwavelet systems   (7.5

- Length of scaling function filter   (5)

    - Compact support wavelet systems
    - Infinite support wavelet systems

- Orthogonality   (5)

    - Orthogonal or Orthonormal wavelet bases
    - Semiorthogonal systems
    - Biorthogonal systems   (7.4)

- Symmetry

    - Symmetric scaling functions and wavelets   (7.4,7.5)
    - Approximately symmetric systems   (6.9)
    - Minimum phase spectral factorization systems   (6)
    - General scaling functions

- Complete and Overcomplete systems   (4,7.6)

    - Frames
    - Tight frames
    - Redundant systems and transforms   (7.6,10.3)
    - Adaptive systems and transforms, pursuit methods   (7.6)

- Discrete and continuous signals and transforms {analogous Fourier method}   (7.8)

    – Discrete Wavelet Transform {Fourier series} (1)
    – Discrete-time Wavelet Transform {Discrete Fourier transforms} (3,8)
    – Continuous-time Wavelet Transform {Fourier transform or integral} (7.8)

- Wavelet design   (6)

    – Max. zero wavelet moments [Daubechies]
    – Max. zero scaling function moments
    – Max. mixture of SF and wavelet moments zero [Coifman]   (6.9)
    – Max. smooth scaling function or wavelet [Heller, Lang, etc.]
    – Min. scaling variation [Gopinath, Odegard, etc.]
    – Frequency domain criteria
        * Butterworth [Daubechies]
        * least-squares, constrained LS, Chebyshev
    – Cosine modulated for M-band systems   (8.5)

- Descriptions   (2)

    – The signal itself
    – The discrete wavelet transform (expansion coefficients)
    – Time functions at various scales or translations
    – Tiling of the time-frequency/scale plane   (7.1)

Chapter 12

References

We would especially recommend five other books that complement this one. An excellent reference for the history, philosophy, and overview of wavelet analysis has been written by Barbara Burke Hubbard [Hub96]. The best source for the mathematical details of wavelet theory is by Ingrid Daubechies [Dau92]. Two good general books which starts with the discrete-time wavelet series and filter bank methods are by Martin Vetterli and Jelena Kovačević [VK95] and by Gilbert Strang and Truong Nguyen [SN96]. P. P. Vaiyanathan has written a good book on general multirate systems as well as filter banks [Vai92].

Much of the recent interest in compactly supported wavelets was stimulated by Daubechies [Dau90, Dau88a, Dau92, CDV93] and S. Mallat [Mal89b, Mal89c, Mal89a] and others [Law90, Law91b]. A powerful point of view has been recently presented by D. L. Donoho, I. M. Johnstone, R. R. Coifman, and others [Don93b, Don95, Don93a, Donar, DJKP95b, DJ95, DJ94a, BDGM94, CD95a, CD95b]. The development in the DSP community using filters has come from Smith and Barnwell [SB86a, SB87], Vetterli [Vet86, Vet87, VG89, VK95], and Vaidyanathan [Vai87a, VM88, Vai92]. Some of the work at Rice is reported in [Gop90, GB95c, SHGB93, GB95b, Gop92, Bur93, GOL*94a, GOL*94b, OGL*95, LGO*95, WOG*95a] [LGOB95, GLOB95, LGO*96, WOG*95b, WB95] Analysis and experimental work was done using the Matlab computer software system [MLB89, MMOP96]. Overview and introductory articles can be found in [GKM89, Mey89, Dau89, Bur93, Coi90, Str89, GB92c, Rus92, BNW94, Bur94, Str94]. [RV91, JS94a, Ngu95b, VD95, BDG96] Two special issues of IEEE Transactions have focused on wavelet methods [DMW92, DFN*93]. Books on wavelets, some of which are edited conference proceedings include [CGT89, KR90, AKM90, Mey90, Mey92a, Chu92a, Chu92b, Dau92, Vai92, AH92, RBC*92, Mal92, Mey92b] [You93, Mey93, BF93, SW93, Koo93, FHV93, Fli94, CMP94, FK94, Kai94] [Wal94, Mas94, VK95, Wic95, SN96, AU96, HW96b, Wor96, Hub96, Tas96] [AH92, Mal97, Tew97, AS96, RW97, Chu97].

Another way to keep up with current research and results on wavelets is to read the *Wavelet Digest* on the world-wide-web at: http://www.math.scarolina.edu/~wavelet edited by Wim Sweldens.

Bibliography

[AAEG89]   F. Argoul, A. Arneodo, J. Elezgaray, and G. Grasseau. Wavelet transform of fractal aggregates. *Physics Letters A.*, 135:327–336, March 1989.

[AAU96]   Akram Aldroubi, Patrice Abry, and Michael Unser. Construction of biorthogonal wavelets starting from any two multiresolutions. *preprint*, 1996.

[AH92]   A. N. Akansu and R. A. Haddad. *Multiresolution Signal Decomposition, Transforms, Subbands, and Wavelets.* Academic Press, San Diego, CA, 1992.

[AKM90]   Louis Auslander, Tom Kailath, and Sanjoy K. Mitter, editors. *Signal Processing, Part I: Signal Processing Theory.* Springer-Verlag, New York, 1990. IMA Volume 22, lectures from IMA program, July 1988.

[Ald96]   A. Aldroubi. *Oblique and Hierarchical Multiwavelet Bases.* Technical Report, National Institutes of Health, December 1996.

[Alp93]   B. Alpert. A class of bases in l^2 for the sparce representation of integral operators. *SIAM J. Math. Analysis*, 24, 1993.

[AS96]   Ali N. Akansu and Mark J. T. Smith. *Subband and Wavelet Transforms, Design and Applications.* Kluwer Academic Publishers, Boston, 1996.

[AU96]   Akram Aldroubi and Michael Unser, editors. *Wavelets in Medicine and Biology.* CRC Press, Boca Raton, 1996.

[Aus89]   P. Auscher. *Ondelettes fractales et applications.* PhD thesis, 1989.

[AWW92]   P. Auscher, G. Weiss, and M. V. Wickerhauser. Local sine and cosine bases of Coifman and Meyer and the construction of smooth wavelets. In C. K. Chui, editor, *Wavelets: A Tutorial in Theory and Applications*, pages 15–51, Academic Press, 1992. Volume 2 in series on Wavelet Analysis and its Applications.

[BBH93]   J. N. Bradley, C. M. Brislawn, and T. Hopper. The FBI wavelet/scalar quantization standard for gray-scale fingerprint image compression. In *Visual Info. Process. II*, SPIE, Orlando, FL, April 1993.

[BBOH96]   C. M. Brislawn, J. N. Bradley, R. J. Onyshczak, and T. Hopper. The FBI compression standard for digitized fingerprint images. In *Proceedings of the SPIE Conference 2847, Applications of Digital Image Processing XIX*, 1996.

[BC92]   S. Basu and C. Chiang. Complete parameterization of two dimensional orthonormal wavelets. In *Proceedings of IEEE-SP Symposium on Time-Frequency and Time-Scale Methods '92, Victoria, BC*, IEEE, 1992.

[BCG94]   Jonathan Berger, Ronald R. Coifman, and Maxim J. Goldberg. Removing noise from music using local trigonometric bases and wavelet packets. *Journal of the Audio Engineering Society*, 42(10):808–817, October 1994.

[BCR91]   G. Beylkin, R. R. Coifman, and V. Rokhlin. Fast wavelet transforms and numerical algorithms I. *Communications on Pure and Applied Mathematics*, 44:141–183, 1991.

[BCR92]   G. Beylkin, R. R. Coifman, and V. Rokhlin. Wavelets in numerical analysis. In M. B. Ruskai, G. Beylkin, I. Daubechies, Y. Meyer, R. Coifman, S. Mallat, and L Raphael, editors, *Wavelets and Their Applications*, pages 181–210, Jones and Bartlett, Boston, 1992. Outgrowth of the NSF/CBMS Conference on Wavelets, Lowell, June 1990.

[BCW90]   T. C. Bell, J. G. Cleary, and I. H. Witten. *Text Compression*. Prentice Hall, N.J., 1990.

[BDG96]   Andrew Brice, David Donoho, and Hong-Ye Gao. Wavelet analysis. *IEEE Spectrum*, 33(10):26–35, October 1996.

[BDGM94]   A. G. Bruce, D. L. Donoho, H.-Y. Gao, and R. D. Martin. Denoising and robust nonlinear wavelet analysis. In *Proceedings of Conference on Wavelet Applications*, pages 325–336, SPIE, Orlando, FL, April 1994.

[BE93]   F. Bao and Nurgun Erdol. On the discrete wavelet transform and shiftability. In *Proceedings of the Asilomar Conference on Signals, Systems and Computers*, pages 1442–1445, Pacific Grove, CA, November 1993.

[BE94]   F. Bao and N. Erdol. The optimal wavelet transform and translation invariance. In *Proceedings of the IEEE International Conference on Acoustics, Speech, and Signal Processing*, pages III:13–16, ICASSP-94, Adelaide, May 1994.

[Bey92]   G. Beylkin. On the representation of operators in bases of compactly supported wavelets. *SIAM Journal on Numerical Analysis*, 29(6):1716–1740, December 1992.

[Bey94]   G. Beylkin. On wavelet-based algorithms for solving differential equations. In John J. Benedetto and Michael W. Frazier, editors, *Wavelets: Mathematics and Applications*, pages 449–466, CRC Press, Boca Raton, 1994.

[Bey97]   Gregory Beylkin. An adaptive pseudo-wavelet approach for solving nonlinear partial differential equations. In W. Dahmen, A. Kurdila, and P. Oswald, editors, *Multiscale Wavelet Methods for Partial Differential Equations*, Academic Press, San Diego, 1997. Volume 6 in the series: Wavelet Analysis and Applications.

[BF93]   John J. Benedetto and Michael W. Frazier, editors. *Wavelets: Mathematics and Applications*. CRC Press, Boca Raton, FL, 1993.

[BK97]   Gregory Beylkin and James M Keiser. On the adaptive nomerical solution of nonlinear partial differential equations in wavelet bases. *Journal of Computational Physics*, 132:233–259, 1997.

[BKV93]   G. Beylkin, J. M. Keiser, and L. Vozovoi. *A New Class of Stabel Time Discretization Schemes for the Solution of Nonlinear PDE's*. Technical Report, Applied Mathematics Program, University of Colorado, Boulder, CO, 1993.

[BM95]   Albert P. Berg and Wasfy B. Mikhael. An efficient structure and algorithm for the mixed transform representation of signals. In *Proceedings of the 29th Asilomar Conference on Signals, Systems, and Computers*, Pacific Grove, CA, November 1995.

[BNW94]   A. Benveniste, R. Nikoukhah, and A. S. Willsky. Multiscale system theory. *IEEE Transactions on Circuits and Systems, I*, 41(1):2–15, January 1994.

[BO96]   C. S. Burrus and J. E. Odegard. Wavelet systems and zero moments. *IEEE Transactions on Signal Processing*, submitted, November 1 1996. Also CML Technical Report, Oct. 1996.

[BO97]   C. Sidney Burrus and Jan E. Odegard. Generalized coiflet systems. In *Proceedings of the International Conference on Digital Signal Processing*, Santorini, Greece, July 1997.

[Boa92]   Boualem Boashash, editor. *Time-Frequency Signal Analysis*. Wiley, Halsted Press, New York, 1992. Result of 1990 Special Converence on Time-Frequency Analysis, Gold Coast, Australia.

[BP85]   C. S. Burrus and T. W. Parks. *DFT/FFT and Convolution Algorithms*. John Wiley & Sons, New York, 1985.

[Bur93]   C. S. Burrus. *Scaling Functions and Wavelets*. first version written in 1989, The Computational Mathematics Lab. and ECE Dept., Rice University, Houston, Tx, 1993.

[Bur94]   Barbara Burke. The mathematical microscope: waves, wavelets, and beyond. In M. Bartusiak, et al, editor, *A Positron Named Priscilla, Scientific Discovery at the Frontier*, chapter 7, pages 196–235, National Academy Press, Washington, DC, 1994.

[BW94]   Michael Burrows and David J. Wheeler. *A Block-Sorting Lossless Data Compression Algorithm*. Technical Report 124, Digital Systems Research Center, Palo Alto, 1994.

[CD94]   Shaobing Chen and David L. Donoho. Basis pursuit. In *Proceedings of the 28th Asilomar Conference on Signals, Systems, and Computers*, pages 41–44, Pacific Grove, CA, November 1994. Also Stanford Statistics Dept. Report, 1994.

[CD95a]   Shaobing Chen and David L. Donoho. *Atomic Decomposition by Basis Pursuit*. Technical Report 479, Statistics Department, Stanford, May 1995. preprint.

[CD95b]   R. R. Coifman and D. L. Donoho. Translation-invariant de-noising. In Anestis Antoniadis, editor, *Wavelets and Statistics*, Springer-Verlag, to appear 1995. Springer Lecture Notes.

[CDF92]   A. Cohen, I. Daubechies, and J. C. Feauveau. Biorthogonal bases of compactly supported wavelets. *Communications on Pure and Applied Mathematics*, 45:485–560, 1992.

[CDM91]   S. Cavaretta, W. Dahmen, and C. A. Micchelli. *Stationary Subdivision*. Volume 93, American Mathematical Society, 1991.

[CDP95]   A. Cohen, I. Daubechies, and G. Plonka. *Regularity of refinable function vectors*. Technical Report 95/16, Universität Rostock, 1995. To appear in: J. Fourier Anal. Appl.

[CDV93]   Albert Cohen, Ingrid Daubechies, and Pierre Vial. Wavelets on the interval and fast wavelet transforms. *Applied and Computational Harmonic Analysis*, 1(1):54–81, December 1993.

[CGT89]   J. M. Combes, A. Grossmann, and P. Tchamitchian, editors. *Wavelets, Time-Frequency Methods and Phase Space*. Springer-Verlag, Berlin, 1989. Proceedings of the International Conference, Marseille, France, December 1987.

[Chu92a]   Charles K. Chui. *An Introduction to Wavelets*. Academic Press, San Diego, CA, 1992. Volume 1 in the series: *Wavelet Analysis and its Applications*.

[Chu92b]   Charles K. Chui. *Wavelets: A Tutorial in Theory and Applications*. Academic Press, San Diego, CA, 1992. Volume 2 in the series: *Wavelet Analysis and its Applications*.

[Chu97]   Charles K. Chui. *Wavelets: A Mathematical Tool for Signal Analysis*. SIAM, Philadilphia, 1997.

[CL96]   C. K. Chui and J. Lian. A study of orthonormal multi-wavelets. *Applied Numerical Mathematics*, 20(3):273–298, March 1996.

[CM94]   Ronald R. Coifman and Fazal Majid. *Adapted Waveform Analysis and Denoising*. Technical Report, Yale University, New Haven, 1994.

[CMP94]   Charles K. Chui, Laura Montefusco, and Luigia Puccio, editors. *Wavelets: Theory, Algorithms, and Applications*. Academic Press, San Diego, 1994. Volume 5 in the series: *Wavelet Analysis and its Applications*.

[CMQW92] R. R. Coifman, Y. Meyer, S. Quake, and M. V. Wickerhauser. Signal processing and compression with wave packets. In Y. Meyer, editor, *Proceedings of the International Conference on Wavelets, 1989 Marseille*, Masson, Paris, 1992.

[CMW92]   R. R. Coifman, Y. Meyer, and M. V. Wickerhauser. Wavelet analysis and signal processing. In M. B. Ruskai et al., editor, *Wavelets and Their Applications*, Jones and Bartlett, Boston, 1992.

[CMX96]   Z. Chen, C. A. Micchelli, and Y. Xu. *The Petrov-Galerkin Method for Second Kind Integral Equations II: Multiwavelet Schemes*. Technical Report, Math. Dept. North Dakota State University, November 1996.

[CNKS96] T. Cooklev, A. Nishihara, M. Kato, and M. Sablatash. Two-channel multifilter banks and multiwavelets. In *IEEE Proc. Int. Conf. Acoust., Speech, Signal Processing*, pages 2769–2772, 1996.

[Coh89]   L. Cohen. Time-frequency distributions - a review. *Proceedings of the IEEE*, 77(7):941–981, 1989.

[Coh92]   A. Cohen. Biorthogonal wavelets. In Charles K. Chui, editor, *Wavelets: A Tutorial in Theory and Applications*, Academic Press, Boca Raton, 1992. Volume 2 in the series: Wavelet Analysis and its Applications.

[Coh95]   Leon Cohen. *Time–Frequency Analysis*. Prentice Hall, Englewood Cliffs, NJ, 1995.

[Coi90]   R. R. Coifman. Wavelet analysis and signal processing. In Louis Auslander, Tom Kailath, and Sanjoy K. Mitter, editors, *Signal Processing, Part I: Signal Processing Theory*, pages 59–68, Springer-Verlag, New York, 1990. IMA vol. 22, lectures from IMA Program, summer 1988.

[Cro96]   Matthew Crouse. *Frame Robustness for De-Noising*. Technical Report, EE 696 Course Report, Rice University, Houston, Tx, May 1996.

[CS93]   A. Cohen and Q. Sun. An arthmetic characterization of the conjugate quadrature filters associated to orthonormal wavelet bases. *SIAM Journal of Mathematical Analysis*, 24(5):1355–1360, 1993.

[CT91]   T. M. Cover and J. A. Thomas. *Elements of Information Theory*. John Wiley $ Sons, N.Y., 1991.

[CW90]   Ronald R. Coifman and M. V. Wickerhauser. *Best-Adapted Wave Packet Bases*. Technical Report, Math Dept., Yale University, New Haven, 1990.

[CW92]   R. R. Coifman and M. V. Wickerhauser. Entropy-based algorithms for best basis selection. *IEEE Transaction on Information Theory*, 38(2):713–718, March 1992.

[Dau88a] Ingrid Daubechies. Orthonormal bases of compactly supported wavelets. *Communications on Pure and Applied Mathematics*, 41:909–996, November 1988.

[Dau88b] Ingrid Daubechies. Time-frequency localization operators: a geometric phase space approach. *IEEE Transactions on Information Theory*, 34(4):605–612, July 1988.

[Dau89]   Ingrid Daubechies. Orthonormal bases of wavelets with finite support – connection with discrete filters. In J. M. Combes, A. Grossman, and Ph. Tchamitchian, editors, *Wavelets, Time-Frequency Methods and Phase Space*, pages 38–66, Springer-Verlag, Berlin, 1989. Proceedings of International Colloquium on Wavelets and Applications, Marseille, France, Dec. 1987.

[Dau90]   Ingrid Daubechies. The wavelet transform, time-frequency localization and signal analysis. *IEEE Transaction on Information Theory*, 36(5):961–1005, September 1990. Also a Bell Labs Technical Report.

[Dau92]   Ingrid Daubechies. *Ten Lectures on Wavelets*. SIAM, Philadelphia, PA, 1992. Notes from the 1990 CBMS-NSF Conference on Wavelets and Applications at Lowell, MA.

[Dau93]   Ingrid Daubechies. Orthonormal bases of compactly supported wavelets II, variations on a theme. *SIAM Journal of Mathematical Analysis*, 24(2):499–519, March 1993.

[Dau96]   Ingrid Daubechies. Where do wavelets comre from? – a personal point of view. *Proceedings of the IEEE*, 84(4):510–513, April 1996.

[DD87]   G. Deslauriers and S. Dubuc. Interpolation dyadique. In G. Cherbit, editor, *Fractals, Dimensions Non Entiérs et Applications*, pages 44–45, Masson, Paris, 1987.

[DDO97]   Wolfgang Dahmen, Andrew Durdila, and Peter Oswald, editors. *Multiscale Wavelet Methods for Partial Differential Equations*. Academic Press, San Diego, 1997.

[DFN*93]   Special issue on wavelets and signal processing. *IEEE Transactions on Signal Processing*, 41(12):3213–3600, December 1993.

[DJ94a]   David L. Donoho and Iain M. Johnstone. Ideal denoising in an orthonormal basis chosen from a library of bases. *C. R. Acad. Sci. Paris, Ser. I*, 319, to appear 1994. Also Stanford Statistics Dept. Report 461, Sept. 1994.

[DJ94b]   David L. Donoho and Iain M. Johnstone. Ideal spatial adaptation via wavelet shrinkage. *Biometrika*, 81:425–455, 1994. Also Stanford Statistics Dept. Report TR-400, July 1992.

[DJ95]   David L. Donoho and Iain M. Johnstone. Adapting to unknown smoothness via wavelet shrinkage. *Journal of American Statist. Assn.*, to appear 1995. Also Stanford Statistics Dept. Report TR-425, June 1993.

[DJJ]   Ingrid Daubechies, Stéphane Jaffard, and Jean-Lin Journé. A simple Wilson orthonormal basis with exponential decay. *preprint*.

[DJKP95a]   David L. Donoho, Iain M. Johnstone, Gérard Kerkyacharian, and Dominique Picard. Discussion of "Wavelet Shrinkage: Asymptopia?". *Journal Royal Statist. Soc. Ser B.*, 57(2):337–369, 1995. Discussion of paper by panel and response by authors.

[DJKP95b]   David L. Donoho, Iain M. Johnstone, Gérard Kerkyacharian, and Dominique Picard. Wavelet shrinkage: asymptopia? *Journal Royal Statistical Society B.*, 57(2):301–337, 1995. Also Stanford Statistics Dept. Report TR-419, March 1993.

[DL91]   Ingrid Daubechies and Jeffrey C. Lagarias. Two-scale difference equations, part I. existence and global regularity of solutions. *SIAM Journal of Mathematical Analysis*, 22:1388–1410, 1991. From an internal report, AT&T Bell Labs, Sept. 1988.

[DL92]   Ingrid Daubechies and Jeffrey C. Lagarias. Two-scale difference equations, part II. local regularity, infinite products of matrices and fractals. *SIAM Journal of Mathematical Analysis*, 23:1031–1079, July 1992. From an internal report, AT&T Bell Labs, Sept. 1988.

[DL93]   R. DeVire and G. Lorentz. *Constructive Approximation*. Springer-Verlag, 1993.

[DM93]   R. E. Van Dyck and T. G. Marshall, Jr. Ladder realizations of fast subband/vq coders with diamond structures. In *Proceedings of IEEE International Symposium on Circuits and Systems*, pages III:177–180, ISCAS, 1993.

[DMW92]   Special issue on wavelet transforms and multiresolution signal analysis. *IEEE Transactions on Information Theory*, 38(2, part II):529–924, March, part II 1992.

[Don93a]   David L. Donoho. Nonlinear wavelet methods for recovery of signals, densities, and spectra from indirect and noisy data. In Ingrid Daubechies, editor, *Different Perspectives on Wavelets, I*, pages 173–205, American Mathematical Society, Providence, 1993. Proceedings of Symposia in Applied Mathematics and Stanford Report 437, July 1993.

[Don93b]   David L. Donoho. Unconditional bases are optimal bases for data compression and for sta-
            tistical estimation. *Applied and Computational Harmonic Analysis*, 1(1):100–115, December
            1993. Also Stanford Statistics Dept. Report TR-410, Nov. 1992.

[Don93c]   David L. Donoho. *Wavelet Skrinkage and W. V. D. – A Ten Minute Tour*. Technical
            Report TR-416, Statistics Department, Stanford University, Stanford, CA, January 1993.
            Preprint.

[Don94]   David L. Donoho. On minimum entropy segmentation. In C. K. Chui, L. Montefusco, and
            L. Puccio, editors, *Wavelets: Theory, Algorithms, and Applications*, Academic Press, San
            Diego, 1994. Also Stanford Tech Report TR-450, 1994; Volume 5 in the series: Wavelet
            Analysis and its Applications.

[Don95]   David L. Donoho. De-noising by soft-thresholding. *IEEE Transactions on Information
            Theory*, 41(3):613–627, May 1995. also Stanford Statistics Dept. report TR-409, Nov. 1992.

[Donar]   David L. Donoho. Interpolating wavelet transforms. *Applied and Computational Harmonic
            Analysis*, to appear. Also Stanford Statistics Dept. report TR-408, Nov. 1992.

[DS52]   R. J. Duffin and R. C. Schaeffer. A class of nonharmonic fourier series. *Transactions of the
            American Mathematical Society*, 72:341–366, 1952.

[DS83]   J. E. Dennis and R. B. Schnabel. *Numerical Methods for Unconstrained Optimization and
            Nonlinear Equations*. Prentice-Hall, Inc., Englewood Cliffs, New Jersey, 1st edition, 1983.

[DS96a]   Ingrid Daubechies and Wim Sweldens. *Factoring Wavelet Transforms into Lifting Steps*.
            Technical Report, Princeton and Lucent Technologies, NJ, September 1996. Preprint.

[DS96b]   T. R. Downie and B. W. Silverman. *The Discrete Multiple Wavelet Transform and Thresh-
            olding Methods*. Technical Report, University of Bristol, November 1996. Submitted to
            IEEE Tran. Signal Processing.

[Dut89]   P. Dutilleux. An implementation of the "algorithme à trou" to compute the wavelet trans-
            form. In J. M. Combes, A. Grossmann, and Ph. Tchamitchian, editors, *Wavelets, Time-
            Frequency Methods and Phase Space*, pages 2–20, Springer-Verlag, Berlin, 1989. Proceedings
            of International Colloquium on Wavelets and Applications, Marseille, France, Dec. 1987.

[DVN88]   Z. Doğanata, P. P. Vaidyanathan, and T. Q. Nguyen. General synthesis procedures for FIR
            lossless transfer matrices, for perfect-reconstruction multirate filter bank applications. *IEEE
            Transactions on Acoustics, Speech, and Signal Processing*, 36(10):1561–1574, October 1988.

[Eir92]   T. Eirola. Sobolev characterization of solutions of dilation equations. *SIAM Journal of
            Mathematical Analysis*, 23(4):1015–1030, July 1992.

[FHV93]   M. Farge, J. C. R. Hunt, and J. C. Vassilicos, editors. *Wavelets, Fractals, and Fourier
            Tranforms*. Clarendon Press, Oxford, 1993. Proceedings of a conference on Wavelets at
            Newnham College, Cambridge, Dec. 1990.

[FK94]   Efi Foufoula-Georgiou and Praveen Kumar, editors. *Wavelets in Geophyics*. Academic
            Press, San Diego, 1994. Volume in the series: Wavelet Analysis and its Applications.

[Fli94]   F. J. Fliege. *Multirate Digital Signal Processing: Multrirate Systems, Filter Banks, and
            Wavelets*. Wiley & Sons, New York, 1994.

[Gab46]   D. Gabor. Theory of communication. *Journal of the Institute for Electrical Engineers*,
            93:429–439, 1946.

[GB90]   R. A. Gopinath and C. S. Burrus. Efficient computation of the wavelet transforms. In *Pro-
            ceedings of the IEEE International Conference on Acoustics, Speech, and Signal Processing*,
            pages 1599–1602, Albuquerque, NM, April 1990.

[GB92a]　　R. A. Gopinath and C. S. Burrus. Cosine–modulated orthonormal wavelet bases. In *Paper Summaries for the IEEE Signal Processing Society's Fifth DSP Workshop*, page 1.10.1, Starved Rock Lodge, Utica, IL, September 13–16, 1992.

[GB92b]　　R. A. Gopinath and C. S. Burrus. On the moments of the scaling function ψ_0. In *Proceedings of the IEEE International Symposium on Circuits and Systems*, pages 963–966, ISCAS-92, San Diego, CA, May 1992.

[GB92c]　　R. A. Gopinath and C. S. Burrus. Wavelet transforms and filter banks. In Charles K. Chui, editor, *Wavelets: A Tutorial in Theory and Applications*, pages 603–655, Academic Press, San Diego, CA, 1992. Volume 2 in the series: Wavelet Analysis and its Applications.

[GB93]　　R. A. Gopinath and C. S. Burrus. Theory of modulated filter banks and modulated wavelet tight frames. In *Proceedings of the IEEE International Conference on Signal Processing*, pages III–169–172, IEEE ICASSP-93, Minneapolis, April 1993.

[GB94a]　　R. A. Gopinath and C. S. Burrus. On upsampling, downsampling and rational sampling rate filter banks. *IEEE Transactions on Signal Processing*, April 1994. Also Tech. Report No. CML TR91-25, 1991.

[GB94b]　　R. A. Gopinath and C. S. Burrus. Unitary FIR filter banks and symmetry. *IEEE Transaction on Circuits and Systems II*, 41:695–700, October 1994. Also Tech. Report No. CML TR92-17, August 1992.

[GB95a]　　R. A. Gopinath and C. S. Burrus. Factorization approach to unitary time-varying filter banks. *IEEE Transactions on Signal Processing*, 43(3):666–680, March 1995. Also a Tech Report No. CML TR-92-23, Nov. 1992.

[GB95b]　　R. A. Gopinath and C. S. Burrus. Theory of modulated filter banks and modulated wavelet tight frames. *Applied and Computational Harmonic Analysis: Wavelets and Signal Processing*, 2:303–326, October 1995. Also a Tech. Report No. CML TR-92-10, 1992.

[GB95c]　　Ramesh A. Gopinath and C. Sidney Burrus. On cosine–modulated wavelet orthonormal bases. *IEEE Transactions on Image Processing*, 4(2):162–176, February 1995. Also a Tech. Report No. CML TR-91-27, March 1992.

[GB96a]　　Haitao Guo and C. Sidney Burrus. Approximate FFT via the discrete wavelet transform. In *Proceedings of SPIE Conference 2825*, Denver, August 6–9 1996.

[GB96b]　　Haitao Guo and C. Sidney Burrus. Convolution using the discrete wavelet transform. In *Proceedings of the IEEE International Conference on Acoustics, Speech, and Signal Processing*, pages III–1291–1294, IEEE ICASSP-96, Atlanta, May 7–10 1996.

[GB96c]　　Haitao Guo and C. Sidney Burrus. Phase-preserving compression of seismic images using the self-adjusting wavelet transform. In *NASA Combined Industry, Space and Earth Science Data Compression Workshop (in conjunction with the IEEE Data Compression Conference, DCC-96), JPL Pub. 96-11*, pages 101–109, Snowbird, Utah, April 4 1996.

[GB97a]　　Haitao Guo and C. Sidney Burrus. Fast approximate Fourier transform via wavelet transforms. *IEEE Transactions on Signal Processing*, submitted, January 1997.

[GB97b]　　Haitao Guo and C. Sidney Burrus. Waveform and image compression with the Burrows Wheeler transform and the wavelet transform. In *Proceedings of the IEEE International Conference on Image Processing*, IEEE ICIP-97, Santa Barbara, October 26-29 1997.

[GB97c]　　Haitao Guo and C. Sidney Burrus. Wavelet transform based fast approximate Fourier transform. In *Proceedings of the IEEE International Conference on Acoustics, Speech, and Signal Processing*, pages III:1973–1976, IEEE ICASSP-97, Munich, April 21–24 1997.

[GGM84a]　　P. Goupillaud, A. Grossman, and J. Morlet. Cyclo-octave and related transforms in seismic signal analysis. *SIAM J. Math. Anal.*, 15:723–736, 1984.

[GGM84b] P. Groupillaud, A. Grossman, and J. Morlet. Cyclo-octave and related transforms in seismic signal analysis. *Geoexploration*, (23), 1984.

[GGT89] S. Ginette, A. Grossmann, and Ph. Tchamitchian. Use of wavelet transforms in the study of propagation of transient acoustic signals across a plane interface between two homogeneous media. In J. M. Combes, A. Grossmann, and Ph. Tchamitchian, editors, *Wavelets: Time-Frequency Methods and Phase Space*, pages 139–146, Springer-Verlag, Berlin, 1989. Proceedings of the International Conference, Marseille, Dec. 1987.

[GHM94] J. S. Geronimo, D. P. Hardin, and P. R. Massopust. Fractal functions and wavelet expansions based on several scaling functions. *Journal of Approximation Theory*, 78:373–401, 1994.

[GKM89] A. Grossman, R. Kronland–Martinet, and J. Morlet. Reading and understanding continuous wavelet transforms. In J. M. Combes, A. Grossmann, and Ph. Tchamitchian, editors, *Wavelets, Time-Frequency Methods and Phase Space*, pages 2–20, Springer-Verlag, Berlin, 1989. Proceedings of International Colloquium on Wavelets and Applications, Marseille, France, Dec. 1987.

[GL93] G. H. Golub and C. F. Van Loan. *Matrix Compuations*. Johns Hopkins University Press, 1993.

[GL94] T. N. T. Goodman and S. L. LEE. Wavelets of multiplicity r. *Tran. American Math. Society*, 342(1):307–324, March 1994.

[GLOB95] H. Guo, M. Lang, J. E. Odegard, and C. S. Burrus. Nonlinear processing of a shift-invariant DWT for noise reduction and compression. In *Proceedings of the International Conference on Digital Signal Processing*, pages 332–337, Limassol, Cyprus, June 26–28 1995.

[GLRT90] R. Glowinski, W. Lawton, M. Ravachol, and E. Tenenbaum. Wavelet solution of linear and nonlinear elliptic, parabolic and hyperbolic problems in one dimension. In *Proceedings of the Ninth SIAM International Conference on Computing Methods in Applied Sciences and Engineering*, Philadelphia, 1990.

[GLT93] T. N. T. Goodman, S. L. Lee, and W. S. Tang. Wavelets in wandering subspaces. *Tran. American Math. Society*, 338(2):639–654, August 1993.

[GOB92] R. A. Gopinath, J. E. Odegard, and C. S. Burrus. On the correlation structure of multiplicity M scaling functions and wavelets. In *Proceedings of the IEEE International Symposium on Circuits and Systems*, pages 959–962, ISCAS-92, San Diego, CA, May 1992.

[GOB94] R. A. Gopinath, J. E. Odegard, and C. S. Burrus. Optimal wavelet representation of signals and the wavelet sampling theorem. *IEEE Transactions on Circuits and Systems II*, 41(4):262–277, April 1994. Also a Tech. Report No. CML TR-92-05, April 1992, revised Aug. 1993.

[GOL*94a] H. Guo, J. E. Odegard, M. Lang, R. A. Gopinath, I. Selesnick, and C. S. Burrus. Speckle reduction via wavelet soft-thresholding with application to SAR based ATD/R. In *Proceedings of SPIE Conference 2260*, San Diego, July 1994.

[GOL*94b] H. Guo, J. E. Odegard, M. Lang, R. A. Gopinath, I. W. Selesnick, and C. S. Burrus. Wavelet based speckle reduction with application to SAR based ATD/R. In *Proceedings of the IEEE International Conference on Image Processing*, pages I:75–79, IEEE ICIP-94, Austin, Texas, November 13-16 1994.

[Gop90] Ramesh A. Gopinath. *The Wavelet Transforms and Time-Scale Analysis of Signals*. Master's thesis, Rice University, Houston, Tx 77251, 1990.

[Gop92] Ramesh A. Gopinath. *Wavelets and Filter Banks – New Results and Applications*. PhD thesis, Rice University, Houston, Tx, August 1992.

[Gop96a] R. A. Gopinath. Modulated filter banks and local trigonometric bases - some connections. Oct 1996. in preparation.

[Gop96b] R. A. Gopinath. Modulated filter banks and wavelets, a general unified theory. In *Proceedings of the IEEE International Conference on Acoustics, Speech, and Signal Processing*, pages 1585–1588, IEEE ICASSP-96, Atlanta, May 7–10 1996.

[GORB96] J. Götze, J. E. Odegard, P. Rieder, and C. S. Burrus. Approximate moments and regularity of efficiently implemented orthogonal wavelet transforms. In *Proceedings of the IEEE International Symposium on Circuits and Systems*, pages II–405–408, IEEE ISCAS-96, Atlanta, May 12-14 1996.

[Gri93] Gustaf Gripenberg. Unconditional bases of wavelets for Sobelov spaces. *SIAM Journal of Mathematical Analysis*, 24(4):1030–1042, July 1993.

[Guo94] Haitao Guo. *Redundant Wavelet Transform and Denoising*. Technical Report CML-9417, ECE Dept and Computational Mathematics Lab, Rice University, Houston, Tx, December 1994.

[Guo95] Haitao Guo. *Theory and Applications of the Shift-Invariant, Time-Varying and Undecimated Wavelet Transform*. Master's thesis, ECE Department, Rice University, April 1995.

[Guo97] Haitao Guo. *Wavelets for Approximate Fourier Transform and Data Compression*. PhD thesis, ECE Department, Rice University, Houston, Tx, May 1997.

[Haa10] Alfred Haar. Zur Theorie der orthogonalen Funktionensysteme. *Mathematische Annalen*, 69:331–371, 1910. Also in PhD thesis.

[HB92] F. Hlawatsch and G. F. Boudreaux-Bartels. Linear and quadratic time-frequency signal representations. *IEEE Signal Processing Magazine*, 9(2):21–67, April 1992.

[Hei93] Henk J. A. M. Heijmans. Descrete wavelets and multiresolution analysis. In Tom H. Koornwinder, editor, *Wavelets: An Elementary Treatment of Theory and Applications*, pages 49–80, World Scientific, Singapore, 1993.

[Hel95] Peter N. Heller. Rank m wavelet matrices with n vanishing moments. *SIAM Journal on Matrix Analysis*, 16:502–518, 1995. Also as technical report AD940123, Aware, Inc., 1994.

[Her71] O. Herrmann. On the approximation problem in nonrecursive digital filter design. *IEEE Transactions on Circuit Theory*, 18:411–413, May 1971. Reprinted in *DSP* reprints, IEEE Press, 1972, page 202.

[HHSM95] A. N. Hossen, U. Heute, O. V. Shentov, and S. K. Mitra. Subband DFT – Part II: accuracy, complexity, and applications. *Signal Processing*, 41:279–295, 1995.

[HKRV92] Cormac Herley, Jelena Kovačević, Kannan Ramchandran, and Martin Vetterli. Time-varying orthonormal tilings of the time-frequency plane. In *Proceedings of the IEEE Signal Processing Society's International Symposium on Time–Frequency and Time–Scale Analysis*, pages 11–14, Victoria, BC, Canada, October 4–6, 1992.

[HKRV93] Cormac Herley, Jelena Kovačević, Kannan Ramchandran, and Martin Vetterli. Tilings of the time-frequency plane: consturction of arbitrary orthogonal bases and fast tiling algorithms. *IEEE Transactions on Signal Processing*, 41(12):3341–3359, December 1993. Special issue on wavelets.

[HPW94] Frédéric Heurtaux, Fabrice Planchon, and Mladen V. Wickerhauser. Scale decomposition in Burgers' equation. In John J. Benedetto and Michael W. Frazier, editors, *Wavelets: Mathematics and Applications*, pages 505–524, CRC Press, Boca Raton, 1994.

[HR96] D. P. Hardin and D. W. Roach. *Multiwavelet Prefilters I: Orthogonal prefilters preserving approximation order $p \leq 2$*. Technical Report, Vanderbilt University, 1996.

[HRW92]   Peter N. Heller, Howard L. Resnikoff, and Raymond O. Wells, Jr. Wavelet matrices and the representation of discrete functions. In Charles K. Chui, editor, *Wavelets: A Tutorial in Theory and Applications*, pages 15–50, Academic Press, Boca Raton, 1992. Volume 2 in the series: Wavelet Analysis and its Applications.

[HSS*95]   P. N. Heller, V. Strela, G. Strang, P. Topiwala, C. Heil, and L. S. Hills. Multiwavelet filter banks for data compression. In *IEEE Proceedings of the International Symposium on Circuits and Systems*, pages 1796–1799, 1995.

[HSS96]   C. Heil, G. Strang, and V. Strela. Approximation by translates of refinable functions. *Numerische Mathematik*, 73(1):75 94, March 1996.

[Hub96]   Barbara Burke Hubbard. *The World According to Wavelets*. A K Peters, Wellesley, MA, 1996.

[HW89]   C. E. Heil and D. F. Walnut. Continuous and discrete wavelet transforms. *SIAM Review*, 31(4):628–666, December 1989.

[HW94]   Peter N. Heller and R. O. Wells. The spectral theory of multiresolution operators and applications. In C. K. Chui, L. Montefusco, and L. Puccio, editors, *Wavelets: Theory, Algorithms, and Applications*, pages 13–31, Academic Press, San Diego, 1994. Also as technical report AD930120, Aware, Inc., 1993; Volume 5 in the series: *Wavelet Analysis and its Applications*.

[HW96a]   Peter N. Heller and R. O. Wells. Sobolev regularity for rank M wavelets. *SIAM Journal on Mathematical Analysis*, submitted, Oct. 1996. Also a CML Technical Report TR9608, Rice University, 1994.

[HW96b]   Eugenio Hernández and Guido Weiss. *A First Course on Wavelets*. CRC Press, Boca Raton, 1996.

[IRP*96]   Plamen Ch. Ivanov, Michael G Rosenblum, C.-K. Peng, Joseph Mietus, Shlomo Havlin, H. Eugene Stanley, and Ary L. Goldberger. Scaling behaviour of heartbeat intervals obtained by wavelet-based time-series analysis. *Nature*, 383:323–327, September 26 1996.

[JB82]   H. W. Johnson and C. S. Burrus. The design of optimal DFT algorithms using dynamic programming. In *Proceedings of the IEEE International Conference on Acoustics, Speech, and Signal Processing*, pages 20–23, Paris, May 1982.

[JCF95]   R. L. Josho, V. J. Crump, and T. R. Fischer. Image subband coding using arithmetic coded trellis coded quantization. *IEEE Transactions on Circuits and Systems*, 515–523, December 1995.

[Jia95]   R. Q. Jia. Subdivision schemes in L_p spaces. *Advances in Computational Mathematics*, 3:309–341, 1995.

[JMNK96]   Bruce R. Johnson, Jason P. Modisette, Peter A. Nordlander, and James L. Kinsey. *Quadrature Integration for Compact Support Wavelets*. Technical Report, Chemistry Dept. Rice University, November 1996. preprint, submitted to *Journal of Computational Physics*.

[JN84]   N. S. Jayant and P. Noll. *Digital Coding of Waveforms*. Prentice-Hall, Inc., Englewood Cliffs, NJ, 1st edition, 1984.

[JRZ96a]   R. Q. Jia, S. D. Riemenschneider, and D. X. Zhou. *Approximation by Multiple Refinable Functions*. Technical Report, University of Alberta, 1996. To appear in: Canadian Journal of Mathematics.

[JRZ96b]   R. Q. Jia, S. D. Riemenschneider, and D. X. Zhou. *Vector Subdivision Schemes and Multiple Wavelets*. Technical Report, University of Alberta, 1996.

[JRZ97]   R. Q. Jia, S. D. Riemenschneider, and D. X. Zhou. *Smoothness of Multiple Refinable Functions and Multiple Wavelets.* Technical Report, University of Alberta, 1997.

[JS94a]   Björn Jawerth and Wim Sweldens. An overview of wavelet based multiresolution analyses. *SIAM Review,* 36:377–412, 1994. Also a University of South Carolina Math Dept. Technical Report, Feb. 1993.

[JS94b]   I. M. Johnstone and B. W. Silverman. *Wavelet Threshold Estimators for Data with Correlated Noise.* Technical Report, Statistics Dept., University of Bristol, September 1994.

[Kai94]   G. Kaiser. *A Friendly Guide to Wavelets.* Birkhäuser, Boston, 1994.

[KMDW95]   H. Krim, S. Mallat, D. Donoho, and A. Willsky. Best basis algorithm for signal enhancement. In *Proceedings of the IEEE International Conference on Acoustics, Speech, and Signal Processing,* pages 1561–1564, IEEE ICASSP-95 Detroit, May 1995.

[Koo93]   Tom H. Koornwinder, editor. *Wavelets: An Elementary Treatment of Theory and Applications.* World Scientific, Singapore, 1993.

[KR90]   G. Kaiser and M. B. Ruskai, editors. *NSF/CMBS Regional Conference on Wavelets.* University of Lowell, MA, June 11 - 15, 1990. Speakers: I.Daubechies, G.Beylkin, R.Coifman, S.Mallat, M.Vetterli, Aware, Inc.

[KS92]   A. A. A. C. Kalker and Imran Shah. Ladder structures for multidimensional linear phase perfect reconstruction filter banks and wavelets. In *Proceedings of SPIE Conference 1818 on Visual Communications and Image Processing,* 1992.

[KT93]   Jaroslav Kautsky and Radko Turcajová. Discrete biorthogonal wavelet transforms as block circulant matrices. *Linear Algegra and its Applications,* submitted 1993.

[KT94a]   Jaroslav Kautsky and Radko Turcajová. A matrix approach to discrete wavelets. In Charles K. Chui, Laura Montefusco, and Luigia Puccio, editors, *Wavelets: Theory, Algorithms, and Applications,* pages 117–136, Academic Press, Boca Raton, 1994. Volume 5 in the series: Wavelet Analysis and its Applications.

[KT94b]   Man Kam Kwong and P. T. Peter Tang. W-matrices, nonorthogonal multiresolution analysis, and finite signals of arbitrary length. *preprint,* 1994.

[KV92]   R. D. Koilpillai and P. P. Vaidyanathan. Cosine modulated FIR filter banks satisfying perfect reconstruction. *IEEE Transactions on Signal Processing,* 40(4):770–783, April 1992.

[Kwo94]   Man Kam Kwong. MATLAB implementation of W-matrix multiresolution analyses. *preprint-MCS-P462-0894,* September 1994.

[Law]   W. Lawton. Private communication.

[Law90]   Wayne M. Lawton. Tight frames of compactly supported affine wavelets. *Journal of Mathematical Physics,* 31(8):1898–1901, August 1990. Also Aware, Inc. Tech Report AD891012.

[Law91a]   Wayne M. Lawton. Multiresolution properties of the wavelet Galerkin operator. *Journal of Mathematical Physics,* 32(6):1440–1443, June 1991.

[Law91b]   Wayne M. Lawton. Necessary and sufficient conditions for constructing orthonormal wavelet bases. *Journal of Mathematical Physics,* 32(1):57–61, January 1991. Also Aware, Inc. Tech. Report AD900402, April 1990.

[Law97]   Wayne M. Lawton. Infinite convolution products & refinable distributions on Lie groups. *Transactions of the American Mathematical Soceity,* submitted 1997.

[LGO*95]   M. Lang, H. Guo, J. E. Odegard, C. S. Burrus, and R. O. Wells, Jr. Nonlinear processing of a shift-invariant DWT for noise reduction. In Harold H. Szu, editor, *Proceedings of SPIE Conference 2491, Wavelet Applications II,* pages 640–651, Orlando, April 17–21 1995.

[LGO*96] M. Lang, H. Guo, J. E. Odegard, C. S. Burrus, and R. O. Wells, Jr. Noise reduction using an undecimated discrete wavelet transform. *IEEE Signal Processing Letters*, 3(1):10–12, January 1996.

[LGOB95] M. Lang, H. Guo, J. E. Odegard, and C. S. Burrus. Nonlinear redundant wavelet methods for image enhancement. In *Proceedings of IEEE Workshop on Nonlinear Signal and Image Processing*, pages 754–757, Halkidiki, Greece, June 20–22 1995.

[LH96] Markus Lang and Peter N. Heller. The design of maximally smooth wavlets. In *Proceedings of the IEEE International Conference on Acoustics, Speech, and Signal Processing*, pages 1463–1466, IEEE ICASSP-96, Atlanta, May 1996.

[Lin95] A. R. Lindsey. *Generalized Orthogonally Multiplexed Communication via Wavelet Packet Bases*. PhD thesis, June 1995.

[LKC*94] R. E. Learned, H. Krim, B. Claus, A. S. Willsky, and W. C. Karl. Wavelet-packet-based multiple access communications. In *Proceedings of SPIE Conference, Wavelet Applications in Signal and Image Processing, Vol. 2303*, pages 246–259, San Diego, July 1994.

[LLK96] K.-C. Lian, J. Li, and C.-C. J. Kuo. Image compression with embedded multiwavelet coding. In *Proceedings of SPIE, Wavelet Application III*, pages 165–176, Orlando, FL, April 1996.

[LLS97a] Wayne M. Lawton, S. L. Lee, and Z. Shen. Convergence of multidimensional cascade algorithm. *Numerische Mathematik*, to appear 1997.

[LLS97b] Wayne M. Lawton, S. L. Lee, and Z. Shen. Stability and orthonormality of multivariate refinable functions. *SIAM Journal of Mathematical Analysis*, to appear 1997.

[LM89] J. L. Larsonneur and J. Morlet. Wavelet and seismic interpretation. In J. M. Combes, A. Grossmann, and Ph. Tchamitchian, editors, *Wavelets: Time-Frequency Methods and Phase Space*, pages 126–131, Springer-Verlag, Berlin, 1989. Proceedings of the International Conference, Marseille, Dec. 1987.

[LP89] G. Longo and B. Picinbono, editors. *Time and Frequency Representation of Signals and Systems*. Springer-Verlag, Wien – New York, 1989. CISM Courses and Lectures No. 309.

[LP94a] Jie Liang and Thomas W. Parks. A translation invariant wavelet representation algorithm with applications. *IEEE Transactions on Signal Processing*, submitted April 1994.

[LP94b] Jie Liang and Thomas W. Parks. A two-dimensional translation invariant wavelet representation and its applications. In *Proceedings of the IEEE International Conference on Image Processing*, pages I:66–70, Austin, November 1994.

[LPT92] J. Liandrat, V. Perrier, and Ph. Tchamitchian. Numerical resolution of nonlinear parital differential equations using the wavelet approach. In M. B. Ruskai, G. Beylkin, I. Daubechies, Y. Meyer, R. Coifman, S. Mallat, and L Raphael, editors, *Wavelets and Their Applications*, pages 181–210, Jones and Bartlett, Boston, 1992. Outgrowth of the NSF/CBMS Conference on Wavelets, Lowell, June 1990.

[LR91] Wayne M. Lawton and Howard L. Resnikoff. *Multidimensional Wavelet Bases*. Aware Report AD910130, Aware, Inc., February 1991.

[LRO97] S. M. LoPresto, K. Ramchandran, and M. T. Orchard. Image coding based on mixture modeling of wavelet coefficients and a fast estimation-quantization framework. *Proc. DCC*, March 1997.

[LSOB94] M. Lang, I. Selesnick, J. E. Odegard, and C. S. Burrus. Constrained FIR filter design for 2-band filter banks and orthonormal wavelets. In *Proceedings of the IEEE Digital Signal Processing Workshop*, pages 211–214, Yosemite, October 1994.

[LV95]   Yuan-Pei Lin and P. P. Vaidyanathan. Linear phase cosine-modulated filter banks. *IEEE Transactions on Signal Processing*, 43, 1995.

[Mal89a]   S. G. Mallat. Multifrequency channel decomposition of images and wavelet models. *IEEE Transactions on Acoustics, Speech and Signal Processing*, 37:2091–2110, December 1989.

[Mal89b]   S. G. Mallat. Multiresolution approximation and wavelet orthonormal bases of L^2. *Transactions of the American Mathematical Society*, 315:69–87, 1989.

[Mal89c]   S. G. Mallat. A theory for multiresolution signal decomposition: the wavelet representation. *IEEE Transactions on Pattern Recognition and Machine Intelligence*, 11(7):674–693, July 1989.

[Mal91]   S. G. Mallat. Zero-crossings of a wavelet transform. *IEEE Transactions on Information Theory*, 37(4):1019–1033, July 1991.

[Mal92]   Henrique S. Malvar. *Signal Processing with Lapped Transforms*. Artech House, Boston, MA, 1992.

[Mal97]   Stéphane Mallat. *Wavelet Signal Processing*. Academic Press, 1997. to appear.

[Mar91]   R. J. Marks II. *Introduction to Shannon Sampling and Interpolation Theory*. Springer-Verlag, New York, 1991.

[Mar92]   T. G. Marshall, Jr. Predictive and ladder realizations of subband coders. In *Proceedings of IEEE Workshop on Visual Signal Processing and Communication*, Raleigh, NC, 1992.

[Mar93]   T. G. Marshall, Jr. A fast wavelet transform based on the eucledean algorithm. In *Proceedings of Conference on Information Sciences and Systems*, Johns Hopkins University, 1993.

[Mas94]   Peter R. Massopust. *Fractal Functions, Fractal Surfaces, and Wavelets*. Academic Press, San Diego, 1994.

[Mau92]   J. Mau. Perfect reconstruction modulated filter banks. In *Proc. Int. Conf. Acoust., Speech, Signal Processing*, pages IV–273, IEEE, San Francisco, CA, 1992.

[Mey87]   Y. Meyer. L'analyses par ondelettes. *Pour la Science*, September 1987.

[Mey89]   Y. Meyer. Orthonormal wavelets. In J. M. Combes, A. Grossmann, and Ph. Tchamitchian, editors, *Wavelets, Time-Frequency Methods and Phase Space*, pages 21–37, Springer-Verlag, Berlin, 1989. Proceedings of International Colloquium on Wavelets and Applications, Marseille, France, Dec. 1987.

[Mey90]   Y. Meyer. *Ondelettes et opérateurs*. Hermann, Paris, 1990.

[Mey92a]   Y. Meyer, editor. *Wavelets and Applications*. Springer-Verlag, Berlin, 1992. Proceedings of the Marseille Workshop on Wavelets, France, May, 1989; Research Notes in Applied Mathematics, RMA-20.

[Mey92b]   Yves Meyer. *Wavelets and Operators*. Cambridge, Cambridge, 1992. Translated by D. H. Salinger from the 1990 French edition.

[Mey93]   Yves Meyer. *Wavelets, Algorithms and Applications*. SIAM, Philadelphia, 1993. Translated by R. D. Ryan based on lectures given for the Spanish Institute in Madrid in Feb. 1991.

[MF97]   Stéphane Mallat and Frédéric Falzon. Understanding image transform codes. In *Proceedings of SPIE Conference, Aerosense*, Orlando, April 1997.

[MLB89]   Cleve Moler, John Little, and Steve Bangert. MATLAB *User's Guide*. The MathWorks, Inc., South Natick, MA, 1989.

[MMOP96]   Michel Misiti, Yves Misiti, Georges Oppenheim, and Jean-Michel Poggi. *Wavelet Toolbox User's Guide*. The MathWorks, Inc., Natick, MA, 1996.

[Mou93]   Pierre Moulin. A wavelet regularization method for diffuse radar-target imaging and speckle-noise reduction. *Journal of Mathematical Imaging and Vision*, 3:123–134, 1993.

[MP89]   C. A. Micchelli and Prautzsch. Uniform refinement of curves. *Linear Algebra, Applications*, 114/115:841–870, 1989.

[MW94a]   Stephen Del Marco and John Weiss. Improved transient signal detection using a wavepacket-based detector with an extended translation-invariant wavelet transform. *IEEE Transactions on Signal Processing*, 43, submitted 1994.

[MW94b]   Stephen Del Marco and John Weiss. M-band wavepacket-based transient signal detector using a translation-invariant wavelet transform. *Optical Engineering*, 33(7):2175–2182, July 1994.

[MWJ94]   Stephen Del Marco, John Weiss, and Karl Jagler. Wavepacket-based transient signal detector using a translation invariant wavelet transform. In *Proceedings of Conference on Wavelet Applications*, pages 792–802, SPIE, Orlando, FL, April 1994.

[MZ93]   S. G. Mallat and Z. Zhang. Matching pursuits with time-frequency dictionaries. *IEEE Transactions on Signal Processing*, 41(12):3397–3415, December 1993.

[NAT96]   Mohammed Nafie, Murtaza Ali, and Ahmed Tewfik. Optimal subset selection for adaptive signal representation. In *Proceedings of the IEEE International Conference on Acoustics, Speech, and Signal Processing*, pages 2511–2514, IEEE ICASSP-96, Atlanta, May 1996.

[Ngu92]   T. Q. Nguyen. A class of generalized cosine-modulated filter banks. In *Proceedings of ISCAS, San Diego, CA*, pages 943–946, IEEE, 1992.

[Ngu94]   Trong Q. Nguyen. Near perfect reconstruction pseudo QMF banks. *IEEE Transactions on Signal Processing*, 42(1):65–76, January 1994.

[Ngu95a]   Trong Q. Nguyen. Digital filter banks design quadratic constrained formulation. *IEEE Transactions on Signal Processing*, 43(9):2103–2108, September 1995.

[Ngu95b]   Truong Q. Nguyen. Aliasing-free reconstruction filter banks. In Wai-Kai Chen, editor, *The Circuits and Filters Handbook*, chapter 85, pages 2682–2717, CRC Press and IEEE Press, Roca Raton, 1995.

[NH96]   Truong Q. Nguyen and Peter N. Heller. Biorthogonal cosine-modulated filter band. In *Proceedings of the IEEE International Conference on Acoustics, Speech, and Signal Processing*, pages 1471–1474, IEEE ICASSP-96, Atlanta, May 1996.

[NK92]   T. Q. Nguyen and R. D. Koilpillai. The design of arbitrary length cosine-modulated filter banks and wavelets satisfying perfect reconstruction. In *Proceedings of IEEE-SP Symposium on Time-Frequency and Time-Scale Methods '92, Victoria, BC*, pages 299–302, IEEE, 1992.

[nRGB91]   D. L. Jones nd R. G. Barniuk. Efficient approximation of continuous wavelet transforms. *Electronics Letters*, 27(9):748–750, 1991.

[NS95]   G. P. Nason and B. W. Silverman. *The Stationary Wavelet Transform and some Statistical Applications*. Technical Report, Department of Mathematics, University of Bristol, Bristol, UK, February 1995. preprint obtained via the internet.

[NV88]   T. Q. Nguyen and P. P. Vaidyanathan. Maximally decimated perfect-reconstruction FIR filter banks with pairwise mirror-image analysis and synthesis frequency responses. *IEEE Transactions on Acoustics, Speech, and Signal Processing*, 36(5):693–706, 1988.

[OB95]   J. E. Odegard and C. S. Burrus. Design of near-orthogonal filter banks and wavelets by Lagrange multipliers. 1995.

[OB96a]   Jan E. Odegard and C. Sidney Burrus. New class of wavelets for signal approximation. In *Proceedings of the IEEE International Symposium on Circuits and Systems*, pages II–189–192, IEEE ISCAS-96, Atlanta, May 12-15 1996.

[OB96b]   Jan E. Odegard and C. Sidney Burrus. Toward a new measure of smoothness for the design of wavelet basis. In *Proceedings of the IEEE International Conference on Acoustics, Speech, and Signal Processing*, pages III–1467–1470, IEEE ICASSP-96, Atlanta, May 7–10 1996.

[OB97]   J. E. Odegard and C. S. Burrus. Wavelets with new moment approximation properties. *IEEE Transactions on Signal Processing*, submitted, January 1997.

[Ode96]   J. E. Odegard. *Moments, smoothness and optimization of wavelet systems*. PhD thesis, Rice University, Houston, TX 77251, USA, May 1996.

[OGB92]   J. E. Odegard, R. A. Gopinath, and C. S. Burrus. Optimal wavelets for signal decomposition and the existence of scale limited signals. In *Proceedings of the IEEE International Conference on Signal Processing*, pages IV 597–600, ICASSP-92, San Francisco, CA, March 1992.

[OGB94]   J. E. Odegard, R. A. Gopinath, and C. S. Burrus. *Design of Linear Phase Cosine Modulated Filter Banks for Subband Image Compression*. Technical Report CML TR94-06, Computational Mathematics Laboratory, Rice University, Houston, TX, February 1994.

[OGL*95]   J. E. Odegard, H. Guo, M. Lang, C. S. Burrus, R. O. Wells, Jr., L. M. Novak, and M. Hiett. Wavelet based SAR speckle reduction and image compression. In *Proceedings of SPIE Conference 2487, Algorithms for SAR Imagery II*, Orlando, April 17–21 1995.

[OS89]   A. V. Oppenheim and R. W. Schafer. *Discrete-Time Signal Processing*. Prentice-Hall, Englewood Cliffs, NJ, 1989.

[OY87]   Özdoğan Yilmaz. *Seismic Data Processing*. Society of Exploration Geophysicists, Tulsa, 1987. Stephen M. Doherty editor.

[P P89]   P. P. Vaidyanathan and Z. Doğanata. The role of lossless systems in modern digital signal processing: a tutorial. *IEEE Transactions on Education*, August 1989.

[Pap77]   Athanasios Papoulis. *Signal Analysis*. McGraw-Hill, New York, 1977.

[PB87]   T. W. Parks and C. S. Burrus. *Digital Filter Design*. John Wiley & Sons, New York, 1987.

[PKC96]   J. C. Pesquet, H. Krim, and H. Carfantan. Time-invariant orthonormal wavelet representations. *IEEE Transactions on Signal Processing*, 44(8):1964–1970, August 1996.

[Plo95a]   G. Plonka. *Approximation order provided by refinable function vectors*. Technical Report 95/1, Universität Rostock, 1995. To appear in: Constructive Approximation.

[Plo95b]   G. Plonka. Approximation properties of multi-scaling functions: a fourier approach. 1995. Rostock. Math. Kolloq. 49, 115–126.

[Plo95c]   G. Plonka. Factorization of refinement masks of function vectors. In C. K. Chui and L. L. Schumaker, editors, *Wavelets and Multilevel Approximation*, pages 317–324, World Scientific Publishing Co., Singapore, 1995.

[Plo97a]   G. Plonka. *Necessary and sufficient conditions for orthonormality of scaling vectors*. Technical Report, Universität Rostock, 1997.

[Plo97b]   G. Plonka. On stability of scaling vectors. In A. Le Mehaute, C. Rabut, and L. L. Schumaker, editors, *Surface Fitting and Multiresolution Methods*, Vanderbilt University Press, Nashville, 1997. Also Technical Report 1996/15, Universität Rostock.

[Polar]   D. Pollen. Daubechies' scaling function on [0,3]. *J. American Math. Soc.*, to appear. Also Aware, Inc. tech. report AD891020, 1989.

[PS95]   G. Plonka and V. Strela. *Construction of multi-scaling functions with approximation and symmetry*. Technical Report 95/22, Universität Rostock, 1995. To appear in: SIAM J. Math. Anal.

[PV96]   See-May Phoong and P. P. Vaidyanathan. A polyphase approach to time-varying filter banks. In *Proc. Int. Conf. Acoust., Speech, Signal Processing*, pages 1554–1557, IEEE, Atlanta, GA, 1996.

[RBC*92]   M. B. Ruskai, G. Beylkin, R. Coifman, I. Daubechies, S. Mallat, Y. Meyer, and L. Raphael, editors. *Wavelets and their Applications*. Jones and Bartlett, Boston, MA, 1992. Outgrowth of NSF/CBMS conference on Wavelets held at the University of Lowell, June 1990.

[RBSB94]   J. O. A. Robertsson, J. O. Blanch, W. W. Symes, and C. S. Burrus. Galerkin wavelet modeling of wave propagation: optimal finite–difference stencil design. *Mathematical and Computer Modelling*, 19(1):31–38, January 1994.

[RC83]   L. Rabiner and D. Crochière. *Multirate Digital Signal Processing*. Prentice-Hall, 1983.

[RD86]   E. A. Robinson and T. S. Durrani. *Geophysical Signal Processing*. Prentice Hall, Englewood Cliffs, NJ, 1986.

[RD92]   Olivier Rioul and P. Duhamel. Fast algorithms for discrete and continuous wavelet transforms. *IEEE Transactions on Information Theory*, 38(2):569–586, March 1992. Special issue on wavelets and multiresolution analysis.

[RD94]   Olivier Rioul and Pierre Duhamel. A Remez exchange algorithm for orthonormal wavelets. *IEEE Transactions on Circuits and Systems II*, 41(8):550–560, August 1994.

[RG95]   Peter Rieder and Jürgen Götze. Algebraic optimizaion of biorthogonal wavelet transforms. *preprint*, 1995.

[RGN94]   Peter Rieder, Jürgen Götze, and Josef A. Nossek. *Algebraic Design of Discrete Wavelet Transforms*. Technical Report TUM-LNS-TR-94-2, Technical University of Munich, April 1994. Also submitted to IEEE Trans on Circuits and Systems.

[Rio91]   O. Rioul. Fast computation of the continuous wavelet transform. In *Proc. Int. Conf. Acoust., Speech, Signal Processing*, IEEE, Toronto, Canada, March 1991.

[Rio92]   Olivier Rioul. Simple regularity criteria for subdivision schemes. *SIAM J. Math. Anal.*, 23(6):1544–1576, November 1992.

[Rio93a]   Olivier Rioul. A discrete-time multiresolution theory. *IEEE Transactions on Signal Processing*, 41(8):2591–2606, August 1993.

[Rio93b]   Olivier Rioul. Regular wavelets: a discrete-time approach. *IEEE Transactions on Signal Processing*, 41(12):3572–3579, December 1993.

[RN96]   P. Rieder and J. A. Nossek. Smooth multiwavelets based on two scaling functions. In *Proc. IEEE Conf. on Time-Frequency and Time-Scale Analysis*, pages 309–312, 1996.

[Ron92]   Amos Ron. *Characterization of Linear Independence and Stability of the Sfifts of a Univariate Refinable Function in Terms of its Refinement Mask*. Technical Report CMS TR 93-3, Computer Science Dept., University of Wisconsin, Madison, September 1992.

[RS83]   L. Rabiner and R. W. Schaefer. *Speech Signal Processing*. Prentice-Hall, Englewood Cliffs, NJ, 1983.

[RT91]   T. A. Ramstad and J. P. Tanem. Cosine modulated analysis synthesis filter bank with critical sampling and perfect reconstruction. In *Proceedings of the IEEE International Conference on Acoustics, Speech, and Signal Processing*, pages 1789–1792, IEEE ICASSP-91, 1991.

[Rus92]   Mary Beth Ruskai. Introduction. In M. B. Ruskai, G. Beylkin, R. Coifman, I. Daubechies, S. Mallat, Y. Meyer, and L Raphael, editors, *Wavelets and their Applications*, Jones and Bartlett, Boston, MA, 1992.

[RV91]   Olivier Rioul and Martin Vetterli. Wavelet and signal processing. *IEEE Signal Processing Magazine*, 8(4):14–38, October 1991.

[RV93]   K. Ramchandran and M. Veterli. Best wavelet packet bases in a rate-distortion sense. *IEEE Transactions on Image Processing*, 2(2):160–175, 1993.

[RW97]   H. L. Resnikoff and R. O. Wells, Jr. *Wavelet Analysis and the Scalable Structure of Information*. Springer-Verlag, New York, 1997. to appear.

[RY90]   K. R. Rao and P. Yip. *Discrete Cosine Transform - Algorithms, Advantages and Applications*. Academic Press, 1990.

[SA92]   E. P. Simoncelli and E. H. Adelson. Subband transforms. In John W. Woods, editor, *Subband Image Coding*, Kluwer, Norwell, MA, to appear 1992. Also, MIT Vision and Modeling Tech. Report No. 137, Sept. 1989.

[Sai94a]   Naoki Saito. *Local Feature Extraction and Its Applications Using a Library of Bases*. PhD thesis, Yale University, New Haven, CN, 1994.

[Sai94b]   Naoki Saito. Simultaneous noise suppression and signal compression using a library of orthonormal bases and the minimum discription length criterion. In E. Foufoula-Georgiou and P. Kumar, editors, *Wavelets in Geophysics*, Academic Press, San Diego, 1994.

[SB86a]   M. J. Smith and T. P. Barnwell. Exact reconstruction techniques for tree-structured subband coders. *IEEE Transactions on Acoustics, Speech, and Signal Processing*, 34:434–441, June 1986.

[SB86b]   M. J. Smith and T. P. Barnwell III. Exact reconstruction techniques for tree-structured subband coders. *IEEE Transactions on Acoustics, Speech, and Signal Processing*, 34:434–441, 1986.

[SB87]   M. J. Smith and T. P. Barnwell. A new filter bank theory for time-frequency representation. *IEEE Transactions on Acoustics, Speech, and Signal Processing*, 35:314–327, March 1987.

[SB93]   H. V. Sorensen and C. S. Burrus. Efficient computation of the DFT with only a subset of input or output points. *IEEE Transactions on Signal Processing*, 41(3):1184–1200, March 1993.

[SC95]   J. A. Storer and M. Cohn, editors. *Proceedings of Data Compression Conference*. IEEE Computer Society Press, Snowbird, Utah, March 1995.

[SC96]   J. A. Storer and M. Cohn, editors. *Proceedings of Data Compression Conference*. IEEE Computer Society Press, Snowbird, Utah, April 1996.

[Sca94]   John A. Scales. *Theory of Seismic Imaging*. Samizat Press, Golden, CO, 1994.

[Sel96]   I. W. Selesnick. *New Techniques for Digital Filter Design*. PhD thesis, Rice University, 1996.

[Sel97]   Ivan W. Selesnick. *Parameterization of Orthogonal Wavelet Systems*. Technical Report, ECE Dept. and Computational Mathematics Laboratory, Rice University, Houston, Tx., May 1997.

[Sha93]   J. M. Shapiro. Embedded image coding using zerotrees of wavelet coefficients. *IEEE Transactions on Signal Processing*, 41(12):3445–3462, December 1993.

[She92]   M. J. Shensa. The discrete wavelet transform: wedding the à trous and Mallat algorithms. *IEEE Transactions on Information Theory*, 40:2464–2482, 1992.

[SHGB93]   P. Steffen, P. N. Heller, R. A. Gopinath, and C. S. Burrus. Theory of regular M-band wavelet bases. *IEEE Transactions on Signal Processing*, 41(12):3497–3511, December 1993. Special Transaction issue on wavelets; Rice contribution also in Tech. Report No. CML TR-91-22, Nov. 1991.

[SHS*96]   V. Strela, P. N. Heller, G. Strang, P. Topiwala, and C. Heil. *The application of multiwavelet filter banks to image processing.* Technical Report, MIT, January 1996. Submitted to IEEE Tran. Image Processing.

[Sie86]   William M. Siebert. *Circuits, Signals, and Systems.* MIT Press and McGraw-Hill, Cambridge and New York, 1986.

[SL96]   M. Sablatash and J. H. Lodge. The design of filter banks with specified minimum stopband attenuation for wavelet packet-based multiple access communications. In *Proceedings of 18th Biennial Symposium on Communications, Queen's University*, Kingston, ON, Canada, June 1996.

[SLB97]   Ivan W. Selesnick, Markus Lang, and C. Sidney Burrus. Magnitude squared design of recursive filters with the Chebyshev norm using a constrained rational Remez algorithm. *IEEE Transactions on Signal Processing*, to appear 1997.

[SMHH95]   O. V. Shentov, S. K. Mitra, U. Heute, and A. N. Hossen. Subband DFT – Part I: definition, interpretation and extensions. *Signal Processing*, 41:261–278, 1995.

[SN96]   Gilbert Strang and T. Nguyen. *Wavelets and Filter Banks.* Wellesley–Cambridge Press, Wellesley, MA, 1996.

[SOB96]   Ivan W. Selesnick, Jan E. Odegard, and C. Sidney Burrus. Nearly symmetric orthogonal wavelets with non-integer DC group delay. In *Proceedings of the IEEE Digital Signal Processing Workshop*, pages 431–434, Loen, Norway, September 2–4 1996.

[SP93]   Wim Sweldens and Robert Piessens. Calculation of the wavelet decomposition using quadrature formulae. In Tom H. Koornwinder, editor, *Wavelets: An Elementary Treatment of Theory and Applications*, pages 139–160, World Scientific, Singapore, 1993.

[SP96a]   A. Said and W. A. Pearlman. A new, fast, and efficient image codec based on set partitioning in hierarchical trees. *IEEE Transactions Cir. Syst. Video Tech.*, 6(3):243–250, June 1996.

[SP96b]   A. Said and W. A. Perlman. An image multiresolution representation for lossless and lossy image compression. *IEEE Transactions on Image Processing*, 5:1303–1310, September 1996.

[SS94]   V. Strela and G. Strang. Finite element multiwavelets. In *Proceedings of SPIE, Wavelet Applications in Signal and Image processing II*, pages 202–213, San Diego, CA, July 1994.

[SS95]   G. Strang and V. Strela. Short wavelets and matrix dilation equations. *IEEE Trans. SP*, 43(1):108–115, January 1995.

[Str86]   Gilbert Strang. *Introduction to Applied Mathematics.* Wellesley-Cambridge Press, Wellesley, MA, 1986.

[Str89]   Gilbert Strang. Wavelets and dilation equations: a brief introduction. *SIAM Review*, 31(4):614–627, 1989. also, MIT Numerical Analysis Report 89-9, Aug. 1989.

[Str94]   Gilbert Strang. Wavelets. *American Scientist*, 82(3):250–255, May 1994.

[Str96a]   G. Strang. Eigenvalues of $(\downarrow 2)H$ and convergence of the cascade algorithm. *IEEE Transactions on Signal Processing*, 44, 1996.

[Str96b]   V. Strela. *Multiwavelets: Theory and Applications.* PhD thesis, Dept. of Mathematics, MIT, June 1996.

[SV93]   A. K. Soman and P. P. Vaidyanathan. On orthonormal wavelets and paraunitary filter banks. *IEEE Transactions on Signal Processing*, 41(3):1170–1183, March 1993.

[SVN93]   A. K. Soman, P. P. Vaidyanathan, and T. Q. Nguyen. Linear phase paraunitary filter banks: theory, factorizations and designs. *IEEE Transactions on Signal Processing*, 41(12):3480–3496, December 1993.

[SW93]   Larry L. Schumaker and Glenn Webb, editors. *Recent Advances in Wavelet Analysis*. Academic Press, San Diego, 1993. Volume in the series: Wavelet Analysis and its Applications.

[SW97]   W. So and J. Wang. Estimating the support of a scaling vector. *SIAM J. Matrix Anal. Appl.*, 18(1):66–73, January 1997.

[Swe95]   Wim Sweldens. *The Lifting Scheme: A Construction of Second Generation Wavelets*. Technical Report TR-1995-6, Math Dept. University of South Carolina, May 1995.

[Swe96a]   Wim Sweldens. The lifting scheme: a custom-design construction of biorthogonal wavelets. *Applied and Computational Harmonic Analysis*, 3(2):186–200, 1996. Also a technical report, math dept. Univ. SC, April 1995.

[Swe96b]   Wim Sweldens. Wavelets: what next? *Proceedings of the IEEE*, 84(4):680–685, April 1996.

[Tas96]   Carl Taswell. *Handbook of Wavelet Transform Algorithms*. Birkhäuser, Boston, 1996.

[Tew97]   Ahmed H. Tewfik. *Wavelets and Multiscale Signal Processing Techniques: Theory and Applications*. to appear, 1997.

[The89]   *The UltraWave Explorer User's Manual*. Aware, Inc., Cambridge, MA, July 1989.

[Tia96]   Jun Tian. *The Mathematical Theory and Applications of Biorthogonal Coifman Wavelet Systems*. PhD thesis, Rice University, February 1996.

[TM94a]   Hai Tao and R. J. Moorhead. Lossless progressive transmission of scientific data using biorthogonal wavelet transform. In *Proceedings of the IEEE Conference on Image Processing*, Austin, ICIP-94, November 1994.

[TM94b]   Hai Tao and R. J. Moorhead. Progressive transmission of scientific data using biorthogonal wavelet transform. In *Proceedings of the IEEE Conference on Visualization*, Washington, October 1994.

[Tur86]   M. Turner. Texture discrimination by Gabor functions. *Biological Cybernetics*, 55:71–82, 1986.

[TVC96]   M. J. Tsai, J. D. Villasenor, and F. Chen. Stack-run image coding. *IEEE Trans. Circ. and Syst. Video Tech.*, 519–521, October 1996.

[TW95]   Jun Tian and Raymond O. Wells, Jr. *Vanishing Moments and Wavelet Approximation*. Technical Report CML TR-9501, Computational Mathematics Lab, Rice University, January 1995.

[TW96]   J. Tian and R. O. Wells. Image compression by reduction of indices of wavelet transform coefficients. *Proc. DCC*, April 1996.

[TWBOar]   J. Tian, R. O. Wells, C. S. Burrus, and J. E. Odegard. Coifman wavelet systems: approximation, smoothness, and computational algorithms. In Jacques Periaux, editor, *Computational Science for the 21st Century*, John Wiley and Sons, New York, 1997 to appear. in honor of Roland Glowinski's 60th birhtday.

[Uns96]   Michael Unser. Approximation power of biorthogonal wavelet expansions. *IEEE Transactions on Signal Processing*, 44(3):519–527, March 1996.

[VA96]   M. J. Vrhel and A. Aldroubi. *Pre-filtering for the Initialization of Multi-wavelet transforms*. Technical Report, National Institutes of Health, 1996.

[Vai87a]   P. P. Vaidyanathan.   Quadrature mirror filter banks, M–band extensions and perfect–reconstruction techniques. *IEEE Acoustics, Speech, and Signal Processing Magazine*, 4(3):4–20, July 1987.

[Vai87b]   P. P. Vaidyanathan.   Theory and design of M-channel maximally decimated quadrature morror filters with arbitrary M, having perfect reconstruction properties. *IEEE Transactions on Acoustics, Speech, and Signal Processing*, 35(4):476–492, April 1987.

[Vai92]   P. P. Vaidyanathan. *Multirate Systems and Filter Banks*. Prentice-Hall, Englewood Cliffs, NJ, 1992.

[VBL95]   J. D. Villasenor, B. Belzer, and J. Liao. Wavelet filter evaluation for image compression. *IEEE Transactions on Image Processing*, 4, August 1995.

[VD89]   P. P. Vaidyanathan and Z. Doğanata. The role of lossless systems in modern digital signal processing: A tutorial. *IEEE Trans. on Education*, 32(3):181–197, August 1989.

[VD95]   P. P. Vaidyanathan and Igor Djokovic. Wavelet transforms. In Wai-Kai Chen, editor, *The Circuits and Filters Handbook*, chapter 6, pages 134–219, CRC Press and IEEE Press, Roca Raton, 1995.

[Vet86]   Martin Vetterli. Filter banks allowing perfect reconstruction. *Signal Processing*, 10(3):219–244, April 1986.

[Vet87]   Martin Vetterli. A theory of multirate filter banks. *IEEE Transactions on Acoustics, Speech, and Signal Processing*, 35(3):356–372, March 1987.

[VG89]   Martin Vetterli and Didier Le Gall. Perfect reconstruction FIR filter banks: some properties and factorizations. *IEEE Transactions on Acoustics, Speech, and Signal Processing*, 37(7):1057–1071, July 1989.

[VH88]   P. P. Vaidyanathan and Phuong-Quan Hoang. Lattice structures for optimal design and robust implementation of two-channel perfect reconstruction QMF banks. *IEEE Transactions on Acoustics, Speech, and Signal Processing*, 36(1):81–93, January 1988.

[VH92]   M. Vetterli and C. Herley. Wavelets and filter banks: theory and design. *IEEE Transactions on Acoustics, Speech, and Signal Processing*, 2207–2232, September 1992.

[VK95]   Martin Vetterli and Jelena Kovačević. *Wavelets and Subband Coding*. Prentice–Hall, Englewood Cliffs, 1995.

[VL89]   M. Vetterli and D. Le Gall. Perfect reconstruction FIR filter banks: some properties and factorizations. *IEEE Transactions on Acoustics, Speech, and Signal Processing*, 37(7):1057–1071, July 1989.

[VLU97]   M. J. Vrhel, C. Lee, and M. Unser. Fast continuous wavelet transform: a least-squares formulation. *Signal Processing*, 57(2):103–120, March 1997.

[VM88]   P. P. Vaidyanathan and S. K. Mitra. Polyphase networks, block digital filtering, LPTV systems, and alias-free QMF banks: a unified approach based on pseudocirculants. *IEEE Transactions on Acoustics, Speech, and Signal Processing*, 36:381–391, March 1988.

[VNDS89]   P. P. Vaidyanathan, T. Q. Nguyen, Z. Doğanata, and T. Saramäki. Improved technique for design of perfect reconstruction FIR QMF banks with lossless polyphase matrices. *IEEE Transactions on Acoustics, Speech, and Signal Processing*, 37(7):1042–1056, July 1989.

[Vol92]   Hans Volkmer. On the regularity of wavelets. *IEEE Transactions on Information Theory*, 38(2):872–876, March 1992.

[Wal94]   Gilbert G. Walter, editor. *Wavelets and Other Orthogonal Systems with Applications*. CRC Press, Boca Raton, FL, 1994.

[WB95]   D. Wei and C. S. Burrus. Optimal soft-thresholding for wavelet transform coding. In *Proceedings of IEEE International Conference on Image Processing*, pages I:610–613, Washington, DC, October 1995.

[WB96]   Dong Wei and Alan C. Bovik. On generalized coiflets: construction, near-symmetry, and optimization. *IEEE Transactions on Circuits and Systems:II*, submitted October 1996.

[WB97]   Dong Wei and Alan C. Bovik. Sampling approximation by generalized coiflets. *IEEE Transactions on Signal Processing*, submitted January 1997.

[Wei95]   Dong Wei. *Investigation of Biorthogonal Wavelets*. Technical Report ECE-696, Rice University, April 1995.

[Wel93]   R. O. Wells, Jr. Parametcrizing smooth compactly supported wavelets. *Transactions of the American Mathematical Society*, 338(2):919–931, 1993. Also Aware tech report AD891231, Dec. 1989.

[Wic92]   M. V. Wickerhauser. Acoustic signal compression with wavelet packets. In Charles K. Chui, editor, *Wavelets: A Tutorial in Theory and Applications*, pages 679–700, Academic Press, Boca Raton, 1992. Volume 2 in the series: Wavelet Analysis and its Applications.

[Wic95]   Mladen Victor Wickerhauser. *Adapted Wavelet Analysis from Theory to Software*. A K Peters, Wellesley, MA, 1995.

[WNC87]   I. Witten, R. Neal, and J. Cleary. Arithmetic coding for data compression. *Communications of the ACM*, 30:520–540, June 1987.

[WO92a]   G. Wornell and A. V. Oppenheim. Estimation of fractal signals from noisy measurements using wavelets. *IEEE Transactions on Acoustics, Speech, and Signal Processing*, 40(3):611–623, March 1992.

[WO92b]   G. W. Wornell and A. V. Oppenheim. Wavelet-based representations for a class of self-similar signals with application to fractal modulation. *IEEE Transactions on Information Theory*, 38(2):785–800, March 1992.

[WOG*95a]   D. Wei, J. E. Odegard, H. Guo, M. Lang, and C. S. Burrus. SAR data compression using best-adapted wavelet packet basis and hybrid subband coding. In Harold H. Szu, editor, *Proceedings of SPIE Conference 2491, Wavelet Applications II*, pages 1131–1141, Orlando, April 17–21 1995.

[WOG*95b]   D. Wei, J. E. Odegard, H. Guo, M. Lang, and C. S. Burrus. Simultaneous noise reduction and SAR image data compression using best wavelet packet basis. In *Proceedings of IEEE International Conference on Image Processing*, pages III:200–203, Washington, DC, October 1995.

[Wor96]   Gregory W. Wornell. *Signal Processing with Fractals, A Wavelet- Based Approach*. Prentice Hall, Upper Saddle River, NJ, 1996.

[WTWB97]   Dong Wei, Jun Tian, Raymond O. Wells, Jr., and C. Sidney Burrus. A new class of biorthogonal wavelet systems for image transform coding. *IEEE Transactions on Image Processing*, to appear 1997.

[WWJ95]   J. Wu, K. M. Wong, and Q. Jin. Multiplexing based on wavelet packets. In *Proceedings of SPIE Conference, Aerosense*, Orlando, April 1995.

[WZ94]   Raymond O. Wells, Jr and Xiaodong Zhou. Wavelet interpolation and approximate solutions of elliptic partial differential equations. In R. Wilson and E. A. Tanner, editors, *Noncompact Lie Groups*, Kluwer, 1994. Also in Proceedings of NATO Advanced Research Workshop, 1992, and CML Technical Report TR-9203, Rice University, 1992.

[XGHS96]   X.-G. Xia, J. S. Geronimo, D. P. Hardin, and B. W. Suter. Design of prefilters for discrete multiwavelet transforms. *IEEE Trans. SP*, 44(1):25–35, January 1996.

[XHRO95]   Z. Xiong, C. Herley, K. Ramchandran, and M. T. Orcgard. Space-frequency quantization for a space-varying wavelet packet image coder. *Proc. Int. Conf. Image Processing*, 1:614–617, October 1995.

[XS96]   X.-G. Xia and B. W. Suter. Vector-valued wavelets and vector filter banks. *IEEE Trans. SP*, 44(3):508–518, March 1996.

[You80]   R. M. Young. *An Introduction to Nonharmonic Fourier Series*. Academic Press, New York, 1980.

[You93]   R. K. Young. *Wavelet Theory and Its Applications*. Kluwer Academic Publishers, Boston, MA, 1993.

[ZT92a]   H. Zou and A. H. Tewfik. Design and parameterization ofm-band orthonormal wavelets. In *Proceedings of the IEEE International Symposium on Circuits and Systems*, pages 983–986, ISCAS-92, San Diego, 1992.

[ZT92b]   H. Zou and A. H. Tewfik. Discrete orthogonal M-band wavelet decompositions. In *Proceedings of the IEEE International Conference on Acoustics, Speech, and Signal Processing*, pages IV–605–608, San Francisco, CA, 1992.

Appendix A.
Derivations for
Chapter 5 on Scaling Functions

This appendix contains outline proofs and derivations for the theorems and formulas given in early part of Chapter 5. They are not intended to be complete or formal, but they should be sufficient to understand the ideas behind why a result is true and to give some insight into its interpretation as well as to indicate assumptions and restrictions.

Proof 1 *The conditions given by (5.10) and (7.7) can be derived by integrating both sides of*

$$\varphi(x) = \sum_n h(n) \sqrt{M} \, \varphi(M\,x - n) \tag{12.1}$$

and making the change of variables $y = Mx$

$$\int \varphi(x) \, dx = \sum_n h(n) \int \sqrt{M} \, \varphi(Mx - n) \, dx \tag{12.2}$$

and noting the integral is independent of translation which gives

$$= \sum_n h(n) \sqrt{M} \int \varphi(y) \frac{1}{M} \, dy. \tag{12.3}$$

With no further requirements other than $\varphi \in L^1$ *to allow the sum and integral interchange and* $\int \varphi(x) \, dx \neq 0$, *this gives (7.7) as*

$$\sum_n h(n) = \sqrt{M} \tag{12.4}$$

and for $M = 2$ *gives (5.10). Note this does not assume orthogonality nor any specific normalization of* $\varphi(t)$ *and does not even assume* M *is an integer.*

This is the most basic necessary condition for the existence of $\varphi(t)$ and it has the fewest assumptions or restrictions.

Proof 2 *The conditions in (5.14) and (7.8) are a down-sampled orthogonality of translates by M of the coefficients which results from the orthogonality of translates of the scaling function given by*

$$\int \varphi(x)\,\varphi(x-m)\,dx = E\,\delta(m) \tag{12.5}$$

in (5.13). The basic scaling equation (12.1) is substituted for both functions in (12.5) giving

$$\int \left[\sum_n h(n)\,\sqrt{M}\,\varphi(Mx-n)\right]\left[\sum_k h(k)\,\sqrt{M}\,\varphi(Mx-Mm-k)\right]dx = E\,\delta(m) \tag{12.6}$$

which, after reordering and a change of variable $y = M\,x$, gives

$$\sum_n\sum_k h(n)\,h(k)\int \varphi(y-n)\,\varphi(y-Mm-k)\,dy = E\,\delta(m). \tag{12.7}$$

Using the orthogonality in (12.5) gives our result

$$\sum_n h(n)\,h(n-Mm) \;=\; \delta(m) \tag{12.8}$$

in (5.14) and (7.8). This result requires the orthogonality condition (12.5), M must be an integer, and any non-zero normalization E may be used.

Proof 3 (Corollary 2) *The result that*

$$\sum_n h(2n) = \sum_n h(2n+1) = 1/\sqrt{2} \tag{12.9}$$

in (5.17) or, more generally

$$\sum_n h(M\,n) = \sum_n h(M\,n+k) = 1/\sqrt{M} \tag{12.10}$$

is obtained by breaking (12.4) for $M = 2$ into the sum of the even and odd coefficients.

$$\sum_n h(n) = \sum_k h(2k) + \sum_k h(2k+1) = K_0 + K_1 = \sqrt{2}. \tag{12.11}$$

Next we use (12.8) and sum over n to give

$$\sum_n\sum_k h(k+2n)h(k) = 1 \tag{12.12}$$

which we then split into even and odd sums and reorder to give:

$$\sum_n \left[\sum_k h(2k+2n)h(2k) + \sum_k h(2k+1+2n)h(2k+1) \right]$$

$$= \sum_k \left[\sum_n h(2k+2n) \right] h(2k) + \sum_k \left[\sum_n h(2k+1+2n) \right] h(2k+1)$$

$$= \sum_k K_0 h(2k) + \sum_k K_1 h(2k+1) = K_0^2 + K_1^2 = 1. \tag{12.13}$$

Solving (12.11) and (12.13) simultaneously gives $K_0 = K_1 = 1/\sqrt{2}$ and our result (5.17) or (12.9) for $M = 2$.

If the same approach is taken with (7.7) and (7.8) for $M = 3$, we have

$$\sum_n x(n) = \sum_n x(3n) + \sum_n x(3n+1) + \sum_n x(3n+2) = \sqrt{3} \tag{12.14}$$

which, in terms of the partial sums K_i, is

$$\sum_n x(n) = K_0 + K_1 + K_2 = \sqrt{3}. \tag{12.15}$$

Using the orthogonality condition (12.8) as was done in (12.12) and (12.13) gives

$$K_0^2 + K_1^2 + K_2^2 = 1. \tag{12.16}$$

Equation (12.15) and (12.16) are simultaneously true if and only if $K_0 = K_1 = K_2 = 1/\sqrt{3}$. This process is valid for any integer M and any non-zero normalization.

Proof 3 *If the support of $\varphi(x)$ is $[0, N-1]$, from the basic recursion equation with support of $h(n)$ assumed as $[N_1, N_2]$ we have*

$$\varphi(x) = \sum_{n=N_1}^{N_2} h(n) \sqrt{2}\, \varphi(2x - n) \tag{12.17}$$

where the support of the right hand side of (12.17) is $[N_1/2, (N-1+N_2)/2)$. Since the support of both sides of (12.17) must be the same, the limits on the sum, or, the limits on the indices of the non zero $h(n)$ are such that $N_1 = 0$ and $N_2 = N$, therefore, the support of $h(n)$ is $[0, N-1]$.

Proof 4 *First define the autocorrelation function*

$$a(t) = \int \varphi(x)\, \varphi(x - t)\, dx \tag{12.18}$$

and the power spectrum

$$A(\omega) = \int a(t)\, e^{-j\omega t}\, dt = \int \int \varphi(x)\, \varphi(x - t)\, dx\; e^{-j\omega t}\, dt \tag{12.19}$$

which after changing variables, $y = x - t$, and reordering operations gives

$$A(\omega) = \int \varphi(x)\, e^{-j\omega x}\, dx \int \varphi(y)\, e^{j\omega y}\, dy \tag{12.20}$$

$$= \Phi(\omega)\, \Phi(-\omega) = |\Phi(\omega)|^2 \tag{12.21}$$

If we look at (12.18) as being the inverse Fourier transform of (12.21) and sample $a(t)$ at $t = k$, we have

$$a(k) = \frac{1}{2\pi} \int_{-\infty}^{\infty} |\Phi(\omega)|^2\, e^{j\omega k}\, d\omega \tag{12.22}$$

$$= \frac{1}{2\pi} \sum_{\ell} \int_0^{2\pi} |\Phi(\omega + 2\pi\ell)|^2\, e^{j\omega k}\, d\omega = \frac{1}{2\pi} \int_0^{2\pi} \left[\sum_{\ell} |\Phi(\omega + 2\pi\ell)|^2 \right] e^{j\omega k}\, d\omega \tag{12.23}$$

but this integral is the form of an inverse discrete-time Fourier transform (DTFT) which means

$$\sum_, a(k)\, e^{j\omega k} = \sum_{\ell} |\Phi(\omega + 2\pi\ell)|^2. \tag{12.24}$$

If the integer translates of $\varphi(t)$ are orthogonal, $a(k) = \delta(k)$ and we have our result

$$\sum_{\ell} |\Phi(\omega + 2\pi\ell)|^2 = 1. \tag{12.25}$$

If the scaling function is not normalized

$$\sum_{\ell} |\Phi(\omega + 2\pi\ell)|^2 = \int |\varphi(t)|^2\, dt \tag{12.26}$$

which is similar to Parseval's theorem relating the energy in the frequency domain to the energy in the time domain.

Proof 6 *Equation (5.20) states a very interesting property of the frequency response of an FIR filter with the scaling coefficients as filter coefficients. This result can be derived in the frequency or time domain. We will show the frequency domain argument. The scaling equation (12.1) becomes (5.51) in the frequency domain. Taking the squared magnitude of both sides of a scaled version of*

$$\Phi(\omega) = \frac{1}{\sqrt{2}} H(\omega/2)\, \Phi(\omega/2) \tag{12.27}$$

gives

$$|\Phi(2\omega)|^2 = \frac{1}{2} |H(\omega)|^2\, |\Phi(\omega)|^2 \tag{12.28}$$

Add $k\pi$ to ω and sum over k to give for the left side of (12.28)

$$\sum_k |\Phi(2\omega + 2\pi k)|^2 = K = 1 \tag{12.29}$$

which is unity from (5.57). Summing the right side of (12.28) gives

$$\sum_k \frac{1}{2} |H(\omega + k\pi)|^2 \, |\Phi(\omega + k\pi)|^2 \tag{12.30}$$

Break this sum into a sum of the even and odd indexed terms.

$$\sum_k \frac{1}{2} |H(\omega + 2\pi k)|^2 \, |\Phi(\omega + 2\pi k)|^2 + \sum_k \frac{1}{2} |H(\omega + (2k+1)\pi)|^2 \, |\Phi(\omega + (2k+1)\pi)|^2 \tag{12.31}$$

$$= \frac{1}{2} |H(\omega)|^2 \sum_k |\Phi(\omega + 2\pi k)|^2 + \frac{1}{2} |H(\omega + \pi)|^2 \sum_k |\Phi(\omega + (2k+1)\pi)|^2$$

which after using (12.29) gives

$$= \frac{1}{2} |H(\omega)|^2 + \frac{1}{2} |H(\omega + \pi)|^2 = 1 \tag{12.32}$$

which gives (5.20). This requires both the scaling and orthogonal relations but no specific normalization of $\varphi(t)$. If viewed as an FIR filter, $h(n)$ is called a quadrature mirror filter (QMF) because of the symmetry of its frequency response about π.

Proof 10 *The multiresolution assumptions in Section 2 require the scaling function and wavelet satisfy (5.1) and (2.24)*

$$\varphi(t) = \sum_n h(n) \sqrt{2} \, \varphi(2t - n), \qquad \psi(t) = \sum_n h_1(n) \sqrt{2} \, \varphi(2t - n) \tag{12.33}$$

and orthonormality requires

$$\int \varphi(t) \, \varphi(t - k) \, dt = \delta(k) \tag{12.34}$$

and

$$\int \psi(t) \, \varphi(t - k) \, dt = 0 \tag{12.35}$$

for all $k \in \mathbf{Z}$. Substituting (12.33) into (12.35) gives

$$\int \sum_n h_1(n) \sqrt{2} \, \varphi(2t - n) \sum_\ell h(\ell) \sqrt{2} \, \varphi(2t - 2k - \ell) \, dt = 0 \tag{12.36}$$

Rearranging and making a change of variables gives

$$\sum_{n,\ell} h_1(n) \, h(\ell) \, \frac{1}{2} \int \varphi(y - n) \, \varphi(y - 2k - \ell) \, dy = 0 \tag{12.37}$$

Using (12.34) gives

$$\sum_{n,\ell} h_1(n) \, h(\ell) \, \delta(n - 2k - \ell) = 0 \tag{12.38}$$

for all $k \in \mathbf{Z}$. Summing over ℓ gives

$$\sum_n h_1(n)\, h(n - 2k) = 0 \tag{12.39}$$

Separating (12.39) into even and odd indices gives

$$\sum_m h_1(2m)\, h(2m - 2k) + \sum_\ell h_1(2\ell + 1)\, h(2\ell + 1 - 2k) = 0 \tag{12.40}$$

which must be true for all integer k. Defining $h_e(n) = h(2n)$, $h_o(n) = h(2n+1)$ and $\widetilde{g}(n) = g(-n)$ for any sequence g, this becomes

$$h_e \star \widetilde{h}_{1e} + h_o \star \widetilde{h}_{1o} = 0. \tag{12.41}$$

From the orthonormality of the translates of φ and ψ one can similarly obtain the following:

$$h_e \star \widetilde{h}_e + h_o \star \widetilde{h}_o = \delta. \tag{12.42}$$

$$h_{1e} \star \widetilde{h}_{1e} + h_{1o} \star \widetilde{h}_{1o} = \delta. \tag{12.43}$$

This can be compactly represented as

$$\begin{bmatrix} h_e & h_o \\ h_{1e} & h_{1o} \end{bmatrix} \star \begin{bmatrix} \widetilde{h}_e & \widetilde{h}_{1e} \\ \widetilde{h}_o & \widetilde{h}_{1o} \end{bmatrix} = \begin{bmatrix} \delta & 0 \\ 0 & \delta \end{bmatrix}. \tag{12.44}$$

Assuming the sequences are finite length (12.44) can be used to show that

$$h_e \star h_{1o} - h_o \star h_{1e} = \pm\delta_k, \tag{12.45}$$

where $\delta_k(n) = \delta(n - k)$. Indeed, taking the Z-transform of (12.44) we get using the notation of Chapter 8 $H_p(z) H_p^T(z^{-1}) = I$. Because, the filters are FIR $H_p(z)$ is a (Laurent) polynomial matrix with a polynomial matrix inverse. Therefore the determinant of $H_p(z)$ is of the form $\pm z^k$ for some integer k. This is equivalent to (12.45). Now, convolving both sides of (12.45) by $\widetilde{h}_e$ we get

$$\begin{aligned}
\pm\widetilde{h}_e \star \delta_k &= [h_e \star h_{1o} - h_o \star h_{1e}] \star \widetilde{h}_e \tag{12.46} \\
&= \left[h_e \star \widetilde{h}_e \star h_{1o} - h_{1e} \star \widetilde{h}_e \star h_o \right] \\
&= \left[h_e \star \widetilde{h}_e \star h_{1o} + h_{1o} \star \widetilde{h}_o \star h_o \right] \\
&= \left[h_e \star \widetilde{h}_e + h_o \star \widetilde{h}_o \right] \star h_{1o} \\
&= h_{1o}.
\end{aligned}$$

Similarly by convolving both sides of (12.45) by $\widetilde{h}_o$ we get

$$\mp\widetilde{h}_o \star \delta_k = h_{1e}. \tag{12.47}$$

Combining (12.47) and (12.47) gives the result

$$h_1(n) = \pm(-1)^n h(-n + 1 - 2k). \tag{12.48}$$

Proof 11 *We show the integral of the wavelet is zero by integrating both sides of (12.33b) gives*

$$\int \psi(t)\, dt = \sum_n h_1(n) \int \sqrt{2}\, \varphi(2t - n)\, dt \tag{12.49}$$

But the integral on the right hand side is A_0, usually normalized to one and from (5.17) or (12.9) and (12.48) we know that

$$\sum_n h_1(n) = 0 \tag{12.50}$$

and, therefore, from (12.49), the integral of the wavelet is zero.

The fact that multiplying in the time domain by $(-1)^n$ is equivalent to shifting in the frequency domain by π gives $H_1(\omega) = H(\omega + \pi)$.

Appendix B.
Derivations for
Section on Properties

In this appendix we develop most of the results on scaling functions, wavelets and scaling and wavelet coefficients presented in Section 5.8 and elsewhere. For convenience, we repeat (5.1), (5.10), (5.13), and (5.15) here

$$\varphi(t) = \sum_n h(n)\,\sqrt{2}\,\varphi(2t - n) \tag{12.51}$$

$$\sum_n h(n) = \sqrt{2} \tag{12.52}$$

$$\int \varphi(t)\,\varphi(t - k)\,dt = E\delta(k) = \begin{cases} E & \text{if } k = 0 \\ 0 & \text{otherwise} \end{cases} \tag{12.53}$$

If normalized

$$\int \varphi(t)\,\varphi(t - k)\,dt = \delta(k) = \begin{cases} 1 & \text{if } k = 0 \\ 0 & \text{otherwise} \end{cases} \tag{12.54}$$

The results in this appendix refer to equations in the text written in bold face fonts.

Equation (5.45) is the normalization of (5.15) and part of the orthonormal conditions required by (12.53) for $k = 0$ and $E = 1$.

Equation (5.53) If the $\varphi(x - k)$ are orthogonal, (12.53) states

$$\int \varphi(x + m)\varphi(x)\,dx = E\,\delta(m) \tag{12.55}$$

Summing both sides over m gives

$$\sum_m \int \varphi(x + m)\varphi(x)\,dx = E \tag{12.56}$$

which after reordering is

$$\int \varphi(x) \sum_m \varphi(x+m)\, dx = E. \tag{12.57}$$

Using (5.50), (12.70), and (12.73) gives

$$\int \varphi(x)\, dx\, A_0 = E \tag{12.58}$$

but $\int \varphi(x)\, dx = A_0$ from (12.68), therefore

$$A_0^2 = E \tag{12.59}$$

If the scaling function is not normalized to unity, one can show the more general result of (5.53). This is done by noting that a more general form of (5.50) is

$$\sum_m \varphi(x+m) = \int \varphi(x)\, dx \tag{12.60}$$

if one does not normalize $A_0 = 1$ in (12.69) through (12.73).

Equation (5.53) follows from summing (12.53) over m as

$$\sum_m \int \varphi(x+m)\varphi(x)\, dx = \int \varphi(x)^2\, dx \tag{12.61}$$

which after reordering gives

$$\int \varphi(x) \sum_m \varphi(x+m)\, dx = \int \varphi(x)^2\, dx \tag{12.62}$$

and using (12.60) gives (5.53).

Equation (5.46) is derived by applying the basic recursion equation to its own right hand side to give

$$\varphi(t) = \sum_n h(n)\,\sqrt{2} \sum_k h(k)\,\sqrt{2}\,\varphi(2(2t-n)-k) \tag{12.63}$$

which, with a change of variables of $\ell = 2n + k$ and reordering of operation, becomes

$$\varphi(t) = \sum_\ell \left[\sum_n h(n)\, h(\ell - 2n) \right] 2\,\varphi(4t - \ell). \tag{12.64}$$

Applying this j times gives the result in (5.46). A similar result can be derived for the wavelet.

Equation (5.48) is derived by defining the sum

$$A_J = \sum_\ell \varphi\!\left(\frac{\ell}{2^J}\right) \tag{12.65}$$

and using the basic recursive equation (12.51) to give

$$A_J = \sum_\ell \sum_n h(n) \sqrt{2} \, \varphi(2\frac{\ell}{2^J} - n). \tag{12.66}$$

Interchanging the order of summation gives

$$A_J = \sqrt{2} \sum_n h(n) \left\{ \sum_\ell \varphi(\frac{\ell}{2^{J-1}} - n) \right\}$$

but the summation over ℓ is independent of an integer shift so that using (12.52) and (12.65) gives

$$A_J = \sqrt{2}\sqrt{2} \sum_n h(n) \left\{ \sum_\ell \varphi(\frac{\ell}{2^{J-1}}) \right\} = 2 \, A_{J-1}. \tag{12.67}$$

This is the linear difference equation

$$A_J - 2 \, A_{J-1} = 0 \tag{12.68}$$

which has as a solution the geometric sequence

$$A_J = A_0 \, 2^J. \tag{12.69}$$

If the limit exists, equation (12.65) divided by 2^J is the Riemann sum whose limit is the definition of the Riemann integral of $\varphi(x)$

$$\lim_{J\to\infty} \left\{ A_J \frac{1}{2^J} \right\} = \int \varphi(x) \, dx = A_0. \tag{12.70}$$

It is stated in (5.57) and shown in (12.56) that if $\varphi(x)$ is normalized, then $A_0 = 1$ and (12.69) becomes

$$A_J = 2^J. \tag{12.71}$$

which gives (5.48).

Equation (12.70) shows another remarkable property of $\varphi(x)$ in that the bracketed term is exactly equal to the integral, independent of J. No limit need be taken!

Equation(5.49) is the "partitioning of unity" by $\varphi(x)$. It follows from (5.48) by setting $J = 0$.

Equation (5.50) is generalization of (5.49) by noting that the sum in (5.48) is independent of a shift of the form

$$\sum_\ell \varphi(\frac{\ell}{2^J} - \frac{L}{2^M}) = 2^J \tag{12.72}$$

for any integers $M \geq J$ and L. In the limit as $M \to \infty$, $\frac{L}{2^M}$ can be made arbitrarily close to any x, therefore, if $\varphi(x)$ is continuous,

$$\sum_\ell \varphi(\frac{\ell}{2^J} - x) = 2^J. \tag{12.73}$$

This gives (5.50) and becomes (5.49) for $J = 0$. Equation (5.50) is called a "partitioning of unity" for obvious reasons.

The first four relationships for the scaling function hold in a generalized form for the more general defining equation (7.4). Only (5.48) is different. It becomes

$$\sum_k \varphi(\frac{k}{M^J}) = M^J \tag{12.74}$$

for M an integer. It may be possible to show that certain rational M are allowed.

Equations (5.51), (5.72), and (5.52) are the recursive relationship for the Fourier transform of the scaling function and are obtained by simply taking the transform (5.2) of both sides of (12.51) giving

$$\Phi(\omega) = \sum_n h(n) \int \sqrt{2}\, \varphi(2t - n)e^{-j\omega t}\, dt \tag{12.75}$$

which after the change of variables $y = 2t - n$ becomes

$$\Phi(\omega) = \frac{\sqrt{2}}{2} \sum_n h(n) \int \varphi(y)e^{-j\omega(y+n)/2}\, dy$$

and using (5.3) gives

$$\Phi(\omega) = \frac{1}{\sqrt{2}} \sum_n h(n)\, e^{-j\omega n/2} \int \varphi(y)e^{-j\omega y/2}\, dy \;=\; \frac{1}{\sqrt{2}} H(\omega/2)\, \Phi(\omega/2)$$

which is (5.51) and (5.72). Applying this recursively gives the infinite product (5.52) which holds for any normalization.

Equation (5.57) states that the sum of the squares of samples of the Fourier transform of the scaling function is one if the samples are uniform every 2π. An alternative derivation to that in Appendix A is shown here by taking the definition of the Fourier transform of $\varphi(x)$, sampling it every $2\pi k$ points and multiplying it times its complex conjugate.

$$\Phi(\omega + 2\pi k)\overline{\Phi(\omega + 2\pi k)} = \int \varphi(x)e^{-j(\omega+2\pi k)x}\, dx \int \varphi(y)e^{j(\omega+2\pi k)y}\, dy \tag{12.76}$$

Summing over k gives

$$\sum_k |\Phi(\omega + 2\pi k)|^2 = \sum_k \int \int \varphi(x)\varphi(y)e^{-j\omega(x-y)}e^{-j2\pi k(x-y)}\, dx\, dy \tag{12.77}$$

$$= \int \int \varphi(x)\varphi(y)e^{j\omega(y-x)} \sum_k e^{j2\pi k(y-x)}\, dx\, dy$$

$$= \int \int \varphi(x)\varphi(x+z)e^{j\omega z} \sum_k e^{j2\pi kz}\, dx\, dz$$

but

$$\sum_k e^{j2\pi kz} = \sum_\ell \delta(z - \ell) \tag{12.78}$$

therefore

$$\sum_k |\Phi(\omega + 2\pi k)|^2 = \int \varphi(x) \sum_\ell \varphi(x + \ell) e^{-j\omega\ell} dx$$

which becomes

$$\sum_\ell \int \varphi(x) \varphi(x + \ell) dx \; e^{j\omega\ell} \tag{12.79}$$

Because of the orthogonality of integer translates of $\varphi(x)$, this is not a function of ω but is $\int |\varphi(x)|^2 dx$ which, if normalized, is unity as stated in (5.57). This is the frequency domain equivalent of (5.13).

Equations (5.58) and (5.59) show how the scaling function determines the equation coefficients. This is derived by multiplying both sides of (12.51) by $\varphi(2x - m)$ and integrating to give

$$\int \varphi(x)\varphi(2x - m) \, dx = \int \sum_n h(n)\varphi(2x - n)\varphi(2x - m) \; dx \tag{12.80}$$

$$= \frac{1}{\sqrt{2}} \sum_n h(n) \int \varphi(x - n)\varphi(x - m) \; dx.$$

Using the orthogonality condition (12.53) gives

$$\int \varphi(x)\varphi(2x - m) \, dx = h(m)\frac{1}{\sqrt{2}} \int |\varphi(y)|^2 \, dy = \frac{1}{\sqrt{2}}h(m)$$

which gives (5.58). A similar argument gives (5.59).

Appendix C.
Matlab Programs

You are free to use these programs or any derivative of them for any scientific purpose but please reference this book. Up-dated versions of these programs and others can be found on our web page at: http://www-dsp.rice.edu/

```
function p = psa(h,kk)
% p = psa(h,kk)    calculates samples of the scaling function
%  phi(t) = p  by  kk  successive approximations from the
%  scaling coefficients  h.  Initial iteration is a constant.
%  phi_k(t) is plotted at each iteration.     csb 5/19/93
%
if nargin==1, kk=11; end;          % Default number of iterations
h2= h*2/sum(h);                    % normalize  h(n)
K = length(h2)-1; S = 128;         % Sets sample density
p = [ones(1,3*S*K),0]/(3*K);       % Sets initial iteration
P = p(1:K*S);                      % Store for later plotting
axis([0 K*S+2 -.5 1.4]);
hu = upsam(h2,S);                  % upsample h(n) by S
for iter = 0:kk                    % Successive approx.
    p = dnsample(conv(hu,p));      % convolve and down-sample
    plot(p); pause;                % plot each iteration
%   P = [P;p(1:K*S)];               % store each iter. for plotting
end
p = p(1:K*S);                      % only the supported part
L = length(p);
x = ([1:L])/(S);
axis([0 3 -.5 1.4]);
plot(x,p);                         % Final plot
title('Scaling Function by Successive Approx.');
ylabel('Scaling Function');
xlabel('x');
```

```
function p = pdyad(h,kk)
% p = pdyad(h,kk)  calculates approx. (L-1)*2^(kk+2) samples of the
%  scaling function  phi(t) = p  by  kk+3  dyadic expansions
%  from the scaling coefficient vector h  where L=length(h).
%  Also plots  phi_k(t)  at each iteration.    csb 5/19/93
%
if nargin==1, kk = 8; end                % Default iterations
h2 = h*2/sum(h);                         % Normalize
N = length(h2); hr = h2(N:-1:1); hh = h2;
axis([0,N-1,-.5,1.4]);
MR = [hr,zeros(1,2*N-2)];                % Generater row for M0
MT = MR; M0 = [];
for k = 1:N-1                            % Generate convolution and
   MR = [0, 0, MR(1:3*N-4)];             %   downsample matrix from h(n)
   MT = [MT; MR];
end
M0 = MT(:,N:2*N-1);                      % M0*p = p  if p samples of phi
MI = M0 - eye(N);
MJ = [MI(1:N-1,:);ones(1,N)];
pp = MJ\[zeros(N-1,1);1];                % Samples of phi at integers
p  = pp(2:length(pp)-1).';
   x = [0:length(p)+1]*(N-1)/(length(p)+1); plot(x,[0,p,0]); pause
p = conv(h2,p);                          % value on half integers
   x = [0:length(p)+1]*(N-1)/(length(p)+1); plot(x,[0,p,0]); pause
y = conv(h2,dnsample(p));                % convolve and downsample
p = merge(y,p);                          % interleave values on Z and Z/2
   x = [0:length(p)+1]*(N-1)/(length(p)+1); plot(x,[0,p,0]); pause
for k = 1:kk
   hh = upsample(hh);                    % upsample coefficients
   y = conv(hh,y);                       % calculate intermediate terms
   p = merge(y,p);                       % insert new terms between old
   x = [0:length(p)+1]*(N-1)/(length(p)+1); plot(x,[0,p,0]); pause;
end
title('Scaling Function by Dyadic Expansion');
ylabel('Scaling Function');
xlabel('x');
axis;
```

```
function [hf,ht] = pf(h,kk)
% [hf,ht] = pf(h,kk) computes and plots hf, the Fourier transform
%  of the scaling function  phi(t)  using the freq domain
%  infinite product formulation with kk iterations from the scaling
%  function coefficients h.  Also calculates and plots ht = phi(t)
%  using the inverse FFT                    csb 5/19/93
if nargin==1, kk=8; end         % Default iterations
L = 2^12; P = L;                % Sets number of sample points
hp = fft(h,L); hf = hp;         % Initializes iteration
plot(abs(hf));pause;            % Plots first iteration
for k = 1:kk                    % Iterations
   hp = [hp(1:2:L), hp(1:2:L)];  % Sample
   hf = hf.*hp/sqrt(2);          % Product
   plot(abs(hf(1:P/2)));pause;   % Plot Phi(omega) each iteration
      P=P/2;                     % Scales axis for plot
end;
ht = real(ifft(hf));            % phi(t) from inverse FFT
ht = ht(1:8*2^kk); plot(ht(1:6*2^kk));   % Plot phi(t)



function hn = daub(N2)
%  hn = daub(N2)
%  Function to compute the Daubechies scaling coefficients from
%  her development in the paper, "Orthonormal bases of compactly
%  supported wavelets", CPAM, Nov. 1988 page 977, or in her book
%  "Ten Lectures on Wavelets", SIAM, 1992 pages 168, 216.
%  The polynomial R in the reference is set to zero and the
%  minimum phase factorization is used.
%  Not accruate for N > 20.  Check results for long h(n).
%    Input:   N2 = N/2, where N is the length of the filter.
%    Output:  hn = h(n) length-N min phase scaling fn   coefficients
%  by rag 10/10/88, csb 3/23/93
a  = 1;   p = 1;   q = 1;      % Initialization of variables
hn = [1 1];                    % Initialize factors of zeros at -1
for j = 1:N2-1,
   hn = conv(hn,[1,1]);         % Generate polynomial for zeros at -1
   a  = -a*0.25*(j+N2-1)/j;     % Generate the binomial coeff. of L
   p  = conv(p,[1,-2,1]);       % Generate variable values for L
   q  = [0 q 0] + a*p;          % Combine terms for L
end;
q  = sort(roots(q));                      % Factor L
hn = conv(hn,real(poly(q(1:N2-1))));      % Combine zeros at -1 and L
hn = hn*sqrt(2)/(sum(hn));                % Normalize
```

```
function h = h246(a,b)
% h = h246(a,b) generates orthogonal scaling function
%   coefficients  h(n)  for lengths 2, 4, and 6 using
%   Resnikoff's parameterization with angles a and b.
%   csb.  4/4/93
if a==b,  h = [1,1]/sqrt(2);              % Length-2
elseif b==0
   h0 = (1 - cos(a) + sin(a))/2;          % Length-4
   h1 = (1 + cos(a) + sin(a))/2;
   h2 = (1 + cos(a) - sin(a))/2;
   h3 = (1 - cos(a) - sin(a))/2;
   h  = [h0 h1 h2 h3]/sqrt(2);
else                                      % Length-6
   h0 = ((1+cos(a)+sin(a))*(1-cos(b)-sin(b))+2*sin(b)*cos(a))/4;
   h1 = ((1-cos(a)+sin(a))*(1+cos(b)-sin(b))-2*sin(b)*cos(a))/4;
   h2 = (1+cos(a-b)+sin(a-b))/2;
   h3 = (1+cos(a-b)-sin(a-b))/2;
   h4 = (1-h0-h2);
   h5 = (1-h1-h3);
   h  = [h0 h1 h2 h3 h4 h5]/sqrt(2);
end


function [a,b] = ab(h)
% [a,b] = ab(h) calculates the parameters a and b from the
%   scaling function coefficient vector h for orthogonal
%   systems of length 2, 4, or 6 only.     csb. 5/19/93
%
h = h*2/sum(h);   x=0;                     % normalization
if length(h)==2, h = [0 0 h 0 0]; x=2; end;
if length(h)==4, h = [0 h 0]; x=4; end;
a = atan2((2*(h(1)^2+h(2)^2-1)+h(3)+h(4)),(2*h(2)*(h(3)-1)-2*h(1)*(h(4)-1)));
b = a - atan2((h(3)-h(4)),(h(3)+h(4)-1));
if x==2, a = 1; b = 1; end;
if x==4, b = 0; end;


function y = upsample(x)
% y = upsample(x) inserts zeros between each term in the row vector x.
%   for example:   [1 0 2 0 3 0] = upsample([1 2 3]).   csb 3/1/93
L = length(x);
y(:) = [x;zeros(1,L)]; y = y.';
y = y(1:2*L-1);
```

```
function y = upsam(x,S)
% y = upsam(x,S) inserts S-1 zeros between each term in the row vector x.
%  for example:   [1 0 2 0 3 0] = upsample([1 2 3]).   csb 3/1/93.
L = length(x);
y(:) = [x;zeros(S-1,L)]; y = y.';
y = y(1:S*L-1);
```

```
function y = dnsample(x)
% y = dnsample(x) samples x by removing the even terms in x.
%  for example:   [1 3] = dnsample([1 2 3 4]).   csb 3/1/93.
L = length(x);
y = x(1:2:L);
```

```
function z = merge(x,y)
% z = merge(x,y) interleaves the two vectors x and y.
% Example [1 2 3 4 5] = merge([1 3 5],[2 4]).
%   csb 3/1/93.
%
z = [x;y,0];
z = z(:);
z = z(1:length(z)-1).';
```

```
function w = wave(p,h)
% w = wave(p,h)   calculates and plots the wavelet  psi(t)
%  from the scaling function  p  and the scaling function
%  coefficient vector h.
%  It uses the definition of the wavelet.  csb. 5/19/93.
%
h2 = h*2/sum(h);
NN = length(h2); LL = length(p); KK = round((LL)/(NN-1));
h1u = upsam(h2(NN:-1:1).*cos(pi*[0:NN-1]),KK);
w  = dnsample(conv(h1u,p)); w  = w(1:LL);
xx = [0:LL-1]*(NN-1)/(LL-1);
axis([1 2 3 4]); axis;
plot(xx,w);
```

```
function g = dwt(f,h,NJ)
% function g = dwt(f,h,NJ); Calculates the DWT of periodic  g
%  with scaling filter  h  and  NJ  scales.   rag & csb 3/17/94.
%
N = length(h);  L = length(f);
c = f;  t = [];
if nargin==2, NJ = round(log10(L)/log10(2)); end; % Number of scales
h0  = fliplr(h);                        % Scaling filter
h1 = h;  h1(1:2:N) = -h1(1:2:N);        % Wavelet filter
for j = 1:NJ                            % Mallat's algorithm
   L = length(c);
   c = [c(mod((-(N-1):-1),L)+1) c];     % Make periodic
   d = conv(c,h1);    d = d(N:2:(N+L-2)); % Convolve & d-sample
   c = conv(c,h0);    c = c(N:2:(N+L-2)); % Convolve & d-sample
   t = [d,t];                           % Concatenate wlet coeffs.
end;
g = [c,t];                              % The DWT


function f = idwt(g,h,NJ)
% function f = idwt(g,h,NJ); Calculates the IDWT of periodic  g
%  with scaling filter  h  and  NJ  scales.   rag & csb 3/17/94.
%
L = length(g);    N = length(h);
if nargin==2, NJ = round(log10(L)/log10(2)); end; % Number of scales
h0 = h;                                 % Scaling filter
h1 = fliplr(h);  h1(2:2:N) = -h1(2:2:N); % Wavelet filter
LJ = L/(2^NJ);                          % Number of SF coeffs.
c  = g(1:LJ);                           % Scaling coeffs.
for j = 1:NJ                            % Mallat's algorithm
   w  = mod(0:N/2-1,LJ)+1;              % Make periodic
   d  = g(LJ+1:2*LJ);                   % Wavelet coeffs.
   cu(1:2:2*LJ+N) = [c c(1,w)];         % Up-sample & periodic
   du(1:2:2*LJ+N) = [d d(1,w)];         % Up-sample & periodic
   c  = conv(cu,h0) + conv(du,h1);      % Convolve & combine
   c  = c(N:N+2*LJ-1);                  % Periodic part
   LJ = 2*LJ;
end;
f = c;                                  % The inverse DWT


function r = mod(m,n)
% r = mod(m,n) calculates r = m modulo n
%
r = m - n*floor(m/n);                   % Matrix modulo n
```

```
function g = dwt5(f,h,NJ)
% function g = dwt5(f,h,NJ)
% Program to calculate the DWT from the L samples of f(t) in
% the vector  f  using  the scaling filter h(n).
%   csb 3/20/94
%
N = length(h);
c = f;   t = [];
if nargin==2
   NJ = round(log10(L)/log10(2));          % Number of scales
end;
h1 = h;   h1(1:2:N) = -h1(1:2:N);          % Wavelet filter
h0  = fliplr(h);                           % Scaling filter
for j = 1:NJ                               % Mallat's algorithm
   L  = length(c);
   d  = conv(c,h1);                        % Convolve
   c  = conv(c,h0);                        % Convolve
   Lc = length(c);
   while Lc > 2*L                          % Multi-wrap?
     d  = [(d(1:L) + d(L+1:2*L)), d(2*L+1:Lc)]; % Wrap output
     c  = [(c(1:L) + c(L+1:2*L)), c(2*L+1:Lc)]; % Wrap output
     Lc = length(c);
   end
   d = [(d(1:N-1) + d(L+1:Lc)), d(N:L)];   % Wrap output
   d = d(1:2:L);                           % Down-sample wlets coeffs.
   c = [(c(1:N-1) + c(L+1:Lc)), c(N:L)];   % Wrap output
   c = c(1:2:L);                           % Down-sample scaling fn c.
   t = [d,t];                              % Concatenate wlet coeffs.
end                                        % Finish wavelet part
g = [c,t];                                 % Add scaling fn coeff.
```

```
function a = choose(n,k)
% a = choose(n,k)
% BINOMIAL COEFFICIENTS
% allowable inputs:
% n : integer, k : integer
% n : integer vector, k : integer
% n : integer, k : integer vector
% n : integer vector, k : integer vector (of equal dimension)
nv = n;
kv = k;
if (length(nv) == 1) & (length(kv) > 1)
nv = nv * ones(size(kv));
elseif (length(nv) > 1) & (length(kv) == 1)
kv = kv * ones(size(nv));
end
a = nv;
for i = 1:length(nv)
   n = nv(i);
   k = kv(i);
   if n >= 0
      if k >= 0
         if n >= k
            c = prod(1:n)/(prod(1:k)*prod(1:n-k));
         else
            c = 0;
         end
      else
         c = 0;
      end
   else
      if k >= 0
         c = (-1)^k * prod(1:k-n-1)/(prod(1:k)*prod(1:-n-1));
      else
         if n >= k
            c = (-1)^(n-k)*prod(1:-k-1)/(prod(1:n-k)*prod(1:-n-1));
         else
            c = 0;
         end
      end
   end
   a(i) = c;
end
```

Index